Computational Group Theory and the Theory of Groups, II

CONTEMPORARY MATHEMATICS

511

Computational Group Theory and the Theory of Groups, II

Computational Group Theory and Cohomology
August 4–8, 2008
Harlaxton College, Grantham, United Kingdom

AMS Special Session
Computational Group Theory
October 17–19, 2008
Western Michigan University
Kalamazoo, MI

Luise-Charlotte Kappe
Arturo Magidin
Robert Fitzgerald Morse
Editors

American Mathematical Society
Providence, Rhode Island

2000 *Mathematics Subject Classification*. Primary 20–06, 20B40, 20B35, 17B55, 18G10, 20F12, 20F18, 20H15, 20J99, 20P05.

The photograph of Harlaxton College appears courtesy of the University of Evansville

---

Library of Congress Cataloging-in-Publication Data

Harlaxton Conference on Computational Group Theory and Cohomology (2008 : Harlaxton College)

Computational group theory and the theory of groups, II : Harlaxton Conference on Computational Group Theory and Cohomology, Harlaxton College, the British Campus of the University of Evansville, Grantham, United Kingdom, August 4–8, 2008 : AMS Special Session on Computational Group Theory, Western Michigan University, Kalamazoo, Michigan, October 17–19, 2008 / Luise-Charlotte Kappe, Arturo Magidin, Robert Fitzgerald Morse, editors.

p. cm. — (Contemporary mathematics ; v. 511)

Includes bibliographical references.

ISBN 978-0-8218-4805-0 (alk. paper)

1. Group theory—Congresses. 2. Computational complexity—Congresses. I. Kappe, Luise-Charlotte. II. Magidin, Arturo. III. Morse, Robert Fitzgerald. IV. AMS Special Session on Computational Group Theory (2008 : Western Michigan University) V. Title.

QA174.H37 2010
512′.2—dc22

2009047805

∞ The paper used in this book is acid-free and falls within the guidelines
established to ensure permanence and durability.
Visit the AMS home page at http://www.ams.org/

10 9 8 7 6 5 4 3 2 1 15 14 13 12 11 10

Contents

Preface

This is a follow-up to the Contemporary Mathematics volume that we edited on the same topic and that was published by the American Mathematical Society in 2008. This volume consists of contributions by researchers who were invited to the Harlaxton Conference on Computational Group Theory and Cohomology, held at Harlaxton College, The British Campus of the University of Evansville, August 4-8, 2008; and to the AMS Special Session on Computational Group Theory held at Western Michigan University, October 17-19, 2008. The Harlaxton Conference was supported financially by the De Brun Centre for Computational Algebra, National University of Ireland, Galway, and the University of Evansville's Institute for Global Enterprise in Indiana.

Both the Conference and the Special Session focused on examples of using CGT to solve problems that arise from many areas of group theory; in this volume we find applications to the enumeration of subgroups of the symmetric group, covering groups by subgroups, the ongoing co-class project for classification of p-groups, construction or computation of homological and cohomological invariants of groups, probabilistic group theory, and the study of free groups, among others. Computational Group Theory plays many roles in these investigations, from exploration that suggests conjectures or proofs, through performing key computations required to establish theorems.

Once again, we present these examples in the hope that they will encourage researchers and graduate students to think about ways in which they can incorporate CGT in their own research by seeing many different applications of CGT to traditional problems in Group Theory.

The Harlaxton Conference was organized by Bettina Eick, Graham Ellis, and Robert F. Morse; we thank them very much for all their work, and we also thank the de Brun Centre and the University of Evansville for their financial support.

The second editor was supported in part by a grant from the Louisiana Board of Regents. The first and third editors thank Arturo for his work managing and editing the submissions. The three of us are grateful to all the participants in the conferences, and to all authors who submitted contributions to this volume. We are also very thankful indeed to the referees who did such an excellent and timely job for both this and the previous volume. Finally, we are also very grateful to the American Mathematical Society for their help in the publication of this volume, particularly to Christine M. Thivierge for her help and patience.

Luise-Charlotte Kappe
Arturo Magidin
Robert Fitzgerald Morse

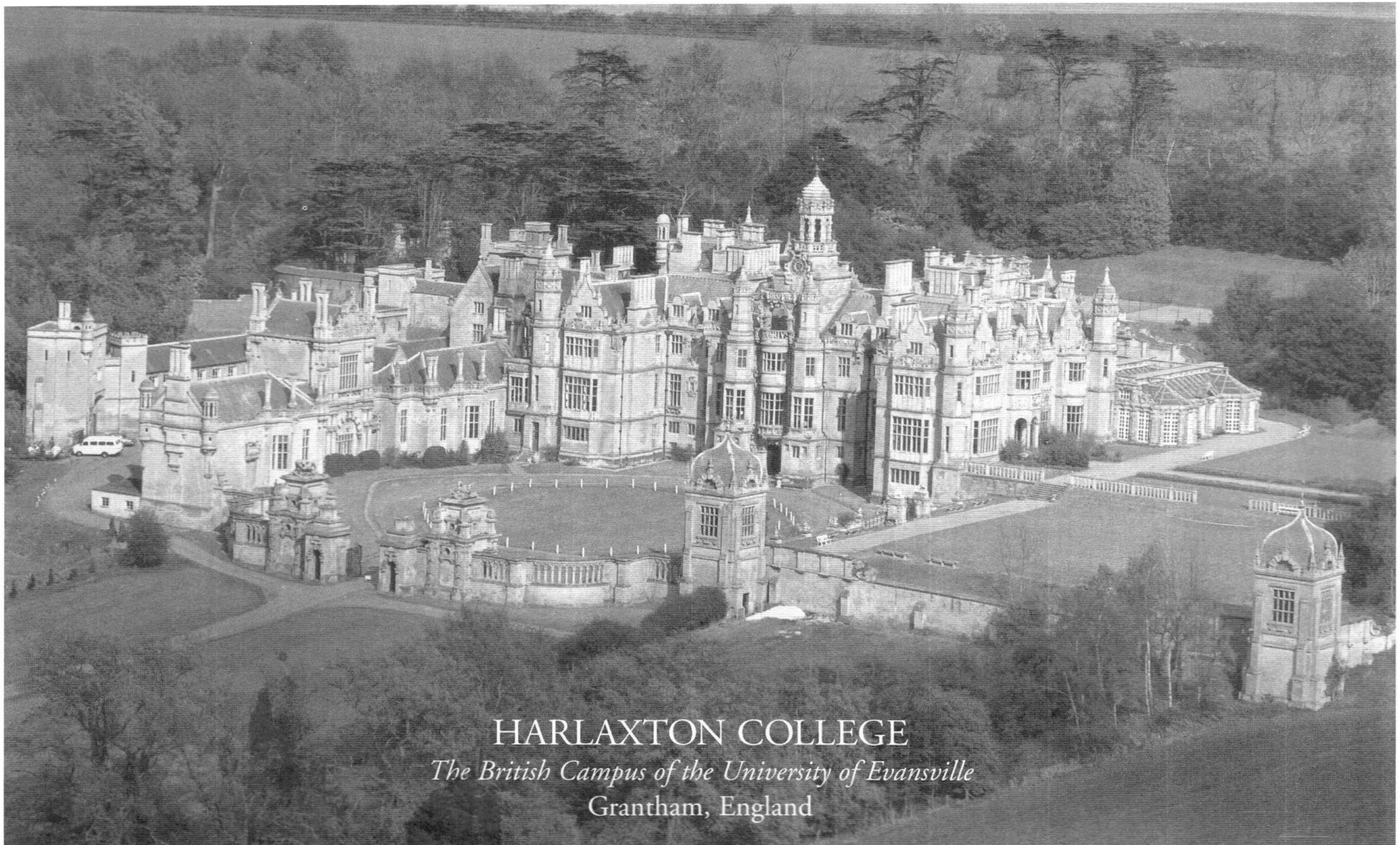
HARLAXTON COLLEGE
The British Campus of the University of Evansville
Grantham, England

Contemporary Mathematics
Volume **511**, 2010

The Probabilistic Zeta Function

Bret Benesh

ABSTRACT. This paper is a summary of results on the $P_G(s)$ function, which is the reciprocal of the probabilistic zeta function for finite groups. This function gives the probability that s randomly chosen elements generate a group G, and information about the structure of the group G is also embedded in it.

1. Introduction and History

Probabilistic group theory has been a growing field of mathematics for the past couple of decades. While other papers have considered this field in greater generality (see [**Di**, **Shal1**, **Shal2**]), we will be focusing on the so-called $P_G(s)$ function, which is the function that gives the probability that s randomly chosen elements (with replacement) of a finite group G generate G.

The study of the $P_G(s)$ function began in 1936, when Philip Hall [**H**] created the Eulerian function $\phi_G(s)$, defined to be the number of s-tuples $(g_1, \ldots, g_s) \in G^s$ such that $\langle g_1, \ldots, g_s \rangle = G$. Hall showed that

$$\phi_G(s) = \sum_{H \leq G} \mu_G(H)|H|^s,$$

where $\mu_G(H)$ is the Möbius function of the subgroup lattice of G, defined inductively as $\mu_G(G) = 1$ and

$$\sum_{H \leq K \leq G} \mu_G(K) = 0$$

if $H < G$.

After Hall, G.E. Wall [**W**] used a variation of the Eulerian function (the Eulerian polynomial) to prove the following theorem.

THEOREM 1.1 (Wall's Theorem). *If G is a finite solvable group, then the number of maximal subgroups in G is less than $|G|$.*

2000 *Mathematics Subject Classification.* 20E34, 20F05, 20P05, 11M41.

Key words and phrases. group theory, zeta functions, Dirichlet series, subgroup lattices, Moebius functions.

The author would like to thank Nigel Boston, Erika Damian, and the referee for their thoughtful comments on the paper.

Wall also conjectured that this result holds for nonsolvable groups, and relevant work has been done to that end in [**LiSh1**] and [**LiPySh**].

The most recent wave of interest in this field began in 1996, when Nigel Boston [**Bo**] and Avinoam Mann [**Ma**] independently defined the $P_G(s)$ function described above. It is clear that $P_G(s) = \frac{\phi_G(s)}{|G|^s}$, or

$$P_G(s) = \sum_{H \leq G} \frac{\mu_G(H)}{|G:H|^s}$$

by using Hall's result, and thus $P_G(s)$ is a Dirichlet series.

A word on the motivation behind this paper: while this topic is interesting in its own right – the author wrote his thesis [**Be**] on a similar subject – it is also a viable topic for undergraduate research. The basic idea is accessible, as only some knowledge of groups and proportions are needed. While many of the ideas below are too advanced for most undergraduates, several of them are not; Boston made a conjecture about the derivative of $P_G(s)$, soon solved by Shareshian in [**Shar**], that a calculus student could understand. Moreover, this topic lends itself well to computational algebra packages like GAP [**G**] and MAGMA [**BoCaPl**]. In fact, Boston's 1996 paper references the use of Cayley, an early version of Magma. Use of a computational algebra package would reduce the amount of background knowledge needed for an undergraduate to begin research, as the student could use simple programs to make conjectures about the $P_G(s)$ function.

2. The Basics of $P_G(s)$

We begin with some basic facts about $P_G(s)$, which are largely from [**Bo**] and [**Ma**]. First, several examples of $P_G(s)$, courtesy of Boston:

Cyclic Groups C_n: $P_{C_n}(s) = \prod_{p|n, p \text{ prime}} \left(1 - \frac{1}{p^s}\right).$

The Alternating Group A_4: $P_{A_4}(s) = \left(1 - \frac{2}{2^s}\right)\left(1 + \frac{2}{2^s}\right)\left(1 - \frac{1}{3^s}\right).$

The Alternating Group A_5: $P_{A_5}(s) = 1 - \frac{5}{5^s} - \frac{6}{6^s} - \frac{10}{10^s} + \frac{20}{20^s} + \frac{60}{30^s} - \frac{60}{60^s}.$

S^n for a Simple Group S: $P_{S^n}(s) = \prod_{i=0}^{n-1} \left(P_S(s) - \frac{i|\text{Aut } S|}{|S|^s}\right).$

By way of motivation, the probability that two integers chosen at random are relatively prime can be solved, rather non-rigorously, by

$$\prod_{\text{primes } p} \left(1 - \frac{1}{p^2}\right) = \frac{1}{\zeta(2)} = \frac{6}{\pi^2},$$

where $\zeta(s)$ is the Riemann zeta function. The left side of the above equation resembles a product of $P_{C_p}(s)$ functions evaluated at $s = 2$ and, we can think of $\frac{1}{P_G(s)}$ as a zeta function of G. We define $\frac{1}{P_G(s)}$ to be the *probabilistic zeta function,* and it is common to label results for $P_G(s)$ as results about the probabilistic zeta function. While it initially only makes sense to consider natural numbers s in the function $P_G(s)$, we will see below that we can gain insight into G by expanding the domain to the complex numbers.

The ring of finite Dirichlet series with coefficients in $\mathbb{Z}$ is a unique factorization domain, so factoring $P_G(s)$ will be of great interest. In fact, if N is a normal subgroup of G, then we can factor $P_G(s)$ as

$$P_G(s) = P_{G/N}(s)P_{G,N}(s),$$

where $P_{G,N}(s)$ is given by the formula

$$P_{G,N}(s) = \sum_{\substack{H\leq G\\ HN=G}} \frac{\mu_G(H)}{|G:H|^s}$$

and interpreted as the conditional probability that a random s-tuple $(g_1,\dots,g_s)$ of G^s generates G given that $\langle g_1,\dots,g_s,N\rangle = G$.

While one might conjecture that $P_{G,N}(s) = P_N(s)$, this is only sometimes true ([**Bo**]). Consider the symmetric group S_5 and its alternating group A_5. It is true that $P_{S_5}(s) = P_{C_2}(s)P_{A_5}(s)$, so that $P_{A_5}(s) = P_{S_5,A_5}(s)$. However, we have

$$P_{S_3}(s) = \left(1-\frac{1}{2^s}\right)\left(1-\frac{3}{3^s}\right),$$

so $P_{S_3,C_3}(s) = 1-\frac{3}{3^s} \neq 1-\frac{1}{3^s} = P_{C_3}(s)$.

Recall that a chief series

$$1 = N_0 \subset N_1 \subset \cdots \subset N_n = G$$

is a collection of normal subgroups N_i of G such that N_{i+1}/N_i is minimal normal in G/N_i. Then the factorization $P_G(s) = P_{G/N}(s)P_{G,N}(s)$ can be used repeatedly on a chief series to obtain factors of $P_G(s)$. Detomi and Lucchini [**DeLu1**] proved that the factorization is independent of the choice of chief series (Gaschütz [**Ga**] had previously proved this independence for solvable groups).

Another immediate result from the above factorization is that if $P_G(s)$ is irreducible, then G is simple. The converse is not true, however. For instance,

$$P_{PSL(2,7)}(s) = \left(1-\frac{2}{2^s}\right)\left(1+\frac{2}{2^s}+\frac{4}{4^s}-\frac{14}{7^s}-\frac{28}{14^s}+\frac{21}{21^s}-\frac{28}{28^s}+\frac{42}{42^s}\right)$$

is reducible. In fact, $P_{PSL(2,p)}(s)$ is always reducible when $t \equiv 3 \pmod 4$ and $p = 2^t-1$ ([**DamLuMo**]).

The Frattini subgroup $\Phi(G)$ of a group G is the intersection of all maximal subgroups and equals the set of non-generators. If we have a normal subgroup N contained in the Frattini subgroup, N contains only non-generators and we obtain $P_G(s) = P_{G/N}(s)$, since $P_{G,N}(s) = 1$.

3. A Motivating Application

An important application of the $P_G(s)$ function is to help determine the minimal number of generators of a group H, denoted $d(H)$. For example, let H be a finite group such that $d(H) = m+1$ for some m, but all proper quotients Q of H have the property that $d(Q) \leq m$. Dalla Volta and Lucchini [**DaLu**] proved such an H must be isomorphic to

$$L_t = \{(l_1,\dots,l_t) \in L^t \mid l_1 \equiv \cdots \equiv l_t \pmod M\},$$

where L is a group with unique minimal normal subgroup M (the group L is called a *primitive monolithic group*) and t is some integer.

The probabilistic zeta function pops up in determining what the integer t is. In the same paper, Dalla Volta and Lucchini proved that if M is nonabelian, then

$$t = 1 + \frac{|M|^m P_L(m)}{|C_{\mathrm{Aut}L}(L/M)| P_{L/M}(m)} = 1 + \frac{|M|^m}{|C_{\mathrm{Aut}L}(L/M)|} P_{L,M}(m).$$

The group L_t is useful in proofs by contradiction that consider minimal counterexamples, as then $P_{L,M}(s)$ can be useful in proving that minimal counterexamples do not exist.

4. What $P_G(s)$ Says About G

As evidenced by the formula

$$P_G(s) = \sum_{H \leq G} \frac{\mu_G(H)}{|G:H|^s},$$

$P_G(s)$ is tied to the subgroup structure of the group G. Because of this, one can think of it as encoding information about the structure of G. This section focuses on examples of information one can gain about a group G solely from knowing the $P_G(s)$ function.

4.1. The primes dividing $|G|$. The first piece of data we can get from $P_G(s)$ is exactly which primes divide $|G|$. Damian and Lucchini [**DamLu1**] proved that if $P_G(s)$ is written as $\sum \frac{a_n}{n^s}$, then:

THEOREM 4.1. *A prime p divides $|G|$ if and only if p divides n for some n with $a_n \neq 0$.*

4.2. The coset poset. Our second example is a case where it is advantageous to view $P_G(s)$ as having a domain greater than the non-negative integers. Brown and Bouc [**Br**] found that letting $s = -1$ gives interesting topological information about the group G. The coset poset $\mathcal{C}(G)$ is the set of cosets xH ($x \in G$ and $H < G$) ordered by inclusion. We can use a simplicial complex $\Delta(\mathcal{C}(G))$ whose simplices are the finite chains in $\mathcal{C}(G)$ to define the Euler characteristic $\chi(\mathcal{C}(G))$. We may then define the reduced Euler characteristic $\tilde{\chi}(\mathcal{C}(G)) = \chi(\mathcal{C}(G)) - 1$. Then Bouc discovered:

THEOREM 4.2. $P_G(-1) = -\tilde{\chi}(\mathcal{C}(G))$.

Moreover, Brown defined an analogue of $P_G(s)$ for finite lattices (instead of groups). Using this analogue, Brown shows that the entire $P_G(s)$ function, not only its value at $s = -1$, can be recreated from a lattice that is defined from the coset poset $\mathcal{C}(G)$.

4.3. Solvability, supersolvability, and nilpotency. Since $P_G(s)$ encodes information about the structure of G, it is natural to wonder whether solvability questions can be answered based solely on $P_G(s)$. Gaschütz [**Ga**] began working on this question in the 1950s, and this question was completely answered by Detomi and Lucchini [**DeLu2**] with the following theorem:

THEOREM 4.3. *G is solvable if and only if $P_G(s)$ is a product of factors of the form $1 - \frac{c_i}{(p_i^{n_i})^s}$, where p_i is a prime.*

A group is supersolvable if it has an invariant normal series where all factors are cyclic. Detomi and Lucchini also describe a condition for supersolvable groups.

THEOREM 4.4. *G is supersolvable if and only if $P_G(s)$ is a product of factors of the form $1 - \frac{c_i}{p_i^s}$, where each p_i is prime and each c_i is positive.*

This begs the question of whether a similar result exists for nilpotent groups, but Gaschütz demonstrated that no such result can exist. Indeed, the functions $P_G(s)$ for $G = C_2 \times C_3 \times C_3$ (nilpotent) and for $G = S_3 \times C_3$ (solvable, but not nilpotent) are both equal to

$$\left(1 - \frac{1}{2^s}\right)\left(1 - \frac{1}{3^s}\right)\left(1 - \frac{3}{3^s}\right).$$

Therefore, it is impossible to determine nilpotency strictly from the $P_G(s)$ function. However, Damian and Lucchini [**DamLu2**] did find the following result on nilpotency. First, define

$$P_G(H, s) = \sum_{H \leq K \leq G} \frac{\mu_G(K)}{|G : K|^s}.$$

Then:

THEOREM 4.5. *A group G is nilpotent if and only if $P_G(H, s)$ divides $P_G(s)$ for all $H \leq G$.*

Finally, suppose that $P_G(s) = \sum \frac{a_n}{n^s}$. Then Detomi and Lucchini [**DeLu2**] proved that G is solvable if and only if $a_{nm} = a_n a_m$ whenever $(n, m) = 1$. Damian and Lucchini [**DamLu1**] were able to generalize this to p-solvable groups:

THEOREM 4.6. *Suppose $P_G(s) = \sum \frac{a_n}{n^s}$. Then G is p-solvable if and only if $a_{p^r d} = a_{p^r} a_d$ whenever $(p, d) = 1$.*

4.4. Simple groups. We now turn our attention to nonsolvable groups, the results of which are found in [**DamLu3, DamLu4, DamLuMo**]. All three papers work toward the same result, which culminates in the following theorem:

THEOREM 4.7. *Let G be a nonabelian finite simple group, let H be a finite group with trivial Frattini subgroup, and assume $P_G(s) = P_H(s)$.*

(1) *If G is an alternating group or a sporadic simple group, then $G \cong H$.*
(2) *If G and H are groups of Lie type defined on a field of characteristic p, then $G \cong H$.*

Finally, Nigel Boston [**Bo**] conjectured that $P'_G(1) = 0$ whenever G is simple nonabelian. This conjecture was proved and generalized by John Shareshian [**Shar**] in the following theorem.

THEOREM 4.8. *$P'_G(1) = 0$ unless $G/O_p(G)$ is cyclic for some prime p.*

5. Computing $P_G(s)$

While $P_G(s)$ can be tedious to compute by hand, computer algebra systems such as MAGMA and GAP can quickly generate the formula and numerical values for $P_G(s)$. The key to computing such a function for a group G is knowing its subgroups, their indices, and the Möbius function. These can all be obtained from MAGMA and GAP, although the subgroups are typically given as conjugacy classes. Because of this, the length of each conjugacy class (that is, the number of subgroups in each class) is also required.

The computer algebra system GAP offers a convenient shortcut for computing examples of the $P_G(s)$ function: it has a command to compute the table of marks. Briefly, the table of marks of a group is a matrix whose entries describe the number of fixed points when a representative of one conjugacy class of subgroups of G acts on another via conjugation. The Möbius function can be determined from the table of marks, as described by Pfeiffer in [**Pf**].

GAP makes it very easy to access this information from the table of marks. In fact, a program—in its entirety—that computes the numerical answer to $P(G, s)$ is:

```
P1:=function(G,s)
return EulerianFunctionByTom(TableOfMarks(G),s)/Order(G)^s;
end;
```

Creating a string that returns the formula is slightly more complicated, although not much more so. Easy access to the Möbius function values, lengths, and orders of the conjugacy classes of subgroups are gotten through the command `TableOfMarks(G)` Below is a very basic program for GAP that returns the formula as a string:

```
P2:=function(G)
local i,tom,mob,ord,len,finalstring;
tom:=TableOfMarks(G);
mob:=MoebiusTom(tom).mu;
ord:=OrdersTom(tom);
len:=LengthsTom(tom);
finalstring:="";

for i in [1..Length(mob)] do
 if IsBound(mob[i]) then
  finalstring:=Concatenation(finalstring,"+",String(len[i]*mob[i]),
   "/",String(Order(G)/ord[i]),"^s);
 fi;
od;

return finalstring;
end;
```

This particular program was designed for simplicity, and the resulting string lacks a certain beauty. Because of this, readers are implored to add additional code to make a more readable output. Additionally, shortcuts can be made for efficiency, such as inserting the line of code `G:=G/FrattiniSubgroup(G);` at the beginning of the program to take advantage of the fact that $P_G(s) = P_{G/\Phi(G)}(s)$.

The logic is similar when using MAGMA; the main difference is that MAGMA does not have a command to access the table of marks, and cannot immediately access the Möbius function of a group. However, the command `SubgroupLattice(G)` contains the lengths and orders of the subgroups of G. Additionally, the command `SubgroupLattice(G)` contains information about the containment of the subgroups, and one can use `SubgroupLattice(G)` and the recursive definition of the Möbius function to create a function that returns the Möbius value of a subgroup.

6. Conjectures and Open Problems

We conclude with several unsolved conjectures and avenues for exploration.

(1) Shareshian [**Shar**] proved that $P'_G(1) = 0$ for simple nonabelian G. It is also true that $P''_{A_6}(1) = 0$. Describe all groups G such that $P''_G(1) = 0$.

(2) Boston [**Bo**] observes that $P_{S_n}(s) = P_{A_n}(s)P_{C_2}(s)$ for $n = 2$, 5, and 6 but not for $n = 3$, 4, 7, 8, and 9. Determine for which n the above equation holds.

(3) A generalization of the previous question: describe the nonabelian finite simple groups S such that $P_S(s) = P_{\mathrm{Aut}\, S,S}(s)$.

(4) If $P_{G,N}(s) \neq P_N(s)$, describe the possibilities for $P_{G,N}(s)$. Detomi and Lucchini [**DeLu1**] gave a partial answer to this challenge in 2003. Let L be a finite group with unique minimal normal subgroup M. Then define the following:
   - $\tilde{P}_{L,1}(s) = P_{L,M}(s)$
   - $\tilde{P}_{L,i}(s) = P_{L,M}(s) - \frac{(1+q_M+\cdots+q_M^{i-2})\gamma_M}{|M|^s}$ if $i > 1$

   where $\gamma_M = |C_{\mathrm{Aut}\, M} L/M|$, $q_M = |\mathrm{End}_L M|$ if M is abelian, and $q_M = 1$ otherwise. Detomi and Lucchini proved that each factor of $P_G(s)$ is equal to $\tilde{P}_{L,i}(s)$ for some primitive monolithic group L and positive integer i, thereby reducing the problem to the study of monolithic groups. Determine the possible values of $P_{L,M}(s)$.

(5) Similar to the earlier result on simple groups, we may conjecture:

   CONJECTURE 6.1. *If G is a simple nonabelian finite group, H a finite group with trivial Frattini subgroup, and $P_G(s) = P_H(s)$, then $G \cong H$.*

   This conjecture would follow if the next conjecture were true. First, some notation. Let

$$a_n(G) = \sum_{n=|G:H|} \mu_G(H),$$

   and let

$$b_n(G,p) = \begin{cases} a_n(G) & \text{if } p\nmid n, \\ 0 & \text{otherwise;} \end{cases}$$

   finally, let $P_G^{(p)}(s) = \displaystyle\sum_{n=1}^{\infty} \frac{b_n(G,p)}{n^s}$. Then:

   CONJECTURE 6.2. *Let G be a group of Lie type. Except for finitely many exceptions, a prime p is the characteristic of the defining field if and only if $\left|P_G^{(p)}(0)\right|$ is a nontrivial p-power.*

   Patassini [**Pat**] has provided some evidence that this conjecture is true.

(6) There have been many theorems of the form "$P_{G,N}(s) > \gamma$ whenever $s \geq f(G)$ for some constant γ and some function f of G" (see for example [**DamLuMo, DeLuMo, LuMo, Pak**]). Improve one of these bounds, or determine similar bounds for $P_G(s)$ (Pak [**Pak**] proves something similar to this).

References

[Be] B. Benesh, *Counting generators in finite groups that are generated by two subgroups of prime power order*, Ph.D. Thesis, University of Wisconsin-Madison, 2005.

[BoCaPl] W. Bosma, J. Cannon, and C. Playoust, *The Magma algebra system. I. The user language*, J. Symbolic Comput. **24** (1997), no. 3-4, 235–265.

[Bo] N. Boston, *A probabilistic generalization of the Riemann zeta function*, Analytic Number Theory, Vol. 1 (Allerton Park, IL, 1995), Progr. Math., vol. 138, Birkhäuser Boston, Boston, MA, 1996, pp. 155–162.

[Br] K. S. Brown, *The coset poset and probabilistic zeta function of a finite group*, J. Algebra **225** (2000), 989–1012.

[DaLu] F. Dalla Volta and A. Lucchini, *Finite groups that need more generators than any proper quotient*, J. Austral. Math. Soc. Ser. A **64** (1998), no. 11, 82–91.

[DaLuFo1] F. Dalla Volta, A. Lucchini, and F. Morini, *On the probability of generating a minimal d-generated group*, J. Austral. Math. Soc. **71** (2001), no. 11, 177–185.

[DaLuFo2] F. Dalla Volta, A. Lucchini, and F. Morini, *Some remarks on the probability of generating an almost simple group*, Glasgow Math. J. **45** (2003), 281–291.

[DamLu1] E. Damian and A. Lucchini, *Finite groups with p-multiplicative probabilistic zeta function*, Comm. Algebra **35** (2007), no. 11, 3451–3472.

[DamLu2] E. Damian and A. Lucchini, *A probabilistic generalization of subnormality*, Journal of Algebra and its Applications **4** (2005), no. 3, 313–323.

[DamLu3] E. Damian and A. Lucchini, *The probabilistic zeta function of finite simple groups*, J. Algebra **313** (2007), 957–971.

[DamLu4] E. Damian and A. Lucchini, *Recognizing the alternating groups from their probabilistic zeta function*, Glasgow Math. J. **46** (2004), 595–599.

[DamLuMo] E. Damian, A. Lucchini, and F. Morini, *Some properties of the probabilistic zeta function on finite simple groups*, Pacific J. Math. **215** (2004), 3–14.

[DeLu1] E. Detomi and A. Lucchini, *Crowns and factorization of the probabilistic zeta function of a finite group*, J. Algebra **265** (2003), 651–668.

[DeLu2] E. Detomi and A. Lucchini, *Recognizing soluble groups from their probabilistic zeta function*, Bull. London Math. Soc. **35** (2003), 659–664.

[DeLuMo] E. Detomi, A. Lucchini, and F. Morini, *How many elements are needed to generate a finite group with good probability?*, Israel J. Math. **132** (2002), 29–44.

[Di] J. Dixon, *Probabilistic group theory*, C.R. Math. Rep. Acad. Sci. Canada **24** (2002), 1–15.

[G] The GAP Group, GAP – *Groups, Algorithms and Programming, Version 4.4.12*, available at `http://www.gap-system.org`.

[Ga] A. W. Gaschütz, *Die Eulersche funktion enlicher auflösbarer Gruppen*, Illinois J. Math. **3** (1959), 469–476.

[H] P. Hall, *The Eulerian functions of a finite group*, Quart. J. Math. **7** (1936), 134–151.

[LiPySh] M. W. Liebech, L. Pyber, and A. Shalev, *On a conjecture of G. E. Wall*, J. Algebra **317** (2007), 184–197.

[LiSh1] M. W. Liebeck and A. Shalev, *Maximal subgroups of symmetric groups*, J. Comb. Th. Ser. A. **75** (1996), 341–352.

[LuMo] A. Lucchini and F. Morini, *On the probability of generating finite groups with a unique minimal normal subgroup*, Pacific J. Math. **203** (2002), no. 2, 429–440.

[Ma] A. Mann, *Positively generated finite groups*, Forum Math. **8** (1996), 429–459.

[Pak] I. Pak, *On probability of generating a finite group* (1999). preprint.

[Pat] M. Patassini, *The probabilistic zeta function of* $\mathrm{PSL}(2,q)$*, of the Suzuki groups* ${}^2B_2(q)$ *and of the Ree groups* ${}^2G_2(q)$, Pacific J. Math. **240** (2009), no. 1, 185–200.

[Pf] G. Pfeiffer, *The subgroups of* M_{24}*, or how to compute the table of marks of a finite group*, Experiment. Math. **6** (1997), 247–270.

[Shal1] A. Shalev, *Asymptotic group theory*, Notices of the Amer. Math. Soc. **48** (2001), 383–389.

[Shal2] A. Shalev, *Simple groups, permutation groups, and probability*, Proceedings of the International Congress of Mathematicians, Vol. II (Berlin, 1998), 1998, pp. 129–137 (electronic).

[**Shar**] J. Shareshian, *On the probabilistic zeta function for finite groups*, J. Algebra **210** (1998), 703–770.

[**W**] G. E. Wall, *Some applications of the Eulerian function of a finite group*, J. Austral. Math. Soc. 2 **8** (1961), 35–59.

Department of Mathematics, College of Saint Benedict/Saint John's University, 37 College Avenue South, Saint Joseph, Minnesota 56374

E-mail address: `bbenesh@csbsju.edu`

Contemporary Mathematics
Volume **511**, 2010

Periodicities for graphs of p-groups beyond coclass

Bettina Eick and Tobias Rossmann

ABSTRACT. We use computational methods to investigate periodic patterns in the graphs $\mathcal{G}(p,(d,w,o))$ associated with the p-groups of rank d, width w, and obliquity o. In the smallest interesting case $\mathcal{G}(p,(3,2,0))$ we obtain a conjectural description of this graph for all $p \geq 3$; in particular, we conjecture that this graph is virtually periodic for all $p \geq 3$. We also investigate other related infinite graphs.

1. Introduction

Which invariants are useful in the classification of p-groups?

The *order* has been considered in many publications, going back to the beginnings of abstract group theory in the 19th century; see [1] for a history. Nowadays, the p-groups of order dividing 2^9 (see [5]) and p^7 (see [16]) are available, but a full classification of the groups of order p^n in general still seems to be out of reach. An important step towards a full classification would be a proof of the famous PORC conjecture [7] which asserts that for fixed n, the number $f(p)$ of p-groups of order p^n is a polynomial on residue classes.

Leedham-Green and Newman [14] suggested using the *coclass* to classify p-groups. Recall that the coclass of a finite p-group G of order p^n and nilpotency class $\mathrm{cl}(G)$ is defined as $\mathrm{cc}(G) = n - \mathrm{cl}(G)$. A first and fundamental idea in classifying all p-groups of a given coclass r is to visualize them in a graph $\mathcal{G}(p,r)$: the vertices of this graph correspond to the isomorphism types of p-groups of coclass r and there is a directed edge $G \to H$ if $G \cong H/\gamma_{\mathrm{cl}(H)}(H)$ holds, where $\gamma_i(H)$ denotes the ith term of the lower central series of H. The classification of all p-groups of coclass r thus translates to an investigation of the infinite graph $\mathcal{G}(p,r)$.

Coclass theory has become a rich and interesting research field in group theory. A highlight in this theory was the complete proof of the coclass-conjectures [14] by Shalev [18] and Leedham-Green [11]. We refer to the book by Leedham-Green and McKay [13] for background and details. Nowadays, the fundamental aim in coclass theory is to prove that every graph $\mathcal{G}(p,r)$ can be constructed from a finite subgraph using certain periodic patterns. This has been proved for $p = 2$ in [3] and [4], but is still open for odd primes. A central problem in the odd prime case is that

2000 *Mathematics Subject Classification.* 20D15.
Key words and phrases. p-groups, rank, width, obliquity, periodicity, coclass.

the graphs $\mathcal{G}(p,r)$ are usually rather thick and thus are often difficult to investigate in detail. As a consequence, only very little detailed experimental evidence on the structure of these graphs is available and explicit conjectures on the nature of any useful periodic patterns are vague at present.

Leedham-Green thus suggested to try other invariants with a similar approach as in coclass theory with the hope of obtaining graphs which have all the nice features of the graphs $\mathcal{G}(p,r)$, but are thinner and thus easier to understand. In particular, Leedham-Green initiated the classification of p-groups by *rank, width and obliquity*; see Chapter 12 of [13] for a discussion. We briefly recall the definitions of these invariants: for any finite or infinite pro-p-group G and a closed subgroup H of G, let $[G:H]_p$ denote the p-logarithm of the index $[G:H]$; further let $\mathrm{d}(G) = [G:\Phi(G)]_p$ be the cardinality of a minimal (topological) generating set of G, and let $\mu_i(G)$ denote the intersection of all closed normal subgroups of G which are not properly contained in $\gamma_i(G)$. Then we define for a pro-p-group G:

- its *rank* $\mathrm{r}(G) = \sup\{\mathrm{d}(U) \mid U \text{ a closed subgroup of } G\}$,
- its *width* $\mathrm{w}(G) = \sup\{[\gamma_i(G):\gamma_{i+1}(G)]_p \mid i \in \mathbb{N}\}$, and
- its *obliquity* $\mathrm{o}(G) = \sup\{[\gamma_i(G):\mu_i(G)]_p \mid i \in \mathbb{N}\}$.

The obliquity of a group determines how restricted its lattice of normal subgroups is. In particular, in a group of obliquity 0 every normal subgroup lies between two consecutive terms of the lower central series.

Let $\tau(G)$ denote the triple $(\mathrm{r}(G), \mathrm{w}(G), \mathrm{o}(G))$ and define the graph $\mathcal{G}(p,(d,w,o))$ similar to the coclass graphs: the vertices of this graph correspond to the isomorphism types of finite p-groups G with $\tau(G) = (d,w,o)$ and there is a directed edge $G \to H$ if $G \cong H/\gamma_{\mathrm{cl}(H)}(H)$ holds. The classification of all p-groups G with $\tau(G) = (d,w,o)$ now translates to understanding the (usually) infinite graph $\mathcal{G}(p,(d,w,o))$.

In this paper we discuss how computational tools can be used to investigate the graphs $\mathcal{G}(p,(d,w,o))$ and we exhibit experimental results for some small and interesting cases. Thus, we give a conjectural description of the graph $\mathcal{G}(p,(3,2,0))$ for $p > 2$ based on our experimental data. It suggests that $\mathcal{G}(p,(3,2,0))$ can be constructed from a finite subgraph using certain periodic patterns and hence $\mathcal{G}(p,(3,2,0))$ seems to have the nice features displayed by the coclass graphs $\mathcal{G}(2,r)$ and, moreover, it is a rather thin graph which can be easily exhibited.

An interesting family of infinite pro-p-groups G with finite $\tau(G)$ are the Sylow pro-p-subgroups of $\mathrm{Aut}(L)$ for simple Lie algebras L of the type $L = \mathfrak{sl}_n(K)$ for $p \geq 3$, where $K/\mathbb{Q}_p$ is a finite extension. The lower central series quotients of such a group G define an infinite path through the graph $\mathcal{G}(p,\tau(G))$. We show how our computational tools can be used to investigate these infinite paths together with certain branches associated with them. Our experiments with these infinite trees indicate that they also exhibit periodic patterns of the same type as $\mathcal{G}(p,(3,2,0))$.

Throughout this paper we assume that p is an odd prime.

2. Preliminaries

There is a correspondence between the infinite paths in $\mathcal{G}(p,(d,w,o))$ and the isomorphism types of infinite pro-p-groups G with $\tau(G) = (d,w,o)$. Hence a first aim in understanding $\mathcal{G}(p,(d,w,o))$ is a classification of the infinite pro-p-groups G with $\tau(G) = (d,w,o)$. In this section, we recall some basic facts about these groups.

Recall that a pro-p-group is *just infinite* if it is infinite but every non-trivial closed normal subgroup has finite index.

LEMMA 2.1. *Let G be an infinite pro-p-group of finite rank, finite width, and finite obliquity. Then G is p-adic analytic and just infinite.*

PROOF. A pro-p-group of finite rank is p-adic analytic by [2], Corollary 8.33. An infinite pro-p-group of finite width and finite obliquity is just infinite, see [10], p. 3. □

If G is a pro-p-group of finite width, then $G/\gamma_i(G)$ is finite for all $i \in \mathbb{N}$. If, moreover, $\tau(G)$ is finite, then there exists a $k \in \mathbb{N}$ with $\tau(G/\gamma_i(G)) = \tau(G)$ for all $i \geq k$ and the groups $G/\gamma_i(G)$ for $i \geq k$ thus define an infinite path through the graph $\mathcal{G}(p, \tau(G))$.

2.1. The solvable case. Let G be an infinite solvable pro-p-group of finite rank, width, and obliquity. Then G is an irreducible p-adic space group, see for example [19], Lemma 8.1. This means that G is an extension of a free p-adic module $\mathbb{Z}_p^d$ for some $d \in \mathbb{N}$ by a finite p-group P which acts faithfully on $\mathbb{Z}_p^d$ and irreducibly on $\mathbb{Q}_p^d$.

The possible dimensions d and point groups P are well-understood, see for example [13]. Excluding the trivial case $d = 1$, irreducible finite p-subgroups of $\mathrm{GL}_d(\mathbb{Z}_p)$ only exist for dimensions d of the form $d = p^{r-1}(p-1)$ for some $r \in \mathbb{N}$. For such d, the group $\mathrm{GL}_d(\mathbb{Z}_p)$ has precisely p^{r-1} conjugacy classes of Sylow p-subgroups with representatives $W_1, \ldots, W_{p^{r-1}}$, say. These groups $W_1, \ldots, W_{p^{r-1}}$ are all conjugate in $\mathrm{GL}_d(\mathbb{Q}_p)$ to an r-fold iterated wreath product $C_p \wr \cdots \wr C_p$ of cyclic groups of order p.

It is well-known that each p-adic space group can be embedded as a subgroup of finite index into a split space group; that is, a space group which is a split extension of $\mathbb{Z}_p^d$ by P, see for example [13], Lemma 10.4.1. This implies the following embedding theorem.

THEOREM 2.2. *Let G be an irreducible p-adic space group of dimension $d = p^{r-1}(p-1)$. Then G embeds as subgroup of finite index into $\mathbb{Z}_p^d \rtimes W_i$ for some $i \in \{1, \ldots, p^{r-1}\}$.*

A simple example of an infinite solvable pro-p-group of finite rank, width, and obliquity is the group $\mathbb{Z}_p^{p-1} \rtimes C_p$. This group has rank $p-1$, width 2, and obliquity 0. It also has finite coclass; in fact, it is the unique infinite pro-p-group of coclass 1.

2.2. The insolvable case. The insolvable infinite pro-p-groups of finite rank, width, and obliquity are discussed in detail in [10]. Here we only recall some of their features. The following theorem yields a characterization of these groups, see Lemma 2.1 and [10], Section IIId.

THEOREM 2.3. *Let G be an insolvable infinite pro-p-group. Then G has finite rank, width, and obliquity if and only if G is p-adic analytic and just infinite.*

Every pro-p-group of finite rank has an associated Lie algebra, see for example [2], Section 9.5. In the case of an insolvable infinite pro-p-group of finite rank, width and obliquity, this Lie algebra is homogeneous of the form $S^{p^k} = S \oplus \cdots \oplus S$ (p^k summands) for a simple finite-dimensional Lie algebra S over $\mathbb{Q}_p$ and some

$k \geq 0$, see [10], Proposition III.6. The following theorem shows how such a pro-p-group can be recovered from its associated Lie algebra; we refer to [10], Proposition III.9 and Corollary III.10 for details and background.

THEOREM 2.4. *Let G be an insolvable infinite pro-p-group of finite rank, width, and obliquity, and let L be its associated Lie algebra over $\mathbb{Q}_p$. Then G embeds as a subgroup of finite index into a Sylow pro-p-subgroup of* $\mathrm{Aut}(L)$.

If $L = S^{p^k}$ for a simple Lie algebra S, then $\mathrm{Aut}(L) = \mathrm{Aut}(S) \wr \mathrm{Sym}(p^k)$ and thus the Sylow pro-p-subgroups of $\mathrm{Aut}(L)$ can be determined from the Sylow pro-p-subgroups of $\mathrm{Aut}(S)$ and $\mathrm{Sym}(p^k)$. The Sylow p-subgroups of $\mathrm{Sym}(p^k)$ are well-understood and can be constructed readily. It is known from [10], Lemma III.16 that the Sylow pro-p-subgroups of $\mathrm{Aut}(L)$ are all conjugate.

Theorem 2.4 suggests an approach for constructing the insolvable infinite pro-p-groups of finite rank, width and obliquity: first classify the finite-dimensional simple Lie algebras over $\mathbb{Q}_p$, then determine the Sylow pro-p-subgroups of their automorphism groups, and finally list the relevant subgroups of finite index in the relevant wreath products based on these Sylow pro-p-subgroups. None of the steps in this approach is straightforward in practice. For example, in [10] the Sylow pro-p-subgroups of the homogeneous Lie algebras of dimension at most 14 are determined and this proved to be highly non-trivial.

Examples of simple Lie algebras over $\mathbb{Q}_p$ are the linear Lie algebras $s\ell_d(K)$ consisting of all $d \times d$ matrices with trace 0 over an extension K of $\mathbb{Q}_p$. If K has degree m over $\mathbb{Q}_p$, then $s\ell_d(K)$ has dimension $m(d^2-1)$ over $\mathbb{Q}_p$. Note that there are only finitely many field extensions K of any given degree over $\mathbb{Q}_p$. The group $\mathrm{Aut}(s\ell_d(K))$ of automorphisms over $\mathbb{Q}_p$ is $(\mathrm{PGL}_d(K) \rtimes D) \rtimes \mathrm{Gal}(K/\mathbb{Q}_p)$, where D is the group of so-called diagram automorphisms. The Sylow pro-p-subgroup of $\mathrm{Aut}(s\ell_d(K))$ is explicitly determined in [10], Lemma XI.4.

2.3. The pro-p-groups of rank 3, width 2, and obliquity 0. A complete classification of the infinite pro-p-groups of rank 3, width 2, and obliquity 0 up to isomorphism is given in [13], Theorem 12.2.3: for every $p > 2$, there are two groups of this type (up to isomorphism). We briefly recall their description.

The first group is the Sylow pro-p-subgroup of $\mathrm{Aut}(L)$, where L is the simple 3-dimensional Lie algebra $s\ell_2(\mathbb{Q}_p)$ of 2×2-matrices with trace 0 over $\mathbb{Q}_p$. As shown in [13], this group can be identified with the subgroup of $\mathrm{SL}_2(\mathbb{Z}_p)$ consisting of the matrices which are upper unitriangular modulo p.

There is exactly one other isomorphism type of simple Lie algebras of dimension 3 over $\mathbb{Q}_p$. To construct a representative, M say, for this isomorphism type, let $\mathbb{Q}_p(a)$ be the unramified extension of $\mathbb{Q}_p$ of degree 2, where a has multiplicative order p^2-1. Let π be the automorphism of $\mathbb{Q}_p(a)$ with $a^\pi = a^p$ and define $K = \mathbb{Q}_p(a,\pi)$. Then K is a division algebra of degree 4 over $\mathbb{Q}_p$. Commutation in K defines a Lie algebra H and $M = [H,H]$ is the desired simple Lie algebra.

The second infinite pro-p-group of rank 3, width 2, and obliquity 0 is the Sylow pro-p-subgroup of $\mathrm{Aut}(M)$. As shown in [13], this group can be identified with the central quotient $U/Z(U)$, where U is the group of 1-units of $\mathbb{Z}_p(a,\pi)$.

3. The graphs $\mathcal{G}(p,(d,w,o))$

For a group G we write $G_i = G/\gamma_i(G)$ for $i \in \mathbb{N}$. If G is an infinite pro-p-group with $\tau(G) = (d,w,o)$, then there exists an integer k such that the quotients

G_i for $i \geq k$ form an infinite path through $\mathcal{G}(p,(d,w,o))$. Conversely, the inverse limit of the groups on an infinite path in $\mathcal{G}(p,(d,w,o))$ is an infinite pro-p-group G with $\tau(G) = (d,w,o)$. Hence the infinite pro-p-groups G with $\tau(G) = (d,w,o)$ parametrize the graph $\mathcal{G}(p,(d,w,o))$ in this sense. Our aim is to investigate the graphs $\mathcal{G}(p,(d,w,o))$ based on this parametrization.

3.1. Descendant trees and τ-trees. For a finite p-group H we denote by $\mathcal{T}(H)$ the *descendant tree* of H: the vertices of $\mathcal{T}(H)$ correspond to the isomorphism types of finite p-groups K such that $K_{\mathrm{cl}(H)+1} \cong H$ and there is an edge $K \to L$ if $L_{\mathrm{cl}(L)} \cong K$. The groups in $\mathcal{T}(H)$ are called *descendants* of H. An *immediate descendant* of H is a descendant of H of class $\mathrm{cl}(H)+1$.

Let G be an infinite pro-p-group with $\tau(G) = (d,w,o)$ and let i be minimal such that $\tau(G_i) = (d,w,o)$ and $G_i \not\cong H_i$ for any infinite pro-p-group $H \not\cong G$ with $\tau(H) = (d,w,o)$. Since there are only finitely many possible isomorphism types for H, see [12], p. 72, such an i exists. The *τ-tree* $\mathcal{T}(G,(d,w,o))$ is defined as the intersection of $\mathcal{T}(G_i)$ with $\mathcal{G}(p,(d,w,o))$; it thus consists of all descendants K of G_i with $\tau(K) = (d,w,o)$. Note that $G_i \to G_{i+1} \to \ldots$ is the unique infinite path in $\mathcal{T}(G,(d,w,o))$ starting at the root G_i of this tree.

3.2. Virtual periodicity of trees. Let $\mathcal{T}$ be a tree with root R having a unique infinite path $R = R_1 \to R_2 \to \ldots$. For each $i \in \mathbb{N}$ let $\mathcal{B}_{R_i}(\mathcal{T})$ be the subtree of $\mathcal{T}$ consisting of the descendants of R_i which are not descendants of R_{i+1}. Then $\mathcal{B}_{R_i}(\mathcal{T})$ is a *branch* of $\mathcal{T}$. We say that $\mathcal{T}$ is *virtually periodic* if there exist l and d such that

$$\mathcal{B}_{R_i}(\mathcal{T}) \cong \mathcal{B}_{R_{i+d}}(\mathcal{T})$$

for every $i \geq l$. We then call the least possible value of such a d the *period* and the corresponding least possible value of l the *defect* of $\mathcal{T}$. Note that a virtually periodic tree $\mathcal{T}$ with defect l and period d can be constructed from its first $l+d-1$ branches $\mathcal{B}_{R_1}(\mathcal{T}), \ldots, \mathcal{B}_{R_{l+d-1}}(\mathcal{T})$.

3.3. Virtual periodicity of $\mathcal{G}(p,(d,w,o))$. We say the graph $\mathcal{G}(p,(d,w,o))$ is *virtually periodic* if all but finitely many groups of $\mathcal{G}(p,(d,w,o))$ are contained in a τ-tree of $\mathcal{G}(p,(d,w,o))$ and if every τ-tree of $\mathcal{G}(p,(d,w,o))$ is virtually periodic. If $\mathcal{G}(p,(d,w,o))$ is virtually periodic, then it can be constructed from a finite subgraph and this would furnish a classification of the p-groups with rank d, width w, and obliquity o. However, the following interesting question is wide open at current.

Question: *For which primes p and which (d,w,o) is $\mathcal{G}(p,(d,w,o))$ virtually periodic?*

It is hoped that for every prime p and every (d,w,o) all but finitely many groups of $\mathcal{G}(p,(d,w,o))$ are contained in a τ-tree of this graph.

It can be deduced from [13], Proposition 3.1.2 and Exercise 3.3(3) that if $p > 2$ and P is a finite p-group of maximal class and order at least p^{p+1}, then $\tau(P) = (p-1,2,0)$ holds. Consequently, all but finitely many groups from the coclass graph $\mathcal{G}(p,1)$ are contained in the graph $\mathcal{G}(p,(p-1,2,0))$. Since it is known that the (unique) maximal infinite subtree of $\mathcal{G}(p,1)$ is not virtually periodic in the sense of Section 3.2 for $p > 3$, it follows that not every graph $\mathcal{G}(p,(d,w,o))$ is virtually periodic.

We conjecture below that the graph $\mathcal{G}(p,(3,2,0))$ is virtually periodic for all primes $p \geq 3$.

4. Computational methods

Many computational methods work with the lower p-series rather than the lower central series: the *lower p-series* of a finite p-group H is defined by $\lambda_1(H) = H$ and $\lambda_{i+1}(H) = [\lambda_i(H), H]\lambda_i(H)^p$ for $i \geq 1$. The length of this series is the *p-class* of H. A finite p-group K is an *immediate p-descendant* of H if $K/\lambda_{d+1}(K) \cong H$, where d is the p-class of H, and if, in addition, the p-class of K is $d+1$.

The ANUPQ [15] program allows the determination of all immediate p-descendants of a given non-trivial finite p-group H up to isomorphism. We want to use this program to determine the immediate descendants of H. For this purpose we investigate under which circumstances the p-descendants coincide with the descendants. We call a finite p-group H *stable* if $\gamma_i(H) = \lambda_i(H)$ holds for all i.

LEMMA 4.1. *Suppose that H is non-abelian and stable. Then E is an immediate descendant of H if and only if it is an immediate p-descendant of H.*

PROOF. It suffices to show that either condition on E implies that E is stable. Thus let E be an immediate descendant of H. As $E/\gamma_2(E) = E/\lambda_2(E)$, it follows from [8], Kapitel III, Satz 2.13b that every lower central factor of E has exponent dividing p. Hence E is stable. Conversely, let E be an immediate p-descendant of H. Then $E/\lambda_2(E) \cong H/\lambda_2(H) = H/H' \cong E/\lambda_{c+1}(E)E'$ and hence $\lambda_2(E) = \lambda_{c+1}(E)E'$. Therefore, $\lambda_2(E)/E' = \lambda_{c+1}(E)E'/E' \cong \lambda_{c+1}(E)/\lambda_{c+1}(E) \cap E'$ has exponent dividing p, whence $\lambda_{c+1}(E) \leq \lambda_3(E) \leq E'$. Thus, $E' = \lambda_2(E)$ and E is stable. □

In this manner the ANUPQ program allows us to construct finite subtrees of a descendant tree $\mathcal{T}(H)$ for a non-abelian and stable p-group H. The resulting descendants are all described in terms of power-commutator (pc) presentations. Such presentations allow effective computations with the groups they define and we can further investigate the resulting groups using the computer algebra system GAP [20].

The width of a group defined by a pc presentation can be read off quite readily: we need to determine the lower central series of the groups under consideration. Even simpler, if the group in question is a descendant determined by ANUPQ, then its lower central series and thus its width can be read off directly.

The rank and the obliquity of a group defined by a pc presentation can be computed, but the available algorithms for this purpose are significantly less effective than those for the computation of the width. We use the formula from [10], p. 74, to determine the obliquity of a finite p-group H. That is, we use the formula

$$\mu_i(H) = \mu_{i-1}(H) \cap \gamma_i(H) \cap \bigcap\{N \lhd H \mid N \not\leq \gamma_i(H) \text{ and } N \leq \gamma_{i-1}(H)\}$$

and then obtain that the obliquity of H is given by

$$\mathrm{o}(H) = \max_i \log_p[\gamma_i(H) : \mu_i(H)].$$

We determine the rank of a finite p-group H by computing a representative for every conjugacy class of subgroups of H and determining the maximum of their minimal generator numbers.

We combine these methods to the following approach for determining finite parts of the τ-tree for an infinite pro-p-group G which has finite rank, width and obliquity and a stable quotient $G/\gamma_2(G)$. Note that the latter condition implies that all lower central quotients $G_i = G/\gamma_i(G)$ are stable. First we choose i large enough such that G_i satisfies $\tau(G_i) = \tau(G) = (d, w, o)$ and $G_i \not\cong H_i$ for any infinite pro-p-group H with $H \not\cong G$ and $\tau(H) = (d, w, o)$. Then we apply the ANUPQ program to start generating the descendant tree of G_i. For every obtained descendant K, we check whether it satisfies $\tau(K) = (d, w, o)$ and if this is not the case, then we discard K. The remaining descendants are groups in $\mathcal{T}(G, (d, w, o))$.

Implementations of our main methods can be found in the Fwtree package [6] for the computer algebra system GAP.

5. The graph $\mathcal{G}(p, (3, 2, 0))$

We have used the algorithms of Section 4 to investigate the graph $\mathcal{G}(p, (3, 2, 0))$ for various primes $p > 2$. This section summarizes our results.

Recall that for every prime $p > 2$ the graph $\mathcal{G}(p, (3, 2, 0))$ contains two τ-trees; these correspond to the two isomorphism types of infinite pro-p-groups of rank 3, width 2, and obliquity 0. Let G and H denote representatives for these isomorphism types, where G corresponds to the Lie algebra $s\ell_2(\mathbb{Q}_p)$.

The following lemma reduces the investigation of $\mathcal{G}(p, (3, 2, 0))$ to an investigation of its τ-trees.

LEMMA 5.1. *If $p > 2$, then almost all groups in $\mathcal{G}(p, (3, 2, 0))$ are contained in a τ-tree.*

PROOF. For $p \geq 5$, this is a consequence of [13], Theorem 12.2.15. For $p = 3$, we can explicitly determine the groups not contained in the τ-trees. If P is a 3-group with $\tau(P) = (3, 2, 0)$, then its class-2 quotient P_3 is isomorphic to the non-abelian group of order 27 and exponent 3, K say; note that $\tau(K) = (2, 2, 0)$. This group K has eleven immediate descendants: four of order 81 and seven others. Among the four of order 81, only one group, Q say, has immediate descendants. The descendants of Q all have coclass 1. As observed in Section 3.3, almost all 3-groups of coclass 1 have rank 2. Hence only finitely many descendants of Q are contained in $\mathcal{G}(3, (3, 2, 0))$. It remains to consider the seven other descendants of K. They all have rank 3, width 2, and obliquity 0, but only two of them have immediate descendants with these properties. These two groups are isomorphic to G_4 and H_4, respectively, which are the roots of the τ-trees of $\mathcal{G}(3, (3, 2, 0))$. □

5.1. A conjectural description of the τ-trees. First we note that G_4 and H_4 are non-isomorphic quotients of G and H which are both contained in $\mathcal{G}(p, (3, 2, 0))$, see [13], Section 12.2. Hence the τ-trees $\mathcal{T}_G := \mathcal{T}(G, (3, 2, 0))$ and $\mathcal{T}_H := \mathcal{T}(H, (3, 2, 0))$ are subtrees of $\mathcal{T}(G_4)$ and $\mathcal{T}(H_4)$, respectively. Each tree $\mathcal{T}_G$ or $\mathcal{T}_H$ consists of its unique infinite path and its branches. To describe the two infinite trees, we describe their branches only. Let $B_j(G)$ and $B_j(H)$ denote the branches of $\mathcal{T}_G$ and $\mathcal{T}_H$ with roots G_j and H_j, respectively.

Figures 1–4 exhibit our conjectural descriptions for all of these branches. Their shapes depend on the underlying group G or H, the class of the root of the branch and the underlying prime p. We use a compact notation to visualize trees: if a vertex v has a number n attached to it (written on the right), then there are n vertices $v = v_1, v_2, \ldots, v_n$ in the tree under consideration and all these v_i have the same ancestor and they all have isomorphic descendant trees.

Thus our experiments suggest the following conjecture.

CONJECTURE 5.2. For every $p > 2$, the graph $\mathcal{G}(p, (3, 2, 0))$ is virtually periodic. More precisely, both of its τ-trees are virtually periodic with period 2 and defect 1.

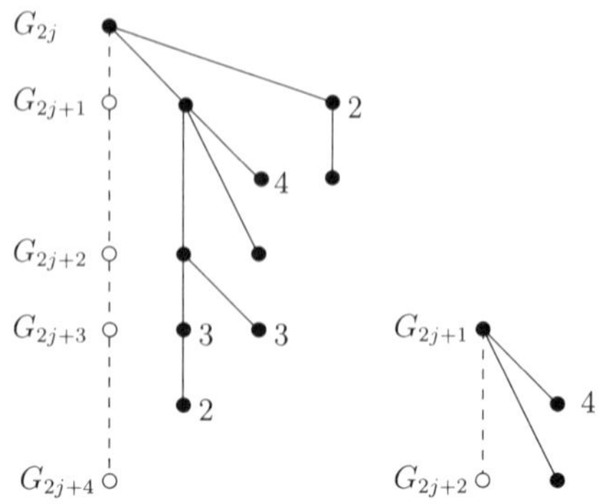

FIGURE 1. The branches $\mathcal{B}_{2j}(G)$ and $\mathcal{B}_{2j+1}(G)$ for $j \geq 2$ and $p = 3$

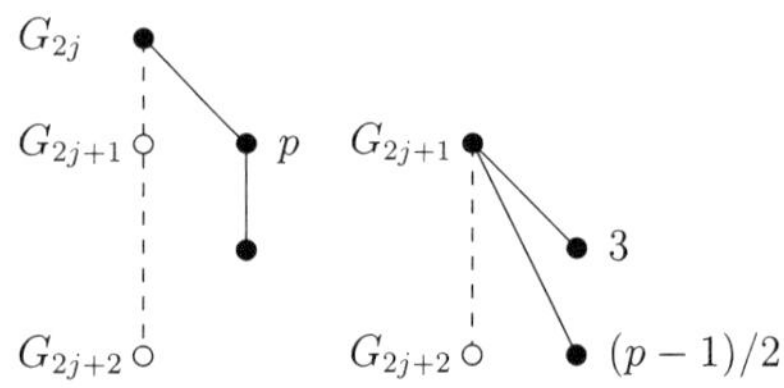

FIGURE 2. The branches $\mathcal{B}_{2j}(G)$ and $\mathcal{B}_{2j+1}(G)$ for $j \geq 2$ and $p \geq 5$

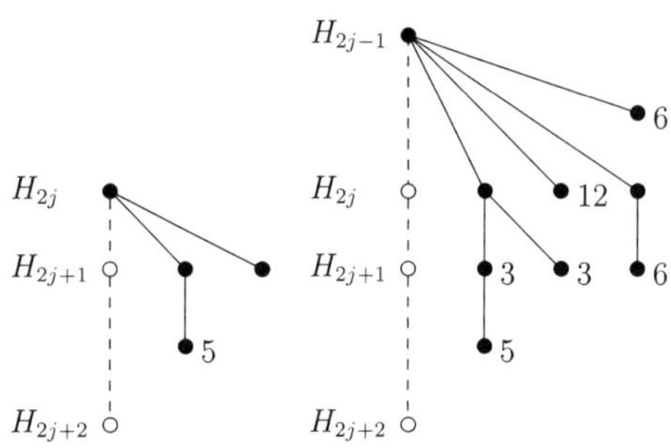

FIGURE 3. The branches $\mathcal{B}_{2j}(H)$ for $j \geq 2$ and $\mathcal{B}_{2j-1}(H)$ for $j \geq 3$ and $p = 3$

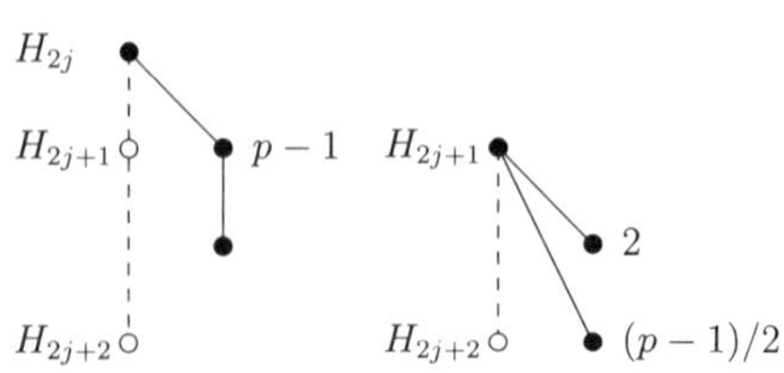

FIGURE 4. The branches $\mathcal{B}_{2j}(H)$ and $\mathcal{B}_{2j+1}(H)$ for $j \geq 2$ and $p \geq 5$

We note that the results of a first investigation of $\mathcal{G}(p, (3, 2, 0))$ for $p \geq 5$ are described in [13], Section 12.2, based on results from [17]. The results obtained there imply that if $p \geq 5$, then the branches of the τ-trees of $\mathcal{G}(p,(3,2,0))$ have depth at most 2.

5.2. Details on our computations. For given p and i, the quotients G_i and H_i can be constructed readily from their definitions, see also [10]. The lower central factors of G and H are known from [13], pp. 273–274. In particular, they are always elementary abelian and thus the quotients G_i and H_i are stable for all i. Hence our methods of Section 4 apply.

We determined the first branches of the τ-trees $\mathcal{T}_G$ and $\mathcal{T}_H$ for all primes $p \leq 13$ and the first branches of the full descendant trees $\mathcal{T}(G_4)$ and $\mathcal{T}(H_4)$ for all primes $p \leq 41$. The numbers of branches that we determined for $p \leq 13$ are given in Table 1.

p	$\mathcal{T}_G$	$\mathcal{T}(G_4)$	$\mathcal{T}_H$	$\mathcal{T}(H_4)$
3	11	26	7	6
5	8	18	5	12
7	6	14	10	16
11	5	8	5	10
13	5	6	5	9

TABLE 1. Numbers of computed branches

The computation of the branches of the full descendant trees is significantly less time-consuming than the corresponding computation for τ-trees, since the determination of the rank and the obliquity for a finite p-group is highly time- and space-consuming. Consequently, we constructed significantly more branches of the full descendants trees than of the corresponding τ-trees. We note that the branches of the descendants trees and the corresponding τ-trees coincided in all cases where we computed both (except for the first branch $\mathcal{B}_4(H)$ in the case $p = 3$).

Our computations were performed on a PC with two 2Ghz processors and 3GB of RAM running under Linux. As an illustration, Table 2 lists approximate runtimes (in seconds) for the construction of the branches $\mathcal{B}_i(G)$ of the τ-tree $\mathcal{T}_G$ and for the corresponding branches $\mathcal{T}(G_i) \setminus \mathcal{T}(G_{i+1})$ of the full descendant trees for $p = 3$.

i	4	5	6	7	8	9	10	11	12	13	14
$\mathcal{B}_i(G)$	59	11	182	32	643	132	2,124	392	6,032	1,149	15,137
$\mathcal{T}(G_i) \setminus \mathcal{T}(G_{i+1})$	11	3	15	3	22	4	31	6	45	9	64

TABLE 2. Approximate runtimes (in seconds) for G and $p = 3$

6. The trees associated with $\mathcal{sl}_n(K)$

Let $\mathrm{G}_n(K)$ be the Sylow pro-p-subgroup of $\mathrm{Aut}(L)$, where L is a simple Lie algebra of the form $L = \mathcal{sl}_n(K)$ for a field extension K of $\mathbb{Q}_p$ with finite degree m. Then $\mathrm{G}_n(K)$ has finite rank, width, and obliquity.

It would be highly interesting to investigate the τ-trees associated with these groups $\mathrm{G}_n(K)$ with a view towards checking whether these trees are virtually periodic. In the smallest case $(n,m) = (2,1)$ this has been done in Section 5 and a complete conjectural outline of the corresponding trees has been obtained. In all larger cases of (n,m) this is less straightforward with our current computational methods and theoretical knowledge. In this section we consider the groups of the form $\mathrm{G}_n(K)$ for some larger values of (n,m) and investigate their descendant trees as a first approximation of their τ-trees.

The groups $\mathrm{G}_n(K)$ are investigated in [10] for the cases with $(n^2-1)m \leq 14$; that is, for the cases $(n,m) = (2,1)$, $(2,2)$, $(2,3)$, $(2,4)$, and $(3,1)$. In each of these cases there may be several groups depending on the number of different fields K. The different fields K can be retrieved from the database [9] for small primes. The following table provides some summary information: it considers the different types of K, the number of corresponding groups $\mathrm{G}_n(K)$ and their parameters $\tau(\mathrm{G}_n(K))$ as far as they are available in [10]. If a $\star$ is listed, then the number of fields (and thus the number of groups) depends on the prime.

(n,m)	type of K	$p=3$ # grps	params	$p \geq 5$ # grps	params
$(3,1)$	$K = \mathbb{Q}_p$	1	$(?,2,5)$	1	$(?,3,?)$
$(2,2)$	K totally ramified	2	$(?,2,0)$	2	$(?,2,?)$
$(2,2)$	K unramified	1	$(?,4,0)$	1	$(?,4,?)$
$(2,3)$	K totally ramified	9	$(?,2,0)$, $(?,3,3)$	$\star$	$(?,2,?)$
$(2,3)$	K unramified	1	$(?,3,4)$	1	$(?,6,?)$
$(2,4)$	K totally ramified	2	$(?,2,0)$	$\star$	$(?,2,?)$
$(2,4)$	K mixed ramified	2	$(?,4,0)$	$\star$	$(?,4,?)$
$(2,4)$	K unramified	1	$(?,8,?)$	1	$(?,8,?)$

Among the infinite pro-p-groups of finite rank, width, and obliquity, the groups $\mathrm{G}_n(K)$ are reasonably well understood. The algorithms of [10] can be used to determine lower central series quotients of these groups and these can then be used as a basis for further computations. Pc presentations of finite quotients of the groups can be obtained from the Fwtree package [6].

From [10] we further obtain that all the groups considered here have stable lower central series quotients so that our algorithms of Section 4 apply and we can compute finite parts of their descendant trees.

6.1. The case $(n,m) = (2,2)$ and $p = 3$. Consider the groups $\mathrm{G}_2(K)$ with K totally ramified of degree 2 over $\mathbb{Q}_3$. There exist two fields K in this case: $\mathbb{Q}_3(\sqrt{\pm 3})$. Denote the corresponding groups by $G = \mathrm{G}_2(\mathbb{Q}_3(\sqrt{-3}))$ and $H = \mathrm{G}_2(\mathbb{Q}_3(\sqrt{3}))$. Our computational evidence suggests the following conjecture.

CONJECTURE 6.1. *The descendant trees $\mathcal{T}(G_7)$ and $\mathcal{T}(H_7)$ are both virtually periodic with period 4 and defect 1.*

In the following, we display the branches with roots $G_7, \ldots, G_{10}$ and with roots $H_7, \ldots, H_{10}$ using the same notation as in Section 5.1. First, the branches with roots G_7 and H_7 and with roots G_9 and H_9 are isomorphic and are given in the following figures.

G_7, H_7

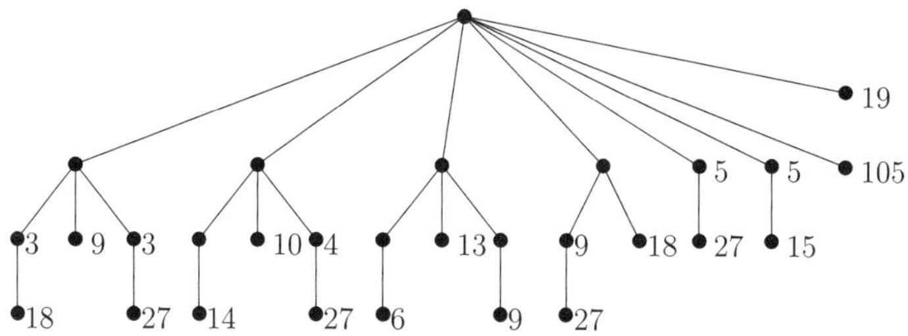

G_9, H_9

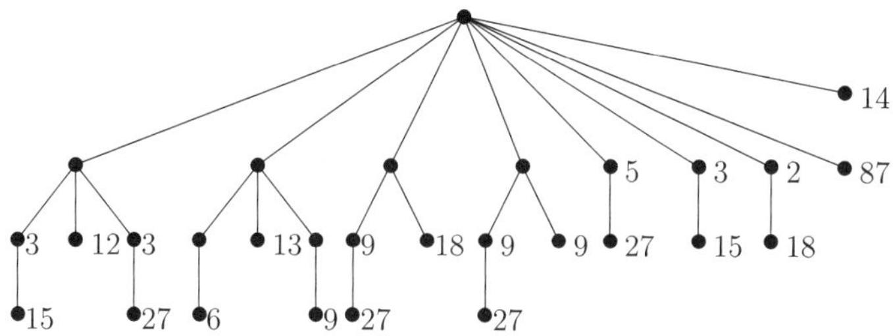

Next, we exhibit the branches with roots G_8 and G_{10}.

G_8

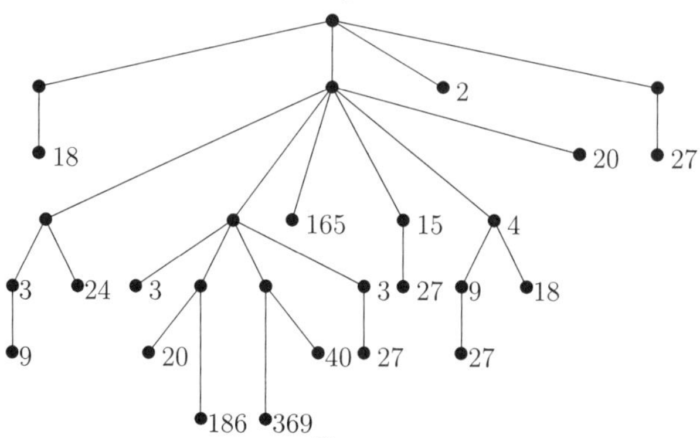

G_{10}

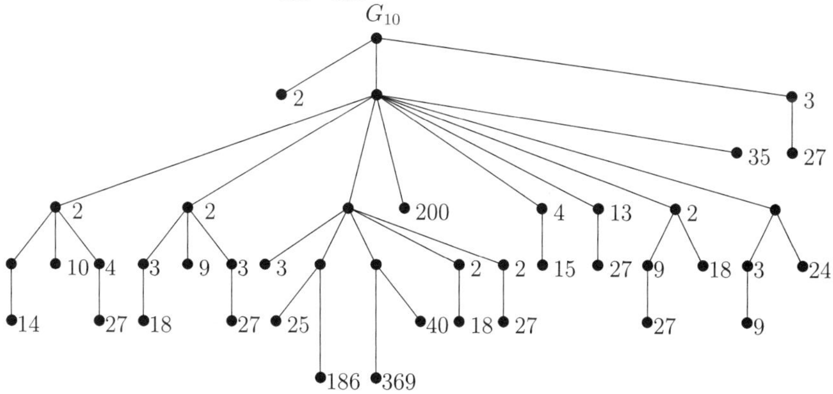

Finally, the branches with roots H_8 and H_{10} are shown in the following figures.

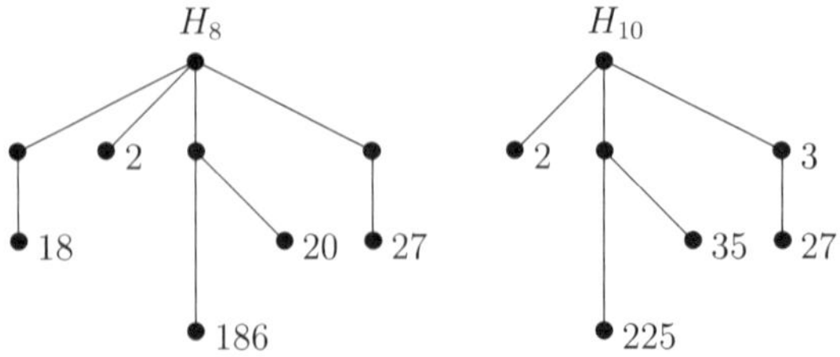

For both choices of the field extension $\mathbb{Q}(\sqrt{\pm 3})$, we have been able to verify five occurrences of the conjectured periodic pattern (consisting of four branches each).

6.2. The case $(n, m) = (2, 2)$ **and** $p > 3$**.** Let $p \geq 5$ and K be a totally ramified extension of degree 2 over $\mathbb{Q}_p$. There are two such fields K to consider and, correspondingly, two groups $\mathrm{G}_2(K)$ exist. Our experimental evidence suggests that there are periodic structures in the descendant trees of the lower central series quotients of these two groups with period a multiple of 4 and defects depending on K.

Let G be either of the two groups $\mathrm{G}_2(K)$. We conjecture that for sufficiently large i the first branch of the descendant tree of G_{2i} is as displayed in the following figure, independent of the field K.

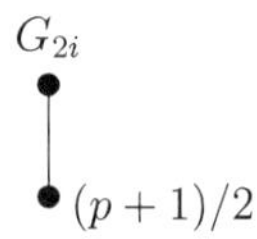

For j odd, we could only properly investigate the branches with root G_j for $p = 5$, and even in this case we have only computed "shaved" versions of these branches. Thus, we have investigated the groups in such a branch of order at most $|G_j|p^3$.

Our conjectural description of the resulting trees in the case $p = 5$ is exhibited in the following figures. Note that, again, these trees do not seem to depend on K.

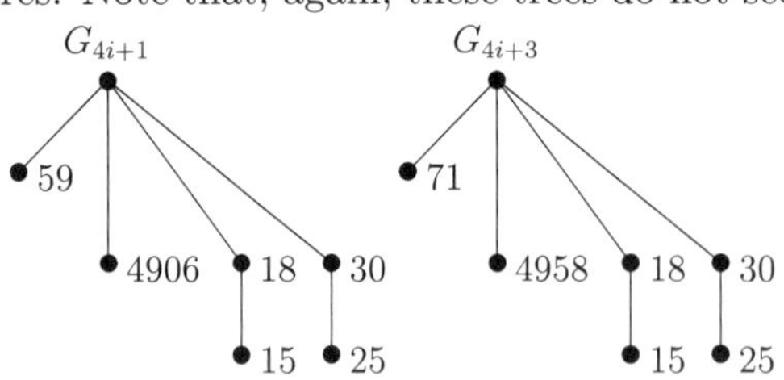

For $p = 5$, we have been able to verify three full and an incomplete fourth occurrence of the above pattern of four "shaved" branches.

6.3. Other cases. We have also experimented with other cases of (n, m), K, and p. The case $p = 3$ proved to be the most accessible one; we include a brief summary of our experiments with this case. Throughout, let $G = \mathrm{G}_n(K)$ and denote by $\mathcal{T}^*(G_j)$ the subtree of $\mathcal{T}(G_j)$ consisting of the groups H in $\mathcal{T}(G_j)$ satisfying the condition that $\gamma_i(H)/\gamma_{i+1}(H) \cong \gamma_i(G)/\gamma_{i+1}(G)$ holds whenever $\gamma_i(H) \neq 1$.

For $(n, m) = (3, 1)$, there is strong evidence that $\mathcal{T}^*(G_{16})$ is virtually periodic with period 6 and defect 1.

For $(n, m) = (2, 3)$, as in the last subsection, we investigated "shaved" versions of $\mathcal{T}^*(G_j)$ consisting only of those groups H in $\mathcal{T}^*(G_j)$ such that $|H| \leq |G_i|p^d$ for some fixed d. For the six ramified non-Galois extensions of degree 3 over $\mathbb{Q}_3$ we obtained that these "shaved" branches seem to be virtually periodic with period 2; the number d we used varied subject to $3 \leq d \leq 6$.

Similarly, for $(n, m) = (2, 4)$, we also investigated "shaved" branches of $\mathcal{T}^*(G_j)$ only. For the two totally ramified extensions of degree 4 of $\mathbb{Q}_3$ we obtained that the resulting "shaved" branches seem to be virtually periodic with period 4.

References

[1] H. U. Besche, B. Eick, and E. A. O'Brien, *A millenium project: constructing small groups*, Internat. J. Algebra Comput. **12** (2002), 623–644.

[2] J. D. Dixon, M. P. F. du Sautoy, A. Mann, and D. Segal, *Analytic pro-p groups*, second edition, Cambridge Studies in Advanced Mathematics, vol. 61, Cambridge University Press, 1999.

[3] M. du Sautoy, *Counting p-groups and nilpotent groups*, Inst. Hautes Etudes Sci. Publ. Math. **92** (2001), 63–112.

[4] B. Eick and C. R. Leedham-Green, *On the classification of prime-power groups by coclass*, Bull. Lond. Math. Soc. **40** (2008), 274–288.

[5] B. Eick and E. A. O'Brien, *Enumerating p-groups*, J. Austral. Math. Soc. **67** (1999), 191–205.

[6] B. Eick and T. Rossmann, *FWtree – Computing trees related to some pro-p-groups of finite width*, available at `http://www.gap-system.org/Packages/Contrib/contrib.html`.

[7] G. Higman, *Enumerating p-groups. II: Problems whose solution is PORC*, Proc. London Math. Soc. **10** (1960), 566–582.

[8] B. Huppert, *Endliche Gruppen I*, Grundlehren der mathematischen Wissenschaften, vol. 134, Springer, 1967.

[9] J. W. Jones and D. P. Roberts, *A database of local fields*, J. Symbolic Comput. **41** (2006), no. 1, 80–97.

[10] G. Klaas, C. R. Leedham-Green, and W. Plesken, *Linear pro-p-groups of finite width*, Lecture Notes in Mathematics, vol. 1674, Springer-Verlag, 1997.

[11] C. R. Leedham-Green, *The structure of finite p-groups*, J. London Math. Soc. (2) **50** (1994), 49–67.

[12] C. R. Leedham-Green and S. McKay, *On the classification of p-groups and pro-p groups*, New horizons in pro-p groups, Progr. Math., vol. 184, Birkhäuser, 2000, pp. 55–74.

[13] C. R. Leedham-Green and S. McKay, *The structure of groups of prime power order*, London Mathematical Society Monographs, Oxford Science Publications, 2002.

[14] C. R. Leedham-Green and M. F. Newman, *Space groups and groups of prime-power order I*, Archiv der Mathematik **35** (1980), 193–203.

[15] E. A. O'Brien, *The p-group generation algorithm*, J. Symb. Comput. **9** (1990), 677–698.

[16] E. A. O'Brien and M. R. Vaughan-Lee, *The groups with order p^7 for odd prime p*, J. Algebra **292** (2005), 243–258.

[17] E. Sanderson, *On strongly l-powerful groups*, PhD Thesis, University of London, 1993.

[18] A. Shalev, *The structure of finite p-groups: Effective proofs of the coclass conjectures*, Invent. Math. **115** (1994), 315–345.

[19] A. Shalev, *Finite p-groups*, Finite and locally finite groups (Istanbul, 1994), NATO Adv. Sci. Inst. Ser. C Math. Phys. Sci., vol. 471, Kluwer Acad. Publ., 1995, pp. 401–450.

[20] The GAP Group, GAP – *Groups, Algorithms and Programming, Version 4.4* (2005), available at `http://www.gap-system.org`.

INSTITUT COMPUTATIONAL MATHEMATICS, TU BRAUNSCHWEIG, POCKELSSTRASSE 14, 38106 BRAUNSCHWEIG, GERMANY.

E-mail address: `beick@tu-bs.de`

SCHOOL OF MATHEMATICS, STATISTICS AND APPLIED MATHEMATICS, NATIONAL UNIVERSITY OF IRELAND, GALWAY; UNIVERSITY ROAD, GALWAY, IRELAND.

E-mail address: `tobias.rossmann@googlemail.com`

Contemporary Mathematics
Volume **511**, 2010

Computing covers of Lie algebras

Graham Ellis, Hamid Mohammadzadeh, and Hamid Tavallaee

ABSTRACT. We describe an algorithm for computing the Lie cover and multiplier of a finite dimensional Lie algebra over a field. A variant of the algorithm can, in certain cases, be used to compute the Leibniz cover of a Lie algebra. The algorithm has been implemented in GAP and the paper presents some computer examples.

1. Introduction

Let L be a finite dimensional Lie algebra over a field k. A *stem extension* of L is a short exact sequence of Lie algebras

$$0 \to M \to C \stackrel{\phi}{\to} L \to 0$$

in which M lies in both the centre $Z(C)$ and derived subalgebra C^2. A stem extension is called a *Lie cover* when the vector space M has dimension equal to that of the Chevalley-Eilenberg homology $H_2(L, k)$. One purpose of this paper is to describe an algorithm for computing the covering homomorphism ϕ; the algorithm provides a new proof of the known fact [6] that Lie covers exist and are unique up to isomorphism. The algorithm has been implemented as part of the HAP package for the GAP computational algebra system. We illustrate this implementation on a solvable (but non-nilpotent) rational Lie algebra introduced in [8].

We should note that the GAP package SOPHUS already contains an efficient implementation of an alternative algorithm for computing covers of nilpotent Lie algebras over finite fields; this is based on the methods in [7] for covering groups of finite p-groups.

The Lie cover of L can be used to decide if there exists some Lie algebra K whose central quotient $K/Z(K)$ is isomorphic to L. Such a K exists if and only if a certain central ideal $Z^*(L)$, which we term the *epicentre* of L, is trivial. The Lie cover has another useful property, namely that all stem extensions of L arise as the quotient of the cover by an ideal in the multiplier. Proofs of these properties were given in [1]; they also follow directly from our algorithm.

2000 *Mathematics Subject Classification.* 17B55.
Key words and phrases. Lie cover, Lie multiplier, algorithm.
The second author was a visitor at National University of Ireland, Galway, during the course of this work.

A second purpose of the paper is to describe how our method can, in certain cases, be used to compute the *Leibniz cover* of a Lie algebra L. This cover is analogous to the Lie cover except that C is only required to be a Leibniz algebra and the central ideal M is required to have the same dimension as that of the Leibniz homology $HL_2(L,k)$. We illustrate this on the Lie algebra of 4×4 rational matrices.

2. Nonabelian tensor squares

For convenience we shall use the language of nonabelian tensor products of Lie algebras introduced in [2,3], though we specialze to the very particular case of the tensor product of a Lie algebra with itself.

Let L and T be Lie algebras over the field k. A bilinear function $\tau\colon L\times L\to T$ is said to be a *Lie pairing* if

$$\begin{aligned}\tau([x,x'],y) &= \tau(x,[x',y])-\tau(x',[x,y]),\\ \tau(x,[y,y']) &= \tau([y',x],y)-\tau([y,x],y'),\\ \tau([y,x],[x',y']) &= -[\tau(x,y),\tau(x',y')],\end{aligned}$$

for all $x,x',y,y'\in L$. The *nonabelian tensor square* $L\otimes L$ arises as a Lie pairing $\otimes\colon L\times L\to L\otimes L$ and is defined up to isomorphism by the following universal property: for any Lie pairing $\tau\colon L\times L\to T$ there exists a unique Lie homomorphism $\psi\colon L\otimes L\to T$ making the following diagram commute.

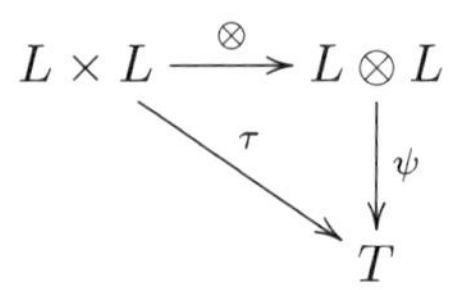

The commutator pairing $L\times L\to L$, $(x,y)\mapsto [x,y]$ induces the homomorphism $\partial\colon L\otimes L\to L$. We denote by $J_2(L)$ the kernel of ∂. One can show [3] that $J_2(L)$ is a central ideal in $L\otimes L$.

Since Lie pairings are bilinear the vector space underlying the nonabelian tensor square is a quotient of the usual tensor product $L\otimes_k L$. The following lemma (whose proof is left to the reader) provides a constructive description of the kernel of this quotient.

LEMMA 1. *Let B be a k-basis for the finite-dimensional Lie algebra L. Let I be the vector subspace of the vector space tensor product $L\otimes_k L$ spanned by the vectors*

$$([x,x']\otimes y)-(x\otimes[x',y])+(x'\otimes[x,y]),$$
$$(x\otimes[y,y'])-([y',x]\otimes y)+([y,x]\otimes y')$$

for $x,x',y,y'\in B$. Let $\mu\colon (L\otimes_k L)\times(L\otimes_k L)\to L\otimes_k L$ be the bilinear map defined on basis elements by $\mu((x\otimes y),(x'\otimes y'))\mapsto -[y,x]\otimes[x',y']$. Then μ induces the structure of a Lie algebra on the quotient vector space $(L\otimes_k L)/I$ and there is a Lie isomorphism $L\otimes L\cong(L\otimes_k L)/I$.

Lemma 1 has been implemented as part of the HAP homological algebra package [4] as a computer function `LieTensorSquare(L)` which inputs a finite-dimensional Lie algebra L represented as a structure constants algebra. It returns the homomorphism $L\otimes L\to L$ as a homomorphism of structure constants algebras; it also returns the Lie pairing function $\otimes\colon L\times L\to L\otimes L$.

A Lie pairing $\tau\colon L \times L \to T$ is said to be an *exterior* Lie pairing if

$$\tau(x,x) = 0$$

for all $x \in L$. The *nonabelian exterior square* $L \wedge L$ is the Lie algebra arising in a universal exterior pairing $\tau\colon L \times L \to L \wedge L$. The commutator pairing

$$\begin{array}{ccc} L \times L & \to & L \\ (x,y) & \mapsto & [x,y] \end{array}$$

again induces a Lie homomorphism $\delta\colon L \wedge L \to L$ whose kernel we denote by $H_2(L)$. This kernel is a central ideal in $L \wedge L$. It was observed in [2] that $H_2(L)$ is isomorphic to the second Chevalley-Eilenberg homology $H_2(L,k)$.

The proof of the following lemma is left to the reader.

LEMMA 2. *Let B be a k-basis for the finite-dimensional Lie algebra L. Let J be the vector subspace of the nonabelian tensor square $L \otimes L$ spanned by the vectors*

$$x \otimes x,$$
$$x \otimes y + y \otimes x = 0$$

for $x, y \in B$. Then J is an ideal and there is a Lie isomorphism $L \wedge L \cong (L \otimes L)/J$.

The computer function `LieExteriorSquare(L)` of [4] implements Lemma 2; the function can be applied to any finite-dimensional Lie algebra L represented as a structure constants algebra.

Example 1. Consider the Lie algebra $L = M_4(\mathbb{Q})$ of 4×4 rational matrices. The following HAP commands show that the second Chevalley-Eilenberg homology $H_2(L,\mathbb{Q})$ is trivial, the second Leibniz homology $HL_2(L,\mathbb{Q})$ has dimension 1, and $J_2(L)$ is isomorphic to $HL_2(L,\mathbb{Q})$. The homology is calculated from implementations in HAP of the standard Chevalley-Eilenberg complex and the standard Leibniz complex. See [5] for a description of these two complexes.

```
gap> L:=MatLieAlgebra(Rationals,4);;
gap> LieAlgebraHomology(L,2);
0
gap> LeibnizAlgebraHomology(L,2);
1
gap> J:=Kernel(LieTensorSquare(L).homomorphism);;
gap> Dimension(J);
1
```

3. A construction of the Lie cover

In a stem extension of Lie algebras

$$0 \to M \to C \stackrel{\phi}{\to} L \to 0 \tag{1}$$

the Lie algebra C is uniquely determined, up to isomorphism, by (i) the derived subalebra C^2, (ii) the restricted homomorphism $\delta = \phi|_{C^2}\colon C^2 \to L$, and (iii) the exterior pairing $\tau\colon L \times L \to C^2$, $(x,y) \mapsto [\tilde{x},\tilde{y}]$, where $\tilde{x} \in C$ denotes an arbitrary lift of $x \in L$. Since M is central the pairing τ is well-defined.

To recover C from the three pieces of data we first construct: a vector space V_{C^2} of dimension $\dim(C^2)$, a vector space $V_{L^{\mathrm{ab}}}$ of dimension $\dim(L^{\mathrm{ab}})$, the direct sum of vector spaces $V = V_{C^2} \oplus V_{L^{\mathrm{ab}}}$, an inclusion of vector spaces $\iota\colon C^2 \to V$ mapping C^2 onto the summand V_{C^2}, and an inclusion of vector spaces $\sigma\colon V_{L^{\mathrm{ab}}} \to L$ whose image has trivial intersection with the derived subalgebra L^2. The vector space V has dimension equal to the dimension of C since $C^{\mathrm{ab}} = L^{\mathrm{ab}}$. There is a linear homomorphism

$$\psi\colon V = V_{C^2} \oplus V_{L^{\mathrm{ab}}} \to L, \quad (x,y) \mapsto \delta(\iota^{-1}(x)) + \sigma(y)$$

and this homomorphism can be composed with the exterior pairing τ and inclusion ι to produce a bilinear bracket operation on V defined by

$$[\ ,\ ]\colon V \times V \to V, \quad (v,w) \mapsto \iota(\tau(\psi(v), \psi(w))).$$

The proof of the following lemma is left to the reader.

LEMMA 3. *The bracket on V is a Lie bracket and the Lie algebra V is Lie isomorphic to C. Modulo this isomorphism the homomorphism ψ corresponds to the stem extension homomorphism ϕ.*

For any finite dimension Lie algebra L the methods of Section 2 can be used to compute the following three pieces of data: (i) the nonabelian exterior square $L \wedge L$, (ii) the Lie homomorphism $\delta\colon L \wedge L \to L$, $(x,y) \mapsto [x,y]$, and (iii) the universal exterior pairing $\wedge\colon L \times L \to L \wedge L$. Since $\ker(\delta) \cong H_2(L,k)$ we can apply the construction of Lemma 3 to this starting data in order to obtain a Lie covering

$$0 \to H_2(L,k) \to L^* \to L \to 0$$

in which the derived subalgebra of L^* is isomorphic to the exterior square $L \wedge L$.

Since the exterior square is uniquely determined by L up to isomorphism we obtain the following.

THEOREM 4. [6] *Any finite dimensional Lie algebra L admits a covering homomorphism $\phi\colon L^* \to L$ and L^* is unique up to Lie isomorphism.*

Theorem 4 is in contrast to the group theoretic situation where finite groups generally admit non-isomorphic covers.

For an arbitrary stem extension (1) the exterior pairing

$$\begin{array}{rcl} \tau\colon L \times L & \longrightarrow & C^2, \\ (x,y) & \longmapsto & [\tilde{x}, \tilde{y}] \end{array}$$

induces a surjection $L \wedge L \to C^2$ whose kernel lies in $\ker(\delta\colon L \wedge L \to L) \cong H_2(L,k)$. We thus obtain the following result.

THEOREM 5. [1] *In any stem extension (1) the Lie algebra C is isomorphic to L^*/I where I is a subspace of the vector space $H_2(L,k)$. Moreover, any subspace $I \leq H_2(L,k)$ gives rise to a stem extension.*

The above construction of the Lie cover has been implemented as a function `LieCoveringHomomorphism(L)` in the HAP package.

Example 2. Let L be a 13-dimensional vector space over $\mathbb{Q}$ with basis

$$\{u, x, y, z, t, e_1, e_2, f_1, f_2, g_1, g_2, h_1, h_2\}.$$

It was observed in [8] that a solvavble Lie algebra structure can be imposed on L by defining a bilinear bracket with

$$
\begin{array}{lll}
[u, a_2] = -a_1, & [u, a_1] = a_2, & \text{for } a = e, f, g, h, \\
[e_1, e_2] = x, & [f_1, f_2] = y, & \\
[e_1, f_2] = -t, & [e_2, f_1] = t, & \\
[x, f_i] = h_i, & [y, e_i] = g_i, & \text{for } i = 1, 2, \\
[t, e_i] = -h_I, & [t, f_i] = -g_i, & \text{for } i = 1, 2, \\
[e_1, g_2] = -\frac{1}{2}z, & [e_2, g_1] = \frac{1}{2}z, & \\
[f_1, h_2] = \frac{1}{2}z, & [f_2, h_1] = -\frac{1}{2}z, & \\
[x, y] = z, & &
\end{array}
$$

and with all other products of generators zero. This Lie algebra can be entered into GAP as a structure constants algebra by using commands

```
gap> SCTL:=EmptySCTable(13,0,"antisymmetric");;
gap> SetEntrySCTable( SCTL, 1, 6, [ 1, 7 ] );;
gap> SetEntrySCTable( SCTL, 1, 8, [ 1, 9 ] );;

    ⋮

gap> L:=LieAlgebraByStructureConstants(Rationals,SCTL);;
```

The GAP commands

```
gap> IsLieSolvable(L);
true
gap> IsLieNilpotent(L);
false
```

show that L is solvable but not nilpotent. The following HAP commands show that the Lie cover L^* is 15-dimensional, has derived length equal to the derived length of L, has trivial Lie multiplier $H_2(L^*, 2) = 0$, and has 6-dimensional second Leibniz homology $HL_2(L^*, k) = k^6$. They also show that L is not isomorphic to a central quotient of Lie algebras $K/Z(K)$.

```
gap> phi:=LieCoveringHomomorphism(L);;
gap> Lstar:=Source(phi);;
gap> Dimension(Lstar);
15
gap> Length(LieDerivedSeries(Lstar))=Length(LieDerivedSeries(L));
true
gap> LieAlgebraHomology(Lstar,2);
0
gap> LeibnizAlgebraHomology(Lstar);
6
gap> LieAlgebraHomology(Lstar,2);
0
gap> Dimension(LieEpicentre(L))>0;
true
```

4. Leibniz algebras

A *Leibniz algebra* is a vector space G over k with a bilinear bracket

$$[\ ,\ ]\colon G \times G \longrightarrow G$$

satisfying the Leibniz identity

$$[x,[y,z]] = [[x,y],z] - [[x,z],y]$$

for all $x, y, z \in G$. Any Lie algebra is a Leibniz algebra. Conversely, a Leibniz algebra is a Lie algebra if $[x,x] = 0$ for all $x \in G$.

The homology $HL_*(G,k)$ of a Leibniz algebra is the homology of the chain complex

$$\cdots \to G^{\otimes n} \xrightarrow{d} G^{\otimes n-1} \to \cdots \to G \to k$$

with boundary map defined by the formula

$$d(x_1 \otimes \cdots \otimes x_n) = \sum_{1 \le i < j \le n} (-1)^j (x_1 \otimes \cdots \otimes x_{i-1} \otimes [x_i, x_j] \otimes x_{i+1} \otimes \cdots \otimes \hat{x}_j \otimes \cdots \otimes x_n).$$

See for example [5].

Given a stem extension $HL_2(G,k) \to G^{\#} \xrightarrow{\phi} G$ of Leibniz algebras we say that $G^{\#}$ is a *Leibniz cover* of G.

Consider now a Lie algebra L. If in our construction of the Lie cover L^* we were to replace the exterior square $L \wedge L$ with the nonabelian tensor square $L \otimes L$ we would get a stem extension

$$0 \to J_2(L) \to C \to L \to 0$$

where C is a Leibniz algebra with derived subalgebra $C^2 \cong L \otimes L$. If $J_2(L)$ happens to equal $HL_2(L,k)$ then C is a Leibniz cover of L.

Example 3. Consider the Lie algebra $L = M_4(\mathbb{Q})$ of 4×4 rational matrices. We saw in Example 1 that $J_2(L) = HL_2(L,k)$. The following HAP commands use the above technique to calculate a Leibniz cover $L^{\#}$ for L and show that $HL_2(L^{\#}, \mathbb{Q}) \cong \mathbb{Q}$.

```
gap> L:=MatLieAlgebra(Rationals,4);;
gap> Lcover:=Source(LeibnizQuasiCoveringHomomorphism(L));;
gap> LeibnizAlgebraHomology(L,1);
1
gap> LeibnizAlgebraHomology(C,2);
1
```

References

[1] P. Batten and E. Stitzinger, *On covers of Lie algebras*, Comm. Algebra **24** (1996), no. 14, 4301–4317.

[2] G. Ellis, *Nonabelian exterior products of Lie algebras and an exact sequence in the homology of Lie algebras*, J. Pure Appl. Algebra **46** (1987), no. 2-3, 111–115.

[3] ———, *A nonabelian tensor product of Lie algebras*, Glasgow Math. J. **33** (1991), no. 1, 101–120.

[4] ———, HAP – *Homological Algebra Programming.* A package for the computer algebra system GAP. `http://www.gap-system.org/Packages/hap.html`.

[5] J.-L. Loday and T. Pirashvili, *Universal enveloping algebras of Leibniz algebras and (co)homology*, Math. Annalen **296** (1993), no. 1, 139–158.

[6] K. Moneyhun, *Isoclinisms in Lie algebras*, Algebras Groups Geom. **11** (1994), no. 1, 9–22.

[7] M. F. Newman, W. Nickel, and A. C. Niemeyer, *Descriptions of groups of prime-power order*, J. Symbolic. Comput. **25** (1998), no. 5, 665–682.
[8] M. Wüstner, *An example of a solvable Lie algebra*, Sem. Sophus Lie **2** (1992), no. 1, 57–58.

MATHEMATICS DEPARTMENT, NATIONAL UNIVERSITY OF IRELAND, GALWAY
E-mail address: graham.ellis@nuigalway.ie

FACULTY OF MATHEMATICAL SCIENCES, IRAN UNIVERSITY OF SCIENCES AND TECHNOLOGY, TEHRAN
E-mail address: h_mohammadzadeh@iust.ac.ir

FACULTY OF MATHEMATICAL SCIENCES, IRAN UNIVERSITY OF SCIENCES AND TECHNOLOGY, TEHRAN

Contemporary Mathematics
Volume **511**, 2010

Enumerating subgroups of the symmetric group

Derek F. Holt

ABSTRACT. We announce our successful computation of a list of representatives of the conjugacy classes of subgroups of S_n for $n \leq 18$, including the 7 274 651 classes of subgroups of S_{18}.

1. Introduction

Early attempts to enumerate complete lists of primitive subgroups and transitive subgroups of the symmetric group S_n for low values of n, up to conjugacy in S_n, began with Ruffini in 1799 and continued until about 1912. We refer the reader to [11] for details and references.

There was little further work on this problem until about 1970 when, with the assistance of computers, Sims [12] compiled a list of primitive permutation groups of degree up to 20. More recently, primitive permutation groups of degree up to 4095 have been enumerated [4,10] as have transitive group of degree up to 32 [6,7]. The lists of groups (currently with primitive groups up to degree 2499 and transitive groups up to degree 31) are available as libraries in GAP [5] and MAGMA [1]. With libraries of this kind containing large numbers of groups, it would be desirable to have some kind of compact storage method that allowed easy reconstruction, but for small-degree permutation groups one cannot do much better than store minimal sized generating sets for each group.

The problem of listing representatives of all conjugacy classes of subgroups of S_n has received relatively little attention. Lists for $n \leq 12$ and also of the subgroups of A_n for $n \leq 13$ are available from the website of Götz Pfeiffer [9], although the subgroups of S_{13} can now be computed routinely in MAGMA in a few hours simply by calling the function `Subgroups(Sym(13))`.

The purpose of this note is to announce the author's enumeration of representatives of the conjugacy classes of S_n for $n \leq 18$. The computations were carried out in MAGMA on a 2.40GHz Intel PC with 4GB of memory. These lists are currently available from the author on request, although it is to be hoped that they will eventually be accessible from GAP and MAGMA.

In the table below, we list the numbers of conjugacy classes of primitive, transitive, and all subgroups of S_n in the first three columns. Note that the number of classes of subgroups of S_n that act fixed-point-freely can be obtained by subtracting

1991 *Mathematics Subject Classification.* Primary 20B35; Secondary 20B40.

Degree	Primitive	Transitive	All (classes)	All (total)
1	1	1	1	1
2	1	1	2	2
3	2	2	4	6
4	2	5	11	30
5	5	5	19	156
6	4	16	56	1455
7	7	7	96	11 300
8	7	50	296	151 221
9	11	34	554	1 694 723
10	9	45	1593	29 594 446
11	8	8	3094	404 126 228
12	6	301	10 723	10 594 925 360
13	9	9	20 832	175 238 308 453
14	4	63	75 154	5 651 774 693 595
15	6	104	159 129	117 053 117 995 400
16	22	1954	686 165	5 320 744 503 742 316
17	10	10	1 466 358	125 889 331 236 297 288
18	4	983	7 274 651	7 598 016 157 515 302 757
19	8	8		
20	4	1117		
21	9	164		
22	4	59		
23	7	7		
24	5	25 000		
25	28	211		
26	7	96		
27	15	2392		
28	14	1854		
29	8	8		
30	4	5712		
31	12	12		
32	7	2 801 324		

the number for S_{n-1} from that of S_n. The final column contains the total number of subgroups of S_n.

It might be possible with a considerable amount of effort and large-scale use of computer power to extend the enumeration to degree 19 or perhaps even to degree 20 in the foreseeable future, but it is doubtful to what extent this would be worthwhile given the very large number of groups that are likely to be involved, and it seems highly unlikely that it could be extended much further than this.

In the following section we describe briefly the methods used to enumerate the subgroups of S_n, and then in the final section we discuss an application that motivated the author to undertake these computations.

2. Methods used

The problem of listing representatives of the conjugacy classes of subgroups of S_n subdivides naturally into a large number of subproblems, each of which can be dealt with independently, so the main problem is highly parallelizable. Since we can assume that the subgroups of S_m are already known for $m < n$, and subgroups of S_n that fix $n-m$ points are conjugate in S_n if and only if they are conjugate in S_m, we can restrict our attention to the fix-point-free subgroups H. The lengths of the orbits of H defines a partition of n in which no component is 1, so we can handle the different partitions individually. The transitive subgroups are already known and available in MAGMA for $n \leq 18$, so we need consider only nontrivial partitions.

Furthermore, for each such partition $n = n_1 + n_2 + \cdots + n_k$ with $k > 1$ and $1 < n_1 \leq n_2 \leq \cdots \leq n_k$, the induced action of H on the i-th orbit is a transitive subgroup of S_{n_i}. So each choice of transitive subgroups H_i of S_{n_i} gives rise to the subproblem of enumerating those H for which the induced action on the i-th orbit is H_i. Of course, if some of the n_i are equal – say $n_j = n_{j+1} = \cdots = n_k$ then, since we are enumerating subgroups up to conjugacy in S_n, we only consider one of the possible orderings of each choice of subgroups H_i for $j \leq i \leq k$.

So we have to solve the following type of subproblem. Given a partition of n, $n = n_1 + n_2 + \cdots + n_k$, and transitive subgroups H_i of S_{n_i} as above, let P be the direct product of the H_i. Then we must find the conjugacy classes in S_n of subdirect products of P; that is, the subgroups of P that project onto each of the factors H_i. The conjugacy test for subgroups can be carried out in an appropriate wreath product of symmetric groups. More precisely, if $H_1 = H_2 = \cdots = H_{i_1}$, $H_{i_1+1} = H_{i_1+2} = \cdots = H_{i_2}$, $\ldots$, $H_{i_{l-1}+1} = \cdots = H_{i_l}$ with no other equalities between the H_i, then the conjugacy test takes place in the direct product of the permutation wreath products $S_{n_{i_j}} \wr S_{m_j}$ for $1 \leq j \leq l$, where $m_j = i_{j+1} - i_j$.

The author tried two methods of finding the subdirect products of P. The more straightforward was to compute maximal subgroups repeatedly using the algorithm described in [3], where we keep only those subgroups that project onto each H_i, and at each stage we remove any subgroups that are conjugate to one that is already on the list. For the second method, we employ the algorithm presented in [2] for finding all subgroups of a permutation group. This involves the initial computation of a series of normal subgroups $1 < P_1 < P_2 < \cdots < P_k < P$ of P in which P_k is the solvable radical of P, and each factor group P_j/P_{j-1} is elementary abelian. We then find the subgroups of P/P_k, $P/P_{k-1}, \ldots, P/P_1$, P successively, where at each stage we remove those subgroups that do not project onto each H_iP_{j-1}/P_{j-1}.

Testing subgroups for conjugacy can be inherently slow in large permutation groups, since all currently known methods involve the use of backtrack searches with potentially exponential complexity. The advantage of the second method is that the lifting process of finding the subgroups of P/P_{j-1} from those of P/P_j involves no explicit conjugacy testing of subgroups. This is replaced by an equivalent orbital computation on the vectors of a vector space over a prime field, which is still theoretically of exponential complexity, but is much faster in practice. So we generally used the first method described above to find the subgroups of the top layer P/P_k and then used the second method for the lifting process.

These techniques proved adequate for finding the subgroups of S_n for $n \leq 17$. The process times in seconds for $n = 13$, 14, 15, 16, and 17 were respectively 105, 653, 1190, 20 234, and 26 640.

Unfortunately, they failed to cope with a few of the partitions of 18 within the memory constraint of 4GB RAM. The problem arose from the direct products P having elementary abelian quotients of order 2^9, which gave rise to inordinately large numbers of subgroups during the lifting process. The difficult cases all involved partitions with $n_1 = 2$, and we devised more specialized techniques for dealing with those.

Consider, for example, the partition $3 \times 2 + 3 \times 4$ of 18. Other partitions were handled in similar fashion, with minor variations. Rather than start with a direct product of six transitive groups, we start with $P = H_1 \times H_2$, where $H_1 \leq S_6$ and $H_2 \leq S_{12}$ are groups from the lists that we have already computed in degrees 6 and 12, and whose orbits lengths form the partitions $2+2+2$ and $4+4+4$ respectively. Since H_1 is elementary abelian, a subdirect product of H_1 and H_2 has the form $\{(h_1, h_2) \mid \phi_1(h_1) = \phi_2(h_2)\}$, where ϕ_1, ϕ_2 are epimorphisms of H_1 and H_2 onto an elementary abelian 2-group of order at most 2^3.

We proceed as follows. Before considering the different possible groups H_1 and H_2, for each e with $0 \leq e \leq 3$, let E be elementary abelian of order 2^e and pre-compute a complete list of subdirect products D of $E \times E$ with $|D| = 2^e$. Then consider each pair H_1, H_2 in turn, and let N_i be the normalizer of H_i in the symmetric group for $i = 1, 2$. Now, for each e with $0 \leq e \leq 3$ and $i = 1, 2$, we find representatives, up to conjugation in N_i, of the normal subgroups K_i of H_i in which H_i/K_i is elementary abelian of order 2^e. For each of our subgroups D in the precomputed list described above, we can now define a subdirect product H of $H_1 \times H_2$ that contains $K_1 \times K_2$ as a subgroup of index 2^e with $H/(K_1 \times K_2) = D$. For fixed K_1 and K_2, we test the resulting list of groups (one for each subgroup D) for conjugacy under the action of $N_{N_1}(K_1) \times N_{N_2}(K_2)$. It is not hard to show that this results in the required list of groups for this partition.

The complete calculation of the 5 808 293 fixed-point-free subgroups of S_{18} using these techniques took approximately four cpu-days.

3. An application

The *support* $s(g)$ of a permutation g is the set of points moved by g. The *minimal degree* $m(G)$ of a permutation group $G \leq S_n$ is defined by:

$$m(G) = \min\{|s(g)| \mid 1 \neq g \in G\}.$$

So, for example, $m(S_n) = 2$ for $n \geq 2$, $m(A_n) = 3$ for $n \geq 3$, and if all elements of G act fixed-point-freely then $m(G) = n$.

Large groups with large minimal degrees are interesting and give rise to potential applications to quantum computing and coding theory. There are, however, a number of results saying that very large groups G (i.e. those whose order is exponential in the degree n) must have small minimal degrees. A recent result of this type, proved in [8], is that if $m \leq \log_2 n$ then $|G| \leq n^{10n/m}$, whereas if $m \geq \log_2 n$ then $|G| \leq 2^{10n}$.

To answer the more specific question, given n and $m \leq n$, what is the largest subgroup G of S_n with $m(G) \geq m$, there appears to be no approach other than to carry out an exhaustive search of all possible G. So the lists described in this paper now enable us to answer this question for $n \leq 18$. We list some examples in which the largest such G is unusually large in the table below. Note that the example $L_2(8){:}3 \times L_2(7){:}2 \leq S_{17}$ is intransitive on its support, which demonstrates that it is not sufficient to restrict our attention to transitive groups.

n	m	Group	Order
8	4	$2^3\!:\!L_3(2)$	1344
9	6	$L_2(8)\!:\!3$	1512
11	8	M_{11}	7920
12	8	M_{12}	95040
16	12	$2^4\!:\!A_7$	40320
17	6	$L_2(8)\!:\!3 \times L_2(7)\!:\!2$	508032

References

[1] Wieb Bosma, John Cannon, and Catherine Playoust, *The* MAGMA *algebra system I: The user language*, J. Symbolic Comput. **24** (1997), 235–265.

[2] J.J. Cannon, B.C. Cox, and D.F. Holt, *Computing the subgroups of a permutation group*, J. Symbolic Comput. **31** (2001), 149–161.

[3] J.J. Cannon and D.F. Holt, *Computing maximal subgroups of finite groups*, J. Symbolic Comput. **37** (2004), 589–609.

[4] H.J. Coutts, M. Quick, and C.M. Roney-Dougal, *The primitive permutation groups of degree less than* 4096. Submitted.

[5] The GAP Group, GAP – *Groups, Algorithms, and Programming, Version 4.4.12* (2008), available at http://www.gap-system.org.

[6] J.J. Cannon and D.F. Holt, *The transitive groups of degree 32*, Experimental Mathematics **17** (2008), 307–314.

[7] A. Hulpke, *Constructing transitive permutation groups*, J. Symbolic Comput. **39** (2005), 1–30.

[8] J. Kempe, L. Pyber, and A. Shalev, *Permutation groups, minimal degrees and quantum computing*, Groups, Geometry, and Dynamics **1** (2007), 553–584.

[9] G. Pfeiffer, available at http://schmidt.nuigalway.ie/subgroups/.

[10] C.M. Roney-Dougal, *The primitive groups of degree less than* 2500, J. Algebra **292** (2005), 154–183.

[11] M.W. Short, *The Primitive Soluble Permutation Groups of Degree Less than* 256, Lecture Notes in Mathematics, vol. 1519, Springer-Verlag, 1992.

[12] Charles C. Sims, *Computational methods in the study of permutation groups*, Computational problems in abstract algebra (J. Leech, ed.), Pergamon Press, Oxford, 1970, pp. 169–183.

MATHEMATICS INSTITUTE, UNIVERSITY OF WARWICK, COVENTRY CV4 7AL, UK
E-mail address: D.F.Holt@warwick.ac.uk

Contemporary Mathematics
Volume **511**, 2010

Weight Five Basic Commutators as Relators

David A. Jackson, Anthony M. Gaglione, and Dennis Spellman

ABSTRACT. Charles Sims has asked whether or not, for a free group F, the lower central subgroup $\gamma_n(F)$ coincides with the normal closure in F of the set of basic commutators of weight n. In this paper, we will show that Sims' question has a positive answer when $n = 5$. The longest part of the proof is showing that this is true when F has rank 3. Once this is known, the results for higher ranks follow fairly quickly.

The authors dedicate this paper to the memory of Hermann V. Waldinger.

CONTENTS

2000 *Mathematics Subject Classification.* 20F05, 20F12, 20F18.
Key words and phrases. basic commutators, nilpotent groups, Sims' question.
The first author thanks the Mathematics Department at the University of Illinois for its hospitality while much of this research was done.
The research of the second author was partially supported by the Naval Academy Research Council.

1. Introduction

We write y^x for $x^{-1}yx$ and $[y,x]$ for the commutator $y^{-1}x^{-1}yx$. If A and B are subsets of any group G, then A^G is the normal closure of A in G and $[A,B]$ is the subgroup of G that is generated by the set of elements $[a,b]$ where $a \in A$ and $b \in B$. We write $[z,x,y]$ for the commutator $[[z,x],y]$.

Following standard notation, we will write $[b, {}_1a]$ for $[b,a]$ as the base step for the inductive definition $[b, {}_ma] = [b, {}_{m-1}a, a]$ for $m \geq 2$. Also standard, $[d,c\,;b,a]$ is an abbreviation for $[[d,c],[b,a]]$ and more generally $[C\,;a_1,a_2,\ldots,a_k]$ abbreviates $[C,[a_1,a_2,\ldots,a_k]]$.

Let F be a free group of rank $r > 1$ on the ordered alphabet $\{x_1,x_2,\ldots,x_r\}$. Let $\mathfrak{C} = x_1,x_2,\ldots,x_r,c_{r+1},c_{r+2},\ldots$ be a basic sequence of commutators which begins with the ordered alphabet X. See, for example, [1, 2, 4–6] for definitions of basic commutators and their weights. Charles Sims [6] has asked whether or not $\gamma_n(F) = (\mathfrak{R}_n)^F$ where $\gamma_n(F)$ is the nth term in the lower central series for F and $\mathfrak{R}_n$ is the set of basic commutators of weight n from $\mathfrak{C}$. In [6], Sims showed, using computer implementations of string rewriting algorithms, that this question has a positive answer when $n = 4$ and for the case $n = 5$, $r = 2$. The principal result of this paper is that $\gamma_5(F) = (\mathfrak{R}_5)^F$.

We repeat here three results from the paper Basic Commutators as Relators, [4].

Basic Lemma . *Suppose that $r > 1$ and that F is the free group on an ordered alphabet $X = \{x_1,x_2,\ldots x_r\}$. Let $\mathfrak{C}$ be any basic sequence of commutators that begins with the ordered alphabet X. Suppose that $n > 1$ and $\gamma_{n-1}(F) = (\mathfrak{R}_{n-1})^F$. If $[y,x] \in (\mathfrak{R}_n)^F$ whenever $y \in \mathfrak{R}_{n-1}$ and $x \in X$, then $\gamma_n(F) = (\mathfrak{R}_n)^F$.*

PROOF. Clearly, $(\mathfrak{R}_n)^F \subseteq \gamma_n(F)$, so we need to show that $\gamma_n(F) \subseteq (\mathfrak{R}_n)^F$. Since $\gamma_{n-1}(F) = (\mathfrak{R}_{n-1})^F$, we have $\gamma_n(F) = [\gamma_{n-1}(F),F] = [(\mathfrak{R}_{n-1})^F,F] = [\mathfrak{R}_{n-1},F]^F$ which is contained in $(\mathfrak{R}_n)^F$ since $[y,x] \in (\mathfrak{R}_n)^F$ whenever $y \in \mathfrak{R}_{n-1}$ and $x \in X$. □

The reader should appreciate that the Basic Lemma provides an obvious divide-and-conquer stategy for proving our principal result. As noted, see [4,6], it is known that $\gamma_4(F) = (\mathfrak{R}_4)^F$. Hence, we will show that Sims' question also has a positive answer in weight five, i.e., $\gamma_5(F) = (\mathfrak{R}_5)^F$, if we show that $[y,x] \in (\mathfrak{R}_5)^F$ whenever y is a basic commutator of weight four and x is a generator from X. This is easily done when $r = 2$. See [4]. Generalizing this to the case $r = 3$ is substantially more difficult and comprises the heart, Section 3, of this paper.

In Appendix A, we list all of the commutators $[y,x]$ on the alphabet $\{a,b,c\}$ where y is basic of weight four and x is a generator. We note where these are shown to be trivial. When F is free on $\{a,b,c\}$, the most challenging cases for showing $[y,x]$ to be trivial are $[c,a\,;b,a\,;c]$, $[c,b\,;c,a\,;c]$ and $[c,b\,;c,a\,;c]$.

Our proof leads us to consider weight 6 basic commutators having patterns [3,3], [2,2,2] and [4,2]. In Appendix B, we list basic commutators on the alphabet $\{a,b,c\}$ which have pattern [3,3] or pattern [2,2,2]. We are also forced naturally to consider nonbasic commutators $[z,x]$ of weight 4 where z is a simple weight 3 commutator and x is a generator. We have included a reference list for transformations of these in Appendix C, which also includes a list of instances of Lemma 2.3.

Perhaps surprisingly, the case $r = 4$ follows with relative ease from the case $r = 3$ and the case $r \geq 5$ follows even more easily from the case $r = 4$. The largest obstacle here is to efficiently deal with the large number of commutators which occur. By using subscripted variables, $\{x_1, x_2, x_3, x_4, x_5\}$ rather than the alphabet $\{a, b, c, d, e\}$, we can state and prove numerous instances of a needed result. Some of the results that we need for the $r = 5$ case also have numerous instances that are needed for the $r = 3$ and $r = 4$ cases. We prove these in Section 2. Many readers will probably find it very useful to list all of the instances of the lemmas in Section 2 that occur for the alphabet $\{x_1 = 1, x_2 = b, x_3 = c\}$ and to retain this list for reading Section 3. The $r = 4$ and $r \geq 5$ cases are treated below in Sections 4 and 5 respectively.

In [3], Havas and Richardson published a wonderfully short proof, due to J.R.J.Groves, that the commutator $[b, a, b, a]$ is in the normal closure of the set of basic commutators of weight 4 on the alphabet $\{a, b\}$. We repeat from [4] our dissection and generalization of Groves' proof. We have expanded Groves' Lemma to include additional conclusions in part (iv).

Groves' Lemma . *Let A, B, and C be elements in any group G.*

(i) $[C, B, A] = [C, B]^{-1}[C, A]^{-1}[B, A, C]^{-1}([C, B][C, A][C, A, B])^{[B,A]}$

(ii) *If $[B, A, C]$, $[C, A, B]$, $[C, A\,; B, A]$, and $[C, B\,; C, A]$ are trivial in G, then $[C, B, A] = [C, B\,; B, A]$.*

(iii) *If $[B, A, C]$, $[C, B, B]$, $[C, A, A]$, $[C, A, B]$, and $[C, A, C]$ are trivial in G, then $[C, B, A]$ and $[C, B\,; B, A]$ are trivial in G.*

(iv.a) *If $[C, B]$, $[C, A, A]$, and $[C, A, B]$ are trivial in G, then so is $[B, A, C]$.*

(iv.b) *If $[C, B]$, $[C, A, A]$, and $[B, A, C]$ are trivial in G, then so is $[C, A, B]$.*

(iv.c) *If $[C, B]$, $[C, A\,; B, A]$, and $[B, A, C]$ are trivial in G, then so is $[C, A, B]$.*

(iv.d) *If $[C, A]$, $[C, B\,; B, A]$, and $[C, B, A]$ are trivial in G, then so is $[B, A, C]$.*

(v) *If $[C, A\,; B, A]$, $[C, B\,; B, A]$, and $[C, B\,; C, A]$ are trivial in G, then when any two of $[C, B, A]$, $[B, A, C]$, and $[C, A, B]$ are trivial in G, the third is also trivial.*

(vi) *If $[C, A, C]$, $[C, B, C]$, $[C, A\,; B, A]$, and $[C, B\,; B, A]$ are trivial in G, then $[C, B, A] = [B, A, C]^{-1}[C, B\,; C, A][C, A, B]^{[B,A]}$.*

PROOF. (i) Note that $BA = AB[B, A]$ and apply the commutator identity $[x, yz] = [x, z][x, y][x, y, z] = [x, z][x, y]^z$ to $[C, BA]$ and $[C, (AB)[B, A]]$. Equate these and solve for $[C, B, A]$.

(ii) With $[B, A, C] = 1$, and $[C, A, B] = 1$, the equation in part (i) simplifies to $[C, B, A] = [C, B]^{-1}[C, A]^{-1}([C, B][C, A])^{[B,A]}$. With $[C, A\,; B, A] = 1$, and $[C, B\,; C, A] = 1$, we can write this as

$$\begin{aligned}[C, B, A] &= [C, B]^{-1}[C, A]^{-1}[C, A][C, B]^{[B,A]} \\ &= [C, B]^{-1}[C, B]^{[B,A]} = [C, B\,; B, A].\end{aligned}$$

(iii) Observe first that with $[C, A, A]$, $[C, A, B]$, and $[C, A, C]$ trivial in G, we do have that $[C, A\,; B, A]$ and $[C, B\,; C, A]$ are trivial in G also, so we have $[C, B, A] = [C, B\,; B, A]$ by part (ii). Write this as $[C, B, A] = [C, B, (B^{-1}A^{-1}B)A]$, and use $[x, yz] = [x, z][x, y]^z$ to rewrite this as $[C, B, A] = [C, B, A][C, B, B^{-1}A^{-1}B]^A$. Multiply this last on the left by $[C, B, A]^{-1}$ and then conclude that $[C, B, B^{-1}A^{-1}B]$ is trivial. It then follows, using the triviality of $[C, B, B]$ that $[C, B, A^{-1}]$ is trivial and hence $[C, B, A]$ is trivial.

(iv.a) Since $[C, B]$ and hence $[C, B, A]$ are trivial, from the first part of Groves' Lemma we have $1 = [C, A]^{-1}[B, A, C]^{-1}([C, A][C, A, B])^{[B,A]}$. Since both $[C, A, A]$

and $[C, A, B]$ are trivial in G, we obtain $1 = ([B, A, C]^{-1})^{[C,A]}$ and the triviality of $[B, A, C]$ follows.

(iv.b) When $[C, B]$ and $[B, A, C]$ are trivial we have from part (i) of Groves' Lemma that $[C, A]^{-1}([C, A][C, A, B])^{[B,A]}$ is trivial. Write $([C, A][C, A, B])^{[B,A]}$ as $[C, A]^{A^{-1}BA}$. Then $1 = [C, A]^{-1}([C, A][C, A, B])^{[B,A]} = [C, A]^{-1}[C, A]^{A^{-1}BA} = [C, A]^{-1}A^{-1}B^{-1}A[C, A]A^{-1}BA$. When $[C, A]$ commutes with A, we can rewrite this as $1 = A^{-1}[C, A]^{-1}B^{-1}[C, A]BA = [C, A, B]^A$ and it follows that $[C, A, B]$ is trivial in G.

(iv.c) As in part (iv.b), we find that $[C, A]^{-1}([C, A][C, A, B])^{[B,A]}$ is trivial, using the triviality of $[C, B]$ and $[B, A, C]$. Since $[C, A; B, A]$ is trivial by hypothesis in this case, we find that $[C, A, B]^{[B,A]}$ and hence $[C, A, B]$ is trivial in G.

(iv.d) With $[C, A] = 1$ and $[C, B, A] = 1$, the equation in part (i) simplifies to $1 = [C, B]^{-1}[B, A, C]^{-1}([C, B])^{[B,A]}$. Since $[C, B; B, A]$ is also assumed to be trivial, this simplifies further to $1 = [C, B]^{-1}[B, A, C]^{-1}[C, B]$ and the conclusion follows.

(v) From part (i), with $[C, B; B, A]$ and $[C, A; B, A]$ trivial, we have $[C, B, A] = [C, B]^{-1}[C, A]^{-1}[B, A, C]^{-1}[C, B][C, A][C, A, B]^{[B,A]}$ If $[B, A, C]$ and $[C, A, B]$ are trivial, then $[C, B, A] = [C, B; C, A]$. If $[C, B, A]$ and $[B, A, C]$ are trivial, then $1 = [C, B; C, A][C, A, B]^{[B,A]}$. If $[C, B, A]$ and $[C, A, B]$ are trivial, then $[B, A, C] = [C, B][C, A][C, B]^{-1}[C, A]^{-1} = [C, B; C, A]^{([C,B][C,A])^{-1}}$. See also Proposition W, below.

(vi) Since $[C, B]$ commutes with both $[B, A]$ and with C, it commutes with $[B, A, C]$. Similarly, $[C, A]$ commutes with $[B, A, C]$. Since $[C, B]^{[B,A]} = [C, B]$ and $[C, A]^{[B,A]} = [C, A]$, the result then follows from the equation of part (i). □

For any group G and elements x, y, and z in G, define $W(x, y, z)$ by

$$W(x, y, z) = [z, y]^{-1}[z, x]^{-1}[y, x]^{-1}[z, y][z, x][y, x].$$

$W(x, y, z)$ can be written in a number of different moderately lengthy forms. We want to be able to make brief reference to this element without immediately committing to any one of the forms.

Proposition W . *If G is any group and x, y, and z are elements of G, then*

$$[z, y, x] = \left([y, x, z]^{-1}\right)^{[z,x][z,y]} W(x, y, z)[z, x, y]^{[y,x]},$$

$$W(x, y, z) = [z, y\,;\, z, x][z, y\,;\, y, x]^{[z,x]}[z, x\,;\, y, x],$$

$$W(x, y, z) = [z, x\,;\, y, x]^{[z,y]}[z, y\,;\, y, x][z, y\,;\, z, x]^{[y,x]}.$$

PROOF. These may be verified by substitution and reduction in the free group on $\{x, y, z\}$. □

2. Preliminary results

The original drafts of this work were concerned only with the case where F has rank 3 and it was convenient to use $\{a, b, c\}$ as the ordered alphabet for F. We will use $\{a, b, c\}$ rather than $\{x_1, x_2, x_3\}$ in Section 3 and we will use the ordered alphabet $\{a, b, c, d\}$ in Section 4. In generalizing the proof to ranks 4 and 5, it became apparent that it was more efficient to present and prove some results using subscripted variables rather than stating and proving multiple instances of the same result on the alphabet $\{a, b, c, d, (e)\}$. In this section, and again in Section 5, we will take F to be free on the ordered alphabet $\{x_i\}_{1\le i\le r}$ to conveniently state and prove results that have multiple instances with a common proof. Here and in Section 5, it is always an implicit hypothesis that subscripts have values between 1 and r.

In the body of this paper and in the appendices, we will write $x \equiv y$ to indicate the equality $x(\mathfrak{N}_n)^F = y(\mathfrak{N}_n)^F$ of cosets in the quotient group $F/(\mathfrak{N}_n)^F$. Similarly, statements about triviality or about commutativity should be regarded as statements about triviality or commutativity in the quotient $F/(\mathfrak{N}_n)^F$. Except for the next lemma n will always be 5.

Lemma 2.1. *Suppose $n \ge 4$, $i < j$, $i \le k$, and $i \le m$. Then*

$$[x_j,{}_{n-3}x_i, x_k, x_m] \equiv 1 \quad \text{and} \quad [x_j,{}_{n-3}x_i, x_k\,;\, x_k, x_m] \equiv 1$$

PROOF. If $k \le m$, then $[x_j,{}_{n-3}x_i, x_k, x_m]$ and $[x_j,{}_{n-3}x_i, x_k, x_k]$ are simple basic commutators of weight n. We may conclude in this case that $[x_j,{}_{n-3}x_i, x_k]$ commutes with $[x_k, x_m]$ also.

For the remaining cases we may assume that $k > m$. We use part (iii) of Groves' Lemma with $C = [x_j,{}_{n-3}x_i]$, $B = x_k$, and $A = x_m$. Then $[B, A, C]^{-1} = [x_k, x_m, [x_j,{}_{n-3}x_i]]^{-1} = [x_j,{}_{n-3}x_i\,;\, x_k, x_m]$ which is basic of weight n. Hence $[B, A, C] \equiv 1$. Since $[C, B, B] = [x_j,{}_{n-3}x_i, x_k, x_k]$, this is a simple basic commutator of weight n, as are the commutators $[C, A, A] = [x_j,{}_{n-3}x_i, x_m, x_m]$ and $[C, A, B] = [x_j,{}_{n-3}x_i, x_m, x_k]$. To show that $[C, A, C]$ is trivial, it will suffice to show that $[C, A, x_i]$ and $[C, A, x_j]$ are both trivial. We consider three cases.

Case **(1)** $m = i$. In this case, the commutators $[C, A, x_i] = [x_j,{}_{n-1}x_i]$ and $[C, A, x_j] = [x_j,{}_{n-2}x_i, x_j]$ are both simple basic commutators of weight n.

Case **(2)** $m > i$ and $j \ge m$. In this case, $[C, A, x_i] = [x_j,{}_{n-3}x_i, x_m, x_i]$ is trivial by case (1), while $[C, A, x_j] = [x_j,{}_{n-3}x_i, x_m, x_j]$ is a simple basic commutator of weight n.

Case **(3)** $m > i$ and $j < m$. Here, $[C, A, x_i]$ is again trivial by case (1), while $[C, A, x_j]$ is trivial by case (2) of this lemma applied to $[x_J,{}_{n-3}x_I, x_K, x_M]$ with $J = j$, $I = i$, $K = m$, and $M = j$. □

When $n = 5$ and $r = 3$, the commutators $[b, a, a, b, a]$, $[b, a, a, c, a]$, $[c, a, a, b, a]$, $[c, a, a, c, a]$, and $[c, b, b, c, b]$ are instances of commutators which are trivial by case (1) of the previous lemma. The commutators $[b, a, a, c, b]$ and $[c, a, a, c, b]$ are instances of commutators which are trivial by case (2) of the lemma. With $n = 5$ and $r = 4$, the commutator $[b, a, a, d, c]$ is trivial by case (3).

Lemma 2.2. *Suppose that $[x_j, x_i, x_\ell]$ and $[x_m, x_k, x_k]$ are basic commutators, so that $i < j$, $i \le \ell$, and $k < m$. If either $k \ge \ell$ or $k \le i$, then*

$$[x_m, x_k, x_k\,;\, x_j, x_i, x_\ell] \equiv 1.$$

PROOF. We use part (iv.a) of Groves' Lemma.

In the case where $k \ge \ell$, let $C = [x_j, x_i, x_\ell]$, $B = [x_m, x_k]$, and $A = x_k$. We then have that $[x_m, x_k, x_k\,;\, x_j, x_i, x_\ell] = [B, A, C]$. The hypotheses guarantee that C is a basic commutator of weight 3 and B is a basic commutator of weight 2, so $[C, B]$ is a basic commutator of weight 5 and hence trivial. Since $k \ge \ell$, $[C, A] = [x_j, x_i, x_\ell, x_k]$ is a basic commutator of weight 4. Then $[C, A, x_k] = [x_j, x_i, x_\ell, x_k, x_k]$ and $[C, A, x_m] = [x_j, x_i, x_\ell, x_k, x_m]$ are basic commutators of weight 5, so $[C, A, A]$ and $[C, A, B]$ are trivial as required.

If $k \le i$, let $C = [x_m, x_k, x_k]$, $B = [x_j, x_i]$, and $A = x_\ell$. Again, $[C, B]$ is a basic commutator of weight 5. By the first paragraph, we may assume that $\ell > k$, so that $[C,A]=[x_m,x_k,x_k,x_\ell]$ is a simple basic commutator of weight 4. It is then clear that $[C,A,A]=[x_m,x_k,x_k,x_\ell,x_\ell]$ is a simple basic commutator of weight 5. If $i=\ell$, then $[C,A,x_i]=[C,A,A]$ is a basic commutator of weight 5. Otherwise, $i<\ell$ and $k\le i$ give us $[C,A,x_i]=[x_m,x_k,x_k,x_\ell,x_i]\equiv 1$ by Lemma 2.1. Similarly, if $j\ge \ell$, then $[C,A,x_j]=[x_m,x_k,x_k,x_\ell,x_j]$ is basic, while if $j<\ell$ (and $k\le i<j$), then $[C,A,x_j]$ is trivial by Lemma 2.1. Since $B=[x_j,x_i]$, we then have that $[C,A,B]$ is trivial also. □

With $r=3$, there are 16 distinct worthwhile instances of Lemma 2.2.

Lemma 2.3. $[x_m,x_k\,;x_j,x_i\,;x_\ell]\equiv[x_m,x_k,x_\ell\,;x_j,x_i,x_\ell]$ *if* $i<j$ *and* $k<m$, *and* $\ell\ge\max\{i,k\}$.

PROOF. We use part (i) of Groves' Lemma with $C=[x_m,x_k]$, $B=[x_j,x_i]$, and $A=x_\ell$. Since $[C,A,B]$ and $[B,A,C]$ are basic commutators of weight 5, we have

$$[x_m,x_k\,;x_j,x_i\,;x_\ell]\equiv[x_m,x_k\,;x_j,x_i]^{-1}[x_m,x_k,x_\ell]^{-1}([x_m,x_k\,;x_j,x_i][x_m,x_k,x_\ell])^{[x_j,x_i,x_\ell]}.$$

Then $[x_m,x_k\,;x_j,x_i]$ commutes obviously with $[x_j,x_i,x_\ell]$ and $[x_m,x_k,x_\ell]^{-1}$. □

Lemma 2.4. *Suppose* $i<j$ *and* $i\le k<m$. *Then* $[x_m,x_k\,;x_j,x_i\,;x_k]$ *is trivial.*

PROOF. We obtain $[x_m,x_k\,;x_j,x_i\,;x_k]\equiv[x_m,x_k,x_k\,;x_j,x_i,x_k]$ by setting $\ell=k$ in Lemma 2.3. The latter commutator is trivial by Lemma 2.2. □

Since we are requiring only an arbitrary order on the basic commutators of weight 2, we haven't decided which of $[x_m,x_k\,;x_j,x_i]$ or $[x_j,x_i\,;x_m,x_k]$ is basic. We could equally well state the conclusion of Lemma 2.4 as $[x_j,x_i\,;x_m,x_k\,;x_k]\equiv 1$ if $k\ge i$. There are similar conclusions if we have instead that $k\le i$.

Lemma 2.5. *If* $i<j$ *and* $i<k$ *then*

$$[x_j,x_i,x_k,x_i]\equiv[x_j,x_i\,;x_k,x_i][x_j,x_i,x_i,x_k].$$

PROOF. We use part (vi) of Groves' Lemma with $C=[x_j,x_i]$, $B=x_k$, and $A=x_i$. Then $[C,B,A]=[x_j,x_i,x_k,x_i]$, $[B,A,C]^{-1}=[x_j,x_i\,;x_k,x_i]$, and $[C,A,B]=[x_j,x_i,x_i,x_k]$. Here $[C,B]$ and $[C,A]$ are basic commutators of weight 3 while C and $[B,A]$ are basic commutators of weight 2, so that $[C,B]$ and $[C,A]$ commute with C and $[B,A]$. We have $[C,B,A]\equiv[B,A,C]^{-1}[C,B\,;C,A][C,A,B]^{[B,A]}$. The factor $[C,B\,;C,A]=[x_j,x_i,x_k\,;x_j,x_i,x_i]$ is then trivial by Lemma 2.2. Since $[C,A,B]=[x_j,x_i,x_i,x_k]$, it commutes obviously with $B=x_k$ and it commutes with $A=x_i$ by Lemma 2.1. □

As noted, $[x_j,x_i,x_i,x_k]$ commutes with x_i, x_j, and x_k, so the factors in the conclusion of Lemma 2.5 commute with each other.

With $j=k$, Lemma 2.5 simplifies to $[x_j,x_i,x_j,x_i]\equiv[x_j,x_i,x_i,x_j]$.

Depending on how we choose to order basic commutators of weight 2, we may prefer to write the equivalence in the conclusion of Lemma 2.5 as $[x_j,x_i,x_k,x_i]\equiv[x_k,x_i\,;x_j,x_i]^{-1}[x_j,x_i,x_i,x_k]$ when $j<k$ but to write this as $[x_j,x_i,x_k,x_i]\equiv[x_j,x_i\,;x_k,x_i][x_j,x_i,x_i,x_k]$ when $k<j$.

Also see Lemma 2.8, below, and the comment following Lemma 4.16.

Lemma 2.6. *If $i < j \le k$ and either $m = i$ or $m \ge j$, then*

$$[x_j, x_i, x_j, x_k, x_m] \equiv 1 \quad \text{and} \quad [x_j, x_i, x_j, x_k\,;\, x_k, x_m] \equiv 1$$

PROOF. If $m \ge k$, then $[x_j, x_i, x_j, x_k, x_m]$ is basic of weight 5. It will suffice to prove the lemma in the three cases: **(1)** $m = i$, **(2)** $m = j$, and **(3)** $j < m < k$.

(1) $m = i$. We use part (iii) of Groves' Lemma with $C = [x_j, x_i, x_j]$, $B = x_k$, and $A = x_i$. Then $[B, A, C]^{-1}$ and $[C, B, B]$ are basic commutators of weight 5. By Lemma 2.5, $[x_j, x_i, x_i, x_j] \equiv [x_j, x_i, x_j, x_i] = [C, A]$. Then $[C, A, A] = [C, A, x_i] \equiv [x_j, x_i, x_i, x_j, x_i]$ which is trivial by Lemma 2.1. The commutators $[C, A, x_j]$ and $[C, A, x_k]$ are equivalent to $[x_j, x_i, x_i, x_j, x_j]$ and $[x_j, x_i, x_i, x_j, x_k]$, respectively, which are simple basic commutators of weight 5. We then see that $[C, A, A]$, $[C, A, B]$, and $[C, A, C]$ are trivial.

(2) $m = j$. We use part (iii) of Groves' Lemma with $C = [x_j, x_i, x_j]$, $B = x_k$, and $A = x_j$. If $k > j$, then $[B, A, C]^{-1}$ is again a basic commutator of weight 5, while if $k = j$, then $[B, A]$ and $[B, A, C]$ are already trivial in the free group. Observe also that $[C, B, B]$ is a simple basic commutator of weight 5. In our current case, $[C, A] = [x_j, x_i, x_j, x_j]$, so $[C, A, A] = [C, A, x_j]$ and $[C, A, B] = [C, A, x_k]$ are both simple basic commutators of weight 5. To show that $[C, A, C]$ is trivial also, it suffices to observe that $[C, A, x_i]$ is trivial by the first case of this lemma.

(3) $j < m < k$. We use part (iii) of Groves' Lemma with $C = [x_j, x_i, x_j]$, $B = x_k$, and $A = x_m$. Again, $[B, A, C]^{-1}$ and $[C, B, B]$ are basic commutators of weight 5. This time, $[C, A] = [x_j, x_i, x_j, x_m]$, so $[C, A, A]$ and $[C, A, B]$ are simple basic commutators of weight 5. Then $[C, A, x_i]$ is trivial by the first case of this lemma and $[C, A, x_j]$ is trivial by the second case, so $[C, A, C]$ is trivial also. □

Recall that the word $W(x, y, z)$ has several forms and is defined by $W(x, y, z) = [z, y]^{-1}[z, x]^{-1}[y, x]^{-1}[z, y][z, x][y, x]$.

Lemma 2.7. *Suppose $i < j < k$ and $w = W(x_i, x_j, x_k)$. Then $[x_k, x_j, x_i]$ is equivalent to each of the forms $[x_j, x_i, x_k]^{-1}[x_k, x_i, x_j]w$, $[x_j, x_i, x_k]^{-1}w[x_k, x_i, x_j]$, and $w[x_j, x_i, x_k]^{-1}[x_k, x_i, x_j]$.*

PROOF. Since basic commutators of weight 2 commute with those of weight 3, this is obvious from the first conclusion of Proposition W. □

Lemma 2.8. *If $i < m$ and $i \le j \le k$, then*

$$[x_m, x_i, x_k, x_j] \equiv [x_m, x_i\,;\, x_k, x_j][x_m, x_i, x_k\,;\, x_m, x_i, x_j][x_m, x_i, x_j, x_k]^{[x_k, x_j]}.$$

PROOF. This follows from part (vi) of Groves' Lemma with $C = [x_m, x_i]$, $B = x_k$, and $A = x_j$. Note that $[C, B]$ and $[C, A]$ are basic commutators of weight 3 which commute with the weight 2 basic commutators C and $[B, A]$. □

Lemma 2.9. *If $i < j$ and $i < k$, then $[x_k, x_i, x_j, x_j\,;\, x_k, x_i] \equiv 1$.*

PROOF. We use part (iv.b) of Groves' Lemma with $C = [x_k, x_i, x_j]$, $B = [x_k, x_i]$, and $A = x_j$. Then $[B, A, C]$ is freely trivial while $[C, B]$ and $[C, A, A]$ are basic commutators of weight 5, so $[C, A, B]$ is trivial. □

Lemma 2.10. *If $i < j < m$ and $i < k$, then $[x_k, x_i, x_j, x_m, x_j] \equiv 1$.*

PROOF. We use part (iii) of Groves' Lemma with $C = [x_k, x_i, x_j]$, $B = x_m$, and $A = x_j$. Then $[B, A, C]^{-1}$, $[C, B, B]$, $[C, A, A]$, and $[C, A, B]$ are basic commutators of weight 5. To show that $[C, A, C]$ is trivial, we note that $[C, A]$ commutes with $[x_k, x_i]$ by Lemma 2.9 and obviously with x_j. □

Lemma 2.11. *If $i < j < k$ and $j \leq m \leq k$, then $[x_k, x_j, x_m\,; x_k, x_i, x_j] \equiv 1$.*

PROOF. We use part (iv.a) of Groves' Lemma with $C = [x_k, x_i, x_j]$, $B = [x_k, x_j]$, and $A = x_m$. Then $[C, B]$ and $[C, A, A]$ are basic commutators of weight 5. To see that $[C, A, B]$ is also trivial, we observe that $[C, A, x_k]$ is a basic commutator of weight 5 and that $[C, A, x_j]$ is either a basic commutator of weight 5, when $j = m$, or else is trivial by Lemma 2.10, when $j < m$. □

3. Rank 3

In this section, F will be the free group on the ordered alphabet $\{a, b, c\}$.

Lemma 3.1. *$[b, a, b]$ commutes with $[b, a, c]$, $[c, a, b]$, $[c, a, c]$, and $[c, b, c]$.*

PROOF. To see that $[b, a, b]$ commutes with $[b, a, c]$, we use part (iv.a) of Groves' Lemma with $C = [b, a, b]$, $B = [b, a]$, and $A = c$. Then $[B, A, C] = [b, a, c\,; b, a, b]$. The commutator $[C, B]$ is a basic commutator of weight 5. Observe that $[C, A]$ is a basic commutator of weight 4 and $[C, A, A]$ is a basic commutator of weight 5. Since $B = [b, a]$, it will now suffice to observe that both $[C, A, b]$ and $[C, A, a]$ are trivial by Lemma 2.6.

To prove that $[b, a, b]$ commutes with $[c, a, b]$, modify the proof in the paragraph above by setting $B = [c, a]$ and $A = b$. Here, $[C, A, c]$ is obviously trivial and $[C, A, a]$ is again trivial by Lemma 2.6. To prove that $[b, a, b]$ commutes with $[c, a, c]$, let $B = [c, a]$ and $A = c$. To prove that $[b, a, b]$ commutes with $[c, b, c]$, let $B = [c, b]$ and $A = c$. In both cases, we can still use Lemma 2.6 to show that $[C, A, B]$ is trivial. □

Lemma 3.2. $[c, a\,; b, a\,; b] \equiv 1$.

PROOF. Since $[c, a, b\,; b, a, b] \equiv 1$ by Lemma 3.1, this follows from Lemma 2.3. □

Lemma 3.3. $[b, a, c, b] \equiv [b, a\,; c, b][b, a, b, c]$.

PROOF. This follows from Lemma 2.8 with $m = j$. Observe that the commutator $[b, a, c\,; b, a, b]$ is trivial by Lemma 3.1. By Lemma 2.6, $[b, a, b, c\,; c, b] \equiv 1$. □

Lemma 3.4. $[c, b, b, c\,; b, a] \equiv 1$.

PROOF. We use part (iv.b) of Groves' Lemma with $C = [c, b, c]$, $B = [b, a]$, and $A = b$. Then $[C, B]$ is a basic commutator of weight 5. By Lemma 2.5, we have $[c, b, b, c] \equiv [c, b, c, b] = [C, A]$. Then $[C, A, B] \equiv [c, b, b, c\,; b, a]$ and $[C, A, A] \equiv [c, b, b, c, b]$ which is trivial by Lemma 2.1. By Lemma 3.1, we also have that the commutator $[B, A, C] = [b, a, b\,; c, b, c]$ is trivial. □

Lemma 3.5. $[c, b, b\,; b, a, c] \equiv 1$.

PROOF. We use part (iv.a) of Groves' Lemma with $C = [c, b, b]$, $B = [b, a]$, and $A = c$. The commutators $[C, B]$ and $[C, A, A]$ are basic commutators of weight 5. The commutator $[C, A, B] = [c, b, b, c\,; b, a]$ is trivial by Lemma 3.4. □

Lemma 3.6. $[c, b\,; b, a\,; b, a] \equiv 1$.

PROOF. By Lemma 3.3, we have $[c,b\,;b,a] \equiv [b,a,b,c][b,a,c,b]^{-1}$. Since a and b commute with $[b,a,b,c]$ by Lemma 2.6, it will suffice to prove that $[b,a,c,b]$ commutes with $[b,a]$. For this, we use part (iv.b) of Groves' Lemma with $C=[b,a,c]$, $B=[b,a]$, and $A=b$. We note that $[C,B]$ is a basic commutator of weight 5 and that $[B,A,C]$ is trivial by Lemma 3.1. To see that $[C,A,A]$ is trivial, we use Lemma 3.3 again to write we $[C,A] \equiv [b,a\,;c,b][b,a,b,c]$. The factor $[b,a\,;c,b]$ commutes with $A=b$ by Lemma 2.4 and $[b,a,b,c]$ commutes with $A=b$ by Lemma 2.6. □

Lemma 3.7. $[c,b,b,c\,;c,a] \equiv 1$.

PROOF. We use part (iv.b) of Groves' Lemma with $C=[c,b,c]$, $B=[c,a]$, and $A=b$. Then $[C,B]$ is a basic commutator of weight 5 and $[B,A,C]$ is trivial by Lemma 2.11. By Lemma 2.5, we have $[C,A] \equiv [c,b,b,c]$. From this, $[C,A,A] \equiv [c,b,b,c,b]$ which is trivial by Lemma 2.1 and $[C,A,B] \equiv [c,b,b,c\,;c,a]$. □

Lemma 3.8. $[c,b,b\,;c,a,c] \equiv 1$.

PROOF. We use part (iv.a) of Groves' Lemma with $C=[c,b,b]$, $B=[c,a]$, and $A=c$. Then $[C,B]$ and $[C,A,A]$ are basic commutators of weight 5 and $[C,A,B]$ is trivial by Lemma 3.7. □

Lemma 3.9. $[b,a,c\,;c,a,b] \equiv [c,b\,;b,a\,;c,a]$.

PROOF. We begin by using part (i) of Groves' Lemma with $C=[b,a,c]$, $B=[c,a]$, and $A=b$, while observing that $[C,B]$ and $[C,B,A]$ are trivial. We have

$$1 \equiv [C,A]^{-1}[B,A,C]^{-1}[C,A]^{B[B,A]}.$$

We may rewrite this equivalence as $[B,A,C]^{-1} \equiv [C,A]([C,A]^{-1})^{B[B,A]}$ or

$$[b,a,c\,;c,a,b] \equiv [b,a,c,b]([b,a,c,b]^{-1})^{[c,a][c,a,b]}.$$

By Lemma 3.3, we may write $[b,a,c,b]$ as $[c,b\,;b,a]^{-1}[b,a,b,c]$, so

$$[b,a,c\,;c,a,b] \equiv [c,b\,;b,a]^{-1}[b,a,b,c]([b,a,b,c]^{-1}[c,b\,;b,a])^{[c,a][c,a,b]}.$$

Since $[b,a,b,c]$ commutes with a and b by Lemma 2.6, this simplifies to

$$[c,b\,;b,a]^{-1}[c,b\,;b,a]^{[c,a][c,a,b]} \equiv [b,a,c\,;c,a,b].$$

Then $[c,a,b]$ commutes with $[c,a]$ and with $[c,b\,;b,a]$. □

Lemma 3.10. $[c,a,c,b] \equiv [c,a,c\,;c,a,b][c,a\,;c,b][c,a,b,c]$.

PROOF. This follows from Lemma 2.8 since $[c,a,b,c,b] \equiv 1$ by Lemma 2.10. □

Throughout this section we will routinely abbreviate $W(a,b,c)$ as w.

Lemma 3.11. $[c,b,a\,;c,a,b] \equiv [b,a,c\,;c,a,b]^{-1}$.

PROOF. We use the commutator identities $[xy,z]=[x,z]^y[y,z]$ and $[x^{-1},z] = ([x,z]^{-1})^{x^{-1}}$. We use Lemma 2.7 to substitute for $[c,b,a]$. We have

$$\begin{aligned}[c,b,a\,;c,a,b] &\equiv [w[b,a,c]^{-1}[c,a,b],[c,a,b]] \equiv [[b,a,c]^{-1}[c,a,b],[c,a,b]] \\ &= [[b,a,c]^{-1},[c,a,b]]^{[c,a,b]} = ([b,a,c\,;c,a,b]^{-1})^{[b,a,c]^{-1}[c,a,b]}.\end{aligned}$$

We then use Lemma 3.9 to observe that $[b,a,c\,;c,a,b]$ commutes with basic commutators of weight 3. □

Lemma 3.12. $[c, b\,; b, a\,; c, b] \equiv 1$.

PROOF. By Lemma 3.3, $[c, b\,; b, a] \equiv [b, a, b, c][b, a, c, b]^{-1}$. The factor $[b, a, b, c]$ commutes obviously with c and commutes with b by Lemma 2.6, so it will suffice to show that $[b, a, c, b]$ also commutes with $[c, b]$. For this, we use part (iv.c) of Groves' Lemma with $C = [b, a, c]$, $B = [c, b]$, and $A = b$. We note that $[C, B]$ is a basic commutator of weight 5 and that $[B, A, C]$ is trivial by Lemma 3.5. To see that $[C, A\,; B, A]$ is trivial, we use Lemma 3.3 again to write $[C, A] \equiv [b, a\,; c, b][b, a, b, c]$. Then $[B, A] = [c, b, b]$ has weight 3 and commutes with the weight 2 factors of $[b, a\,; c, b]$. Since $[b, a, b, c]$ commutes obviously with c and commutes with b by Lemma 2.6, it also commutes with $[B, A]$. □

Lemma 3.13. $[c, b\,; c, a\,; c, b] \equiv 1$.

PROOF. By Lemma 3.10, $[c, b\,; c, a] \equiv [c, a, b, c][c, a, c, b]^{-1}[c, a, c\,; c, a, b]$. Since $[c, b]$ commutes with $[c, a, b, c]$ using Lemma 2.10 and commutes obviously with $[c, a, c\,; c, a, b]$, it will suffice to show that $[c, b]$ commutes with $[c, a, c, b]$. For this we use part (iv.c) of Groves' Lemma, with $C = [c, a, c]$, $B = [c, b]$, and $A = b$. Then $[C, B]$ is a basic commutator of weight 5 and $[B, A, C]$ is trivial by Lemma 3.8. To show that $[C, A\,; B, A]$ is trivial, we use Lemma 3.10 again to write $[C, A] = [c, a, c, b] \equiv [c, a, c\,; c, a, b][c, a\,; c, b][c, a, b, c]$. Since $[B, A]$ commutes with $[c, a, c]$ by Lemma 3.8 and with $[c, a, b]$ by Lemma 2.2, $[B, A]$ commutes with the first factor. Since $[B, A]$ is a basic commutator of weight 3, it commutes obviously with $[c, a\,; c, b]$. The third factor, $[c, a, b, c]$, commutes obviously with c and commutes with b by Lemma 2.10, hence with $[B, A]$. □

Lemma 3.14. $[c, a\,; b, a]$ *commutes with* $[b, a]$ *and* $[c, a]$.

PROOF. By Lemma 2.5, we have $[c, a\,; b, a] \equiv [c, a, b, a][c, a, a, b]^{-1}$. Since the commutator $[c, a, a, b]$ commutes with a, b, and c by Lemma 2.1, it will suffice to show that $[c, a, b, a]$ commutes with $[b, a]$ and $[c, a]$. To show that $[c, a, b, a]$ commutes with $[b, a]$, we use part (iv.c) of Groves' Lemma with $C = [c, a, b]$, $B = [b, a]$, and $A = a$. We note that $[C, B]$ is a basic commutator of weight 5 and that $[B, A, C]$ is trivial by Lemma 2.2. By Lemma 2.5, we have $[C, A] \equiv [c, a\,; b, a][c, a, a, b]$. The factor $[c, a\,; b, a]$ is a product of weight 2 basic commutators and their inverses, so it commutes with the weight 3 basic commutator $[B, A]$. The factor $[c, a, a, b]$ commutes with a, b, and c by Lemma 2.1, so it also commutes with $[B, A]$. To prove that $[c, a, b, a]$ commutes with $[c, a]$ replace $B = [b, a]$ by $B = [c, a]$ in this proof. □

Lemma 3.15. $[b, a, c\,; c, a, b\,; a] \equiv 1$.

PROOF. We use part (v) of Groves' Lemma with $C = [b, a, c]$, $B = [c, a, b]$, and $A = a$ to show that $[C, B, A]$ is trivial. By Lemma 2.5, $[C, A]$ is equivalent to the product $[c, a\,; b, a]^{-1}[b, a, a, c]$ and $[B, A]$ is equivalent to the product $[c, a\,; b, a][c, a, a, b]$. For both of these, the first factor commutes obviously with basic commutators of weight 3 (e.g. B and C) and the second factor commutes with a, b and c by Lemma 2.1. We will be done when we show that $[C, A\,; B, A]$, $[C, B\,; B, A]$, and $[C, B\,; C, A]$ are all trivial.

First, $[C, A\,; B, A] \equiv [[c, a\,; b, a]^{-1}[b, a, a, c], [c, a\,; b, a][c, a, a, b]]$. We use Lemma 2.1 again to argue that $[b, a, a, c]$ and $[c, a, a, b]$ commute with all other factors and deduce that $[C, A\,; B, A] \equiv [[c, a\,; b, a]^{-1}, [c, a\,; b, a]]$ which is obviously trivial.

Next, $[C, B; B, A] \equiv [[b, a, c; c, a, b], [c, a; b, a][c, a, a, b]]$. This is trivial since the basic commutators of weight 3 commute with the basic commutators of weight 2, and $[c, a, a, b]$ commutes with all of the occurring variables. The argument for the triviality of $[C, B; C, A]$ is similar. □

Lemma 3.16. $[c, a; b, a; c, b] \equiv [b, a, c; c, a, b]^{-1}$.

PROOF. We begin by using part (i) of Groves' Lemma with $C = [c, a, b]$, $B = [c, b]$, and $A = a$. Since $[C, B]$ is trivial, $[C, A]^{-1}[B, A, C]^{-1}([C, A][C, A, B])^{[B,A]} \equiv 1$. Here, it will be convenient to express this as $[B, A, C] \equiv [C, A]^{B[B,A]}[C, A]^{-1}$. We apply Lemma 3.11 to $[B, A, C]$ and apply Lemma 2.5 to $[C, A]$ to rewrite $[B, A, C] \equiv [C, A]^{B[B,A]}[C, A]^{-1}$ as

$$[b, a, c; c, a, b]^{-1} \equiv ([c, a; b, a][c, a, a, b])^{a^{-1}[c,b]a}[c, a, a, b]^{-1}[c, a; b, a]^{-1}.$$

Since $[c, a, a, b]$ commutes with a, b, and c by Lemma 2.1 and $[c, a; b, a]$ commutes with basic commutators of weight 3, we may simplify to

$$[b, a, c; c, a, b]^{-1} \equiv [c, a; b, a]^{-1}[c, a; b, a]^{a^{-1}[c,b]a}.$$

Conjugate both sides of this equivalence by a^{-1}. The conclusion then follows since a commutes with $[c, a; b, a]$ by Lemma 2.4 and with $[b, a, c; c, a, b]$ by Lemma 3.15. □

Lemma 3.17. $[[c, b; b, a], [c, a; b, a]] \equiv 1$.

PROOF. We use part (iv.d) of Groves' Lemma with $C = [c, b; b, a]$, $B = [c, a]$, and $A = [b, a]$. Observe that $[C, A]$ is trivial by Lemma 3.6. By Lemma 3.9, $[C, B]$ is equivalent to $[b, a, c; c, a, b]$, so $[C, B]$ commutes with all basic commutators of weight 2 and hence with A and $[B, A]$. □

Lemma 3.18. $[[c, b; c, a], [c, a; b, a]] \equiv 1$.

PROOF. We use part (iv.d) of Groves' Lemma with $C = [c, a; b, a]$, $B = [c, b]$, and $A = [c, a]$. Observe that $[C, A]$ is trivial by Lemma 3.14. By Lemma 3.16, $[C, B]$ is equivalent to $[b, a, c; c, a, b]^{-1}$, so $[C, B]$ commutes with all basic commutators of weight 2 and hence with A and $[B, A]$. □

Lemma 3.19. $[[c, b; c, a], [c, b; b, a]] \equiv 1$.

PROOF. We use part (iv.d) of Groves' Lemma with $C = [c, b; b, a]$, $B = [c, a]$, and $A = [c, b]$. Observe that $[C, A]$ is trivial by Lemma 3.12. By Lemma 3.9, $[C, B]$ is equivalent to $[b, a, c; c, a, b]$, so $[C, B]$ commutes with all basic commutators of weight 2 and hence with A and $[B, A]$. Note that $[B, A] = [c, a; c, b]$ here but that $C = [c, b; b, a]$ commutes also with $[B, A]^{-1} = [c, b; c, a]$. □

Lemma 3.20. $[c, b; b, a; a] \equiv 1$.

PROOF. We begin by using part (i) of Groves' Lemma with $C = [b, a, c]$, $B = b$, and $A = a$. Since $[B, A, C]$ is obviously trivial, we may write $[C, A][C, B]^A \equiv ([C, B][C, A]^B)^{[B,A]}$ or $[b, a, c, a][b, a, c, b]^a \equiv ([b, a, c, b][b, a, c, a]^b)^{[b,a]}$. Use Lemma 2.5 to write $[b, a, c, a]$ as $[b, a; c, a][b, a, a, c]$ and use Lemma 3.3 to write $[b, a, c, b]$ as $[b, a; c, b][b, a, b, c]$. By Lemmas 2.1 and 2.6, $[b, a, a, c]$ and $[b, a, b, c]$ commute with a, b, and c, so we obtain $[b, a; c, a][b, a; c, b]^a \equiv ([b, a; c, b][b, a; c, a]^b)^{[b,a]}$ which we invert to obtain $[c, b; b, a]^a[c, a; b, a] \equiv ([c, a; b, a]^b[c, b; b, a])^{[b,a]}$. Then $[c, a; b, a]$

commutes with b and $[b,a]$ by Lemmas 3.2 and 3.14, while $[c,b\,;b,a]$ commutes with $[b,a]$ by Lemma 3.6, so we may simplify again to obtain

$$[c,b\,;b,a]^a[c,a\,;b,a] \equiv [c,a\,;b,a][c,b\,;b,a].$$

Since $[c,a\,;b,a]$ commutes with $[c,b\,;b,a]$ by Lemma 3.17, the conclusion follows. □

Recall that we are using w as an abbreviation for $W(a,b,c)$ in this section.

Lemma 3.21. $[w,[b,a]] \equiv 1$.

PROOF. We show first that $[w,[b,a]] \equiv [c,b,a\,;b,a]$ and then that $[c,b,a\,;b,a] \equiv 1$. By Lemma 2.7, $[c,b,a] \equiv [b,a,c]^{-1}[c,a,b]w$, so

$$\begin{aligned}[c,b,a\,;b,a] &\equiv [[b,a,c]^{-1}[c,a,b]w,[b,a]]\\ &= [[b,a,c]^{-1}[c,a,b],[b,a]]^w[w,[b,a]] \equiv [w,[b,a]].\end{aligned}$$

To show that $[c,b,a\,;b,a]$ is trivial, we use part (v) of Groves' Lemma with $C=[c,b]$, $B=[b,a]$, and $A=a$ to show that $[C,A,B]$ is trivial. Observe that $[B,A,C]$ is a basic commutator of weight 5 and that $[C,B,A]$ is trivial by Lemma 3.20. The commutator $[B,A]=[b,a,a]$ is basic of weight 3, so it commutes obviously with basic commutators of weight 2. Hence $[C,B\,;B,A]$ is trivial.

By Lemma 2.7, we may write $[C,A]=[c,b,a]$ as $[b,a,c]^{-1}[c,a,b]w$ where $w=[c,b]^{-1}[c,a]^{-1}[b,a]^{-1}[c,b][c,a][b,a]$. Then $[B,A]$ commutes obviously with w and commutes with $[b,a,c]$ and $[c,a,b]$ by Lemma 2.2, so $[C,A\,;B,A]$ is trivial.

To show that $[C,B\,;C,A]$ is trivial, we note that $[C,B]=[c,b\,;b,a]$ commutes obviously with $[b,a,c]$ and $[c,a,b]$, so it will suffice to show that $[C,B]$ commutes with w. We have $w=[c,b\,;c,a][c,b\,;b,a][c,b\,;b,a\,;c,a][c,a\,;b,a]$ by the second part of Proposition W. Then $[C,B]$ commutes with the first of these four factors by Lemma 3.19, trivially with the second factor and with the final factor by Lemma 3.17. By Lemma 3.9, we may write the third factor as $[b,a,c\,;c,a,b]$ which commutes obviously with $[C,B]$. □

Lemma 3.22. $[c,b\,;c,a\,;b,a] \equiv 1$.

PROOF. We have $[c,b\,;c,a] \equiv w[c,a\,;b,a]^{-1}[b,a,c\,;c,a,b]^{-1}[c,b\,;b,a]^{-1}$ using the second part of Proposition W and Lemma 3.9. Then $[b,a]$ commutes with w by Lemma 3.21, with $[c,a\,;b,a]$ by Lemma 3.14, obviously with $[b,a,c\,;c,a,b]$ and with $[c,b\,;b,a]$ by Lemma 3.6. □

Lemma 3.23. $[c,a,b,c\,;b,a] \equiv 1$.

PROOF. We begin by using part (i) of Groves' Lemma with $C=[c,a,c]$, $B=[b,a]$, and $A=b$. Since $[C,B]$ and $[C,B,A]$ are obviously trivial and $[B,A,C]$ is trivial by Lemma 3.1 we have $[C,A] \equiv ([C,A]^B)^{[B,A]}$ or $[c,a,c,b] \equiv [c,a,c,b]^{[b,a][b,a,b]}$. We use Lemma 3.10 to substitute $[c,a,c\,;c,a,b][c,b\,;c,a]^{-1}[c,a,b,c]$ for the occurrences of $[c,a,c,b]$ and obtain

$$[c,a,c\,;c,a,b][c,b\,;c,a]^{-1}[c,a,b,c] \equiv ([c,a,c\,;c,a,b][c,b\,;c,a]^{-1}[c,a,b,c])^{[b,a][b,a,b]}.$$

Since $[b,a]$ commutes with $[c,b\,;c,a]$ by Lemma 3.22 and $[b,a,b]$ commutes with the factors $[c,a,c]$ and $[c,a,b]$ of $[c,a,c\,;c,a,b]$ by Lemma 3.1, we simplify first to

$$[c,a,c\,;c,a,b][c,b\,;c,a]^{-1}[c,a,b,c] \equiv [c,a,c\,;c,a,b][c,b\,;c,a]^{-1}[c,a,b,c]^{[b,a][b,a,b]}$$

and then to $[c,a,b,c] \equiv [c,a,b,c]^{b^{-1}[b,a]b}$. Since b commutes with $[c,a,b,c]$ by Lemma 2.10, the conclusion follows. □

Lemma 3.24. $[c,a,b\,;\,b,a,c] \equiv 1$. *As consequences,* $[c,b\,;\,b,a\,;\,c,a] \equiv 1$ *and* $[c,a\,;\,b,a\,;\,c,b] \equiv 1$.

PROOF. We use part (iv.a) of Groves' Lemma with $C = [c,a,b]$, $B = [b,a]$, and $A = c$. Then $[C,B]$ and $[C,A,A]$ are basic commutators of weight 5, while $[C,A,B]$ is trivial by Lemma 3.23. The consequences follow from Lemmas 3.9 and 3.16. □

Recall that we are using w as an abbreviation for $W(a,b,c)$ in this section.

Lemma 3.25. $w \equiv [c,b\,;\,c,a][c,b\,;\,b,a][c,a\,;\,b,a]$, *where the three factors on the right commute with each other.*

PROOF. The form for w follows from the second part of Proposition W, since we now know by Lemma 3.24 that $[c,b\,;\,b,a]$ commutes with $[c,a]$. The commutativity of the three factors is given in Lemmas 3.17, 3.18, and 3.19. □

Lemma 3.26. $[c,a,c\,;\,b,a,c\,;\,c,a,c] \equiv 1$.

PROOF. We use part (i) of Groves' Lemma with $C = [c,a\,;\,b,a]$, $B = [c,a]$, and $A = c$. Then $[C,B]$ is trivial by Lemma 3.14 and $[B,A,C]$ is obviously trivial. Since $[C,A] \equiv [c,a,c\,;\,b,a,c]$ by Lemma 2.3, $[C,A,B]$ is also obviously trivial. We then have $1 \equiv [C,A]^{-1}[C,A]^{[B,A]} \equiv [c,a,c\,;\,b,a,c]^{-1}[c,a,c\,;\,b,a,c]^{[c,a,c]}$. □

Lemma 3.27. $[c,b,c\,;\,b,a,c\,;\,b,a,c] \equiv 1$.

PROOF. The proof is similar to the proof of Lemma 3.26, but here we use $C = [c,b\,;\,b,a]$ and $B = [b,a]$. Here, $[C,B]$ is trivial by Lemma 3.6. □

Lemma 3.28. $[c,b,c\,;\,b,a,c\,;\,c,b,c] \equiv 1$.

PROOF. The proof is similar to the proof of Lemma 3.26, but here we use $C = [c,b\,;\,b,a]$ and $B = [c,b]$. Here, $[C,B]$ is trivial by Lemma 3.12. □

Lemma 3.29. $[c,b,c\,;\,c,a,c\,;\,c,b,c] \equiv 1$.

PROOF. The proof is similar to the proof of Lemma 3.26, but here we use $C = [c,b\,;\,c,a]$ and $B = [c,b]$. Here, $[C,B]$ is trivial by Lemma 3.13. □

Lemma 3.30. $[c,b,c\,;\,c,a,c\,;\,b,a,c] \equiv 1$.

PROOF. The proof is similar to the proof of Lemma 3.26, but here we use $C = [c,b\,;\,c,a]$ and $B = [b,a]$. Here, $[C,B]$ is trivial by Lemma 3.22. □

Lemma 3.31. $[c,a,c\,;\,b,a,c\,;\,b] \equiv 1$.

PROOF. We use part (i) of Groves' Lemma with $C = [c,a\,;\,b,a]$, $B = c$, and $A = b$. Then $[B,A,C]$ is trivial as one of the consequences of Lemma 3.24, while $[C,A]$ is trivial by Lemma 3.2. The equation from part (i) of Groves' Lemma simplifies to $[C,B]^A \equiv [C,B]^{[B,A]}$. Since $[C,B]$ is equivalent to $[c,a,c\,;\,b,a,c]$ by Lemma 2.3, $[C,B]$ commutes with $[B,A]$, and the conclusion follows. □

Lemma 3.32. $[c,b,c\,;\,c,a,c\,;\,b] \equiv 1$.

PROOF. We use part (i) of Groves' Lemma with $C = [c, b\,;\, c, a]$, $B = c$, and $A = b$. Then $[B, A, C]$ is trivial by Lemma 3.13, while $[C, A]$ is trivial by Lemma 2.4. The equation from part (i) of Groves' Lemma simplifies to $[C, B]^A \equiv [C, B]^{[B,A]}$. Since $[C, B]$ is equivalent to $[c, b, c\,;\, c, a, c]$ by Lemma 2.3, $[C, B]$ commutes with $[B, A]$ and the conclusion follows. □

Lemma 3.33. $[c, b, a\,;\, c, a, c] \equiv 1$.

PROOF. We use part (iv.a) of Groves' Lemma with $C = [c, a, c]$, $B = [c, b]$, and $A = a$. Then $[C, B]$ is a basic commutator of weight 5. By Lemma 2.5, we have $[C, A] \equiv [c, a, a, c]$. Since $[c, a, a, c]$ commutes with a and b by Lemma 2.1, $[C, A, A]$ and $[C, A, B]$ are trivial. □

Lemma 3.34. $[c, a, b, c]$ *commutes with* $[c, a\,;\, b, a]$, $[c, b\,;\, b, a]$, $[c, b\,;\, c, a]$, *and* w.

PROOF. First observe that $[c, a\,;\, b, a]$, $[c, b\,;\, b, a]$, and $[c, b\,;\, c, a]$ commute with $[c, a, c, b]$ since these commute obviously with $[c, a, c]$ and they commute with b by Lemmas 3.2 and 2.4.

Using Lemma 3.10, we have $[c, a, b, c] \equiv [c, a\,;\, c, b]^{-1}[c, a, c\,;\, c, a, b]^{-1}[c, a, c, b]$. Then $[c, a\,;\, b, a]$, $[c, b\,;\, b, a]$, and $[c, b\,;\, c, a]$ commute with the third factor by the above paragraph and they commute obviously with the second factor. $[c, a\,;\, b, a]$ commutes with the first factor by Lemma 3.18 and $[c, b\,;\, b, a]$ commutes with the first factor by Lemma 3.19. By Lemma 3.25, w also commutes with $[c, a, b, c]$. □

Lemma 3.35. $[c, a, c\,;\, b, a, c\,;\, c, a, b] \equiv 1$ *and* $[c, b, c\,;\, c, a, c\,;\, c, a, b] \equiv 1$.

PROOF. Both $[c, a, c\,;\, b, a, c]$ and $[c, b, c\,;\, c, a, c]$ commute obviously with $[c, a]$ and these commute with b by Lemmas 3.31 and 3.32. □

Lemma 3.36. $[c, b, a, b] \equiv [b, a, b, c]^{-1}[c, b\,;\, b, a][c, a, b, b]$.

PROOF. Observe first that w commutes with b, since the three factors of w given in Lemma 3.25 commute with b by Lemmas 2.4 and 3.2. By Lemma 2.7, we have $[c, b, a] \equiv w[b, a, c]^{-1}[c, a, b]$. Using the commutator identity $[xy, z] = [x, z]^y[y, z]$ with $x = w$, $y = [b, a, c]^{-1}[c, a, b]$, and $z = b$, we find $[c, b, a, b] \equiv [[b, a, c]^{-1}[c, a, b], b]$ since $[w, b] \equiv 1$. Using $[xy, z] = [x, z]^y[y, z]$ again with $x = [b, a, c]^{-1}$ and $y = [c, a, b]$ followed by $[[b, a, c]^{-1}, b] = ([b, a, c, b]^{-1})^{[b,a,c]^{-1}}$ we find $[c, b, a, b]$ is equivalent to $([b, a, c, b]^{-1})^{[b,a,c]^{-1}[c,a,b]}[c, a, b, b]$. We use Lemma 3.3 to replace $[b, a, c, b]^{-1}$ by $[b, a, b, c]^{-1}[c, b\,;\, b, a]$ and find $[c, b, a, b] \equiv ([b, a, b, c]^{-1}[c, b; b, a])^{[b,a,c]^{-1}[c,a,b]}[c, a, b, b]$. Then $[c, b\,;\, b, a]$ commutes obviously with the weight 3 exponents. Since $[b, a, b, c]$ commutes with a and with b by Lemma 2.6 and commutes obviously with c, the factor $[b, a, b, c]^{-1}$ also commutes with these exponents. □

Lemma 3.37. $[c, a, b, b, a] \equiv 1$.

PROOF. We use part (iii) of Groves' Lemma with $C = [c, a, b]$, $B = b$, and $A = a$. Then $[B, A, C]^{-1}$ and $[C, B, B]$ are basic commutators of weight 5. By Lemma 2.5, we have $[C, A] \equiv [c, a\,;\, b, a][c, a, a, b]$. The factor $[c, a\,;\, b, a]$ commutes with $A = a$ by Lemma 2.4, with $B = b$ by Lemma 3.2 and obviously with C. The factor $[c, a, a, b]$ commutes with a by Lemma 2.1 and hence with A, B and C. □

Lemma 3.38. $[c, b, b, a] \equiv [b, a, b, c]^{-1}[c, a, b, b][c, b\,;\, b, a]^2$.

PROOF. We use part (i) of Groves' Lemma with $C = [c, b]$, $B = b$, and $A = a$ to see $[c, b, b, a] = [c, b, b]^{-1}[c, b, a]^{-1}[c, b\,; b, a]([c, b, b][c, b, a][c, b, a, b])^{[b,a]}$. We use Lemma 2.7 to substitute above for $[c, b, a]$. Since w commutes with $[b, a]$ by Lemma 3.21, $[c, b, b, a]$ is equivalent to the product

$$[c, b, b]^{-1}w^{-1}[c, a, b]^{-1}[b, a, c][c, b\,; b, a][c, b, b][b, a, c]^{-1}[c, a, b]w[c, b, a, b]^{[b,a]}.$$

Then $[c, b, b]$ commutes obviously with w and $[c, b\,; b, a]$, with $[c, a, b]$ by Lemma 2.2, and with $[b, a, c]$ by Lemma 3.5. Since $[c, b\,; b, a]$ commutes obviously with $[b, a, c]$ and $[c, a, b]$ and is one of the three commuting factors of w in Lemma 3.25, we have $[c, b, b, a] \equiv [c, b\,; b, a][c, b, a, b]^{[b,a]}$. Next, we use Lemma 3.36 to replace $[c, b, a, b]$ by $[b, a, b, c]^{-1}[c, b\,; b, a][c, a, b, b]$ to obtain

$$[c, b, b, a] \equiv [c, b\,; b, a]([b, a, b, c]^{-1}[c, b\,; b, a][c, a, b, b])^{[b,a]}.$$

Then $[b, a, b, c]$ commutes with a and b by Lemma 2.6, $[c, a, b, b]$ commutes with a by Lemma 3.37 and $[c, b\,; b, a]$ commutes with $[b, a]$ by Lemma 3.6, so we may remove the exponent $[b, a]$ and permute the factors. □

Lemma 3.39. $[c, b, b, b, a] \equiv 1$.

PROOF. We use part (iii) of Groves' Lemma with $C = [c, b, b]$, $B = b$, and $A = a$. Then $[C, B, B]$ and $[B, A, C]^{-1}$ are basic commutators of weight 5. We need to show that $[C, A]$ commutes with A, B, and C.

By Lemma 3.38, we have $[C, A] \equiv [b, a, b, c]^{-1}[c, a, b, b][c, b\,; b, a]^2$. The first factor commutes with a and b by Lemma 2.6 and the second factor commutes with a by Lemma 3.37. Finally, $[c, b\,; b, a]$ commutes with $A = a$ by Lemma 3.20, with $B = b$ by Lemma 2.4 and obviously with C. □

Lemma 3.40. *$[b, a, c, c]$ commutes with $[b, a, c]$ and $[c, a, b]$.*

PROOF. $[b, a, c, c]$ commutes with $[b, a]$ by Lemma 2.9 and obviously with c, so it commutes with $[b, a, c]$.

To see that $[b, a, c, c]$ commutes with $[c, a, b]$, we use part (iv.a) of Groves' Lemma with $C = [c, a, b]$, $B = [b, a, c]$, and $A = c$. Then $[C, B]$ is trivial by Lemma 3.24 and $[C, A, A]$ is a basic commutator of weight 5. Since $[C, A]$ commutes with $[b, a]$ by Lemma 3.23 and obviously with c, $[C, A, B]$ is also trivial. □

Lemma 3.41. $[c, a, c\,; b, a, c] \equiv [c, a, c\,; c, a, b]$.

PROOF. Using Lemmas 2.7 and 3.24, we have $[b, a, c] \equiv [c, b, a]^{-1}w[c, a, b]$. Then $[c, a, c]$ commutes with $[c, b, a]$ by Lemma 3.33 and obviously with w, so we have

$$[c, a, c\,; b, a, c] \equiv [[c, a, c], ([c, b, a]^{-1}w)[c, a, b]] \equiv [c, a, c\,; c, a, b].$$

□

Lemma 3.42. $[c, a, b, c, a] \equiv [c, a, b, c\,; c, a][c, a, c\,; b, a, c]$.

PROOF. We begin by using part (i) of Groves' Lemma with $C = [c, a, b]$, $B = c$, and $A = a$. Since $[B, A, C]^{-1}$ is a basic commutator of weight 5, we obtain

$$[c, a, b, c, a] \equiv [c, a, b, c]^{-1}[c, a, b, a]^{-1}([c, a, b, c][c, a, b, a][c, a, b, a, c])^{[c,a]}.$$

By Lemma 2.5, we have $[c, a, b, a] \equiv [c, a\,; b, a][c, a, a, b]$, where $[c, a, a, b]$ commutes with a, b, and c by Lemma 2.1. Using the obvious triviality of $[c, a, a, b, c]$, we have

$[c,a,b,a,c] \equiv [c,a\,;b,a\,;c]$ and this last commutator is equivalent to $[c,a,c\,;b,a,c]$ by Lemma 2.3. We substitute for $[c,a,b,a]$ and $[c,a,b,a,c]$ above to obtain

$$[c,a,b,c,a] \equiv [c,a,b,c]^{-1}[c,a,a,b]^{-1}[c,a\,;b,a]^{-1} \\ ([c,a,b,c][c,a\,;b,a][c,a,a,b][c,a,c\,;b,a,c])^{[c,a]}.$$

Since $[c,a,a,b]$ commutes with a, b, and c by Lemma 2.1, we may cancel the factors $[c,a,a,b]$ above. The parenthesized factors $[c,a,b,c]$ and $[c,a\,;b,a]$ on the second line above commute by Lemma 3.34, and then $[c,a\,;b,a]$ commutes with $[c,a]$ by Lemma 3.14. Obviously, $[c,a,c\,;b,a,c]$ commutes with $[c,a]$, so we may simplify the display above to $[c,a,b,c,a] \equiv [c,a,b,c]^{-1}[c,a,b,c]^{[c,a]}[c,a,c\,;b,a,c]$. □

Lemma 3.43. $[c,a,c\,;c,a,b\,;c,a,b,c] \equiv 1$.

PROOF. By Lemma 3.41, $[c,a,c\,;c,a,b] \equiv [c,a,c\,;b,a,c]$, so it will suffice to show that $[c,a,c\,;b,a,c\,;c,a,b,c] \equiv 1$. To do this, we begin by using part (i) of Groves' Lemma with $C=[c,a,b]$, $B=[c,a\,;b,a]$, and $A=c$. Then $[C,B]$ is obviously trivial, while $[C,A,B]$ is trivial by Lemma 3.34. Since $[B,A]$ is equivalent to $[c,a,c\,;b,a,c]$ by Lemma 2.3, $[B,A,C]$ is trivial by Lemma 3.35. The equation from part (i) of Groves' Lemma reduces to $1 \equiv [C,A]^{-1}[C,A]^{[B,A]}$. We again use that $[B,A]$ is equivalent to $[c,a,c\,;b,a,c]$ by Lemma 2.3. □

Lemma 3.44. $[c,a,c,b\,;c,a,b] \equiv 1$.

PROOF. We use part (v) of Groves' Lemma with $C=[c,a,b]$, $B=[c,a,c]$, and $A=b$ to show that $[B,A,C]$ is trivial. Since $[C,A]$ commutes obviously with b and c and commutes with a by Lemma 3.37 we have that $[C,A,B]$, $[C,A\,;B,A]$, and $[C,A\,;C,B]$ are trivial. Since $[C,B]^{-1}$ is equivalent to $[c,a,c\,;b,a,c]$ by Lemma 3.41, $[C,B]$ commutes with $A=b$ by Lemma 3.31 and commutes with B by Lemma 3.26. □

Lemma 3.45. $[c,b\,;c,a\,;c,a] \equiv [c,a,c\,;b,a,c][c,a,b,c\,;c,a]$, *where the two factors on the right commute.*

PROOF. We begin by using part (i) of Groves' Lemma with $C=[c,a,c]$, $B=[c,a]$, and $A=b$. Since $[C,B]$ is a basic commutator of weight 5, we obtain $[C,A] \equiv [B,A,C]^{-1}[C,A]^{B[B,A]}$ or $[c,a,c,b] \equiv [c,a,c\,;c,a,b][c,a,c,b]^{[c,a][c,a,b]}$. Since $[c,a,b]$ commutes obviously with $[c,a]$ and commutes with $[c,a,c,b]$ by Lemma 3.44, we can simplify this equivalence to $[c,a,c,b] \equiv [c,a,c\,;c,a,b][c,a,c,b]^{[c,a]}$. We use Lemma 3.10 to substitute $[c,a,c\,;c,a,b][c,a\,;c,b][c,a,b,c]$ for $[c,a,c,b]$ and then divide on the left by $[c,a,c\,;c,a,b]$ to obtain

$$[c,a\,;c,b][c,a,b,c] \equiv ([c,a,c\,;c,a,b][c,a\,;c,b][c,a,b,c])^{[c,a]}.$$

Next, $[c,a,b,c]$ commutes with $[c,b\,;c,a]$ by Lemma 3.34 and $[c,a,c\,;c,a,b]$ commutes obviously with $[c,a]$ so we can rewrite the equivalence above as

$$[c,a,b,c][c,b\,;c,a]^{-1} \equiv [c,a,c\,;c,a,b][c,a,b,c]^{[c,a]}([c,b\,;c,a]^{-1})^{[c,a]}.$$

Now multiply both sides of this equivalence on the left by $[c,a,b,c]^{-1}$ and on the right by $[c,b\,;c,a]^{[c,a]}$ to obtain

$$[c,b\,;c,a]^{-1}[c,b\,;c,a]^{[c,a]} \equiv [c,a,b,c]^{-1}[c,a,c\,;c,a,b][c,a,b,c]^{[c,a]}.$$

We then obtain $[c,b\,;c,a\,;c,a] \equiv [c,a,c\,;c,a,b][c,a,b,c\,;c,a]$ since $[c,a,c\,;c,a,b]$ commutes with $[c,a,b,c]$ by Lemma 3.43. The two factors on the right commute since $[c, a, c\,;\, c, a, b]$ commutes with commutes with $[c, a, b, c]$ by Lemma 3.43 and obviously with $[c, a]$. By Lemma 3.41, the factor $[c, a, c\,;\, c, a, b]$ is equivalent to $[c, a, c\,;\, b, a, c]$ in either position. □

Lemma 3.46. $[c, b\,;\, c, a\,;\, c, a\,;\, c] \equiv [c, a, b, c\,;\, c, a]^{-1}$.

PROOF. We begin by using part (i) of Groves' Lemma with $C = [c, a, b, c]$, $B = c$, and $A = a$. Since $[C, B]$ is a basic commutator of weight 5, we obtain $[C, A] \equiv [B, A, C]^{-1}[C, A]^{B[B,A]}$ or

$$[c, a, b, c, a] \equiv [c, a, b, c\,;\, c, a][c, a, b, c, a]^{c[c,a]}.$$

By Lemma 3.42, $[c, a, b, c, a]$ is equivalent to $[c, a, b, c\,;\, c, a][c, a, c\,;\, b, a, c]$. Using both Lemmas 3.42 and 3.45, $[c, a, b, c, a]$ is also equivalent to $[c, b\,;\, c, a\,;\, c, a]$. Substitute the product $[c, a, b, c\,;\, c, a][c, a, c\,;\, b, a, c]$ for $[c, a, b, c, a]$ on the left side above and substitute $[c, b\,;\, c, a\,;\, c, a]$ for $[c, a, b, c, a]$ on the right side above to obtain

$$[c, a, b, c\,;\, c, a][c, a, c\,;\, b, a, c] \equiv [c, a, b, c\,;\, c, a][c, b\,;\, c, a\,;\, c, a]^{c[c,a]}.$$

Divide both sides of this equivalence on the left by $[c, a, b, c\,;\, c, a]$ and then conjugate both sides of the result by $[c, a]^{-1}$ to obtain $[c, a, c\,;\, b, a, c] \equiv [c, b\,;\, c, a\,;\, c, a]^{c}$. We use Lemma 3.45 to replace $[c, a, c\,;\, b, a, c]$ by $[c, b\,;\, c, a\,;\, c, a][c, a, b, c\,;\, c, a]^{-1}$ and the conclusion follows. □

Lemma 3.47. $[c, a, b, c\,;\, c, a\,;\, c, a, c] \equiv 1$.

PROOF. We have $[c, a, b, c\,;\, c, a] \equiv [c, a, c\,;\, b, a, c]^{-1}[c, b\,;\, c, a\,;\, c, a]$ by Lemma 3.45. The conclusion follows since $[c, a, c]$ commutes obviously with $[c, b\,;\, c, a\,;\, c, a]$ and commutes with $[c, a, c\,;\, b, a, c]$ by Lemma 3.26. □

Lemma 3.48. $[c, a, b, c\,;\, c, a, c] \equiv [c, a, c\,;\, b, a, c][c, a, b, c\,;\, c, a]^{-1}$.

PROOF. We use part (i) of Groves' Lemma with $C = [c, a, b]$, $B = [c, a]$, and $A = c$. Since $[C, B]$ is a basic commutator of weight 5, we have

$$1 \equiv [c, a, b, c]^{-1}[c, a, b\,;\, c, a, c][c, a, b, c]^{[c,a][c,a,c]}.$$

Then $[c, a, b\,;\, c, a, c]$ commutes with $[c, a, b, c]$ by Lemma 3.43 and is equivalent to $[c, a, c\,;\, b, a, c]^{-1}$ by Lemma 3.41, so we rewrite and calculate

$$\begin{aligned}[c, a, c\,;\, b, a, c] &\equiv [c, a, b, c]^{-1}[c, a, b, c]^{[c,a][c,a,c]} = [[c, a, b, c], [c, a][c, a, c]] \\ &= [c, a, b, c\,;\, c, a, c][c, a, b, c\,;\, c, a]^{[c,a,c]}.\end{aligned}$$

We are done since $[c, a, c]$ commutes with $[c, a, b, c\,;\, c, a]$ by Lemma 3.47. □

Lemma 3.49. $[c, a, b, c\,;\, c, a\,;\, c] \equiv [c, a, c\,;\, b, a, c][c, a, b, c\,;\, c, a]^{-1}$.

PROOF. We use part (i) of Groves' Lemma with $C = [c, a, b, c]$, $B = [c, a]$, and $A = c$. Then $[C, A]$ is a basic commutator of weight 5 and we have

$$[c, a, b, c\,;\, c, a\,;\, c] \equiv [c, a, b, c\,;\, c, a]^{-1}[c, a, b, c\,;\, c, a, c][c, a, b, c\,;\, c, a]^{[c,a,c]}.$$

Since $[c, a, b, c\,;\, c, a]$ commutes with $[c, a, c]$ by Lemma 3.47 and $[c, a, b, c\,;\, c, a, c]$ is equivalent to $[c, a, c\,;\, b, a, c][c, a, b, c\,;\, c, a]^{-1}$ by Lemma 3.48, we have

$$\begin{aligned}[c, a, b, c\,;\, c, a\,;\, c] &\equiv [c, a, b, c\,;\, c, a]^{-1}([c, a, c\,;\, b, a, c][c, a, b, c\,;\, c, a]^{-1})[c, a, b, c\,;\, c, a] \\ &= [c, a, b, c\,;\, c, a]^{-1}[c, a, c\,;\, b, a, c].\end{aligned}$$

We are done since $[c, a, c\,;\, b, a, c]$ commutes with $[c, a, b, c\,;\, c, a]$ by Lemma 3.45. □

Theorem 3.50. $[c, a, c\,; b, a, c] \equiv 1$.

PROOF. It will suffice to show that $[c, a, c\,; b, a, c\,; c] \equiv [c, a, c\,; b, a, c]^{-1}$, since from $[c, a, c\,; b, a, c]^{-1}c^{-1}[c, a, c\,; b, a, c]c \equiv [c, a, c\,; b, a, c]^{-1}$ we first obtain the triviality of $[c, a, c\,; b, a, c]^c$ and then the triviality of $[c, a, c\,; b, a, c]$.

We will use the equivalence $[c, a, c\,; b, a, c] \equiv [c, b\,; c, a\,; c, a][c, a, b, c\,; c, a]^{-1}$ from Lemma 3.45 to compute $[c, a, c\,; b, a, c\,; c]$.

$$\begin{aligned}[c, a, c\,; b, a, c\,; c] &\equiv [[c, b\,; c, a\,; c, a][c, a, b, c\,; c, a]^{-1}, c] \\ &= [c, b\,; c, a\,; c, a\,; c]^{[c,a,b,c\,;c,a]^{-1}}[[c, a, b, c\,; c, a]^{-1}, c] \\ &= ([c, b\,; c, a\,; c, a\,; c][c, a, b, c\,; c, a\,; c]^{-1})^{[c,a,b,c\,;c,a]^{-1}}.\end{aligned}$$

We may substitute $[c, a, b, c\,; c, a]^{-1}$ for $[c, b\,; c, a\,; c, a\,; c]$ using Lemma 3.46 and substitute $[c, a, c\,; b, a, c][c, a, b, c\,; c, a]^{-1}$ for $[c, a, b, c\,; c, a\,; c]$ using Lemma 3.49 to obtain

$$\begin{aligned}[c, a, c\,; b, a, c\,; c] &\equiv ([c, a, b, c\,; c, a]^{-1}([c, a, b, c\,; c, a][c, a, c\,; b, a, c]^{-1}))^{[c,a,b,c\,;c,a]^{-1}} \\ &= ([c, a, c\,; b, a, c]^{-1})^{[c,a,b,c\,;c,a]^{-1}}.\end{aligned}$$

We are done since $[c, a, c\,; b, a, c]$ commutes with $[c, a, b, c\,; c, a]$ by Lemma 3.45. □

Corollary 3.51. $[c, a, c\,; c, a, b] \equiv 1$.

PROOF. This follows from Theorem 3.50 and Lemma 3.41. □

Corollary 3.52. $[c, a\,; b, a\,; c] \equiv 1$.

PROOF. This follows from Theorem 3.50 and Lemma 2.3. □

Corollary 3.53. $[b, a, c, c, a] \equiv 1$.

PROOF. We use part (iii) of Groves' Lemma with $C = [b, a, c]$, $B = c$, and $A = a$. Then $[B, A, C]^{-1}$ and $[C, B, B]$ are basic commutators of weight 5. By Lemma 2.5, $[C, A]$ is equivalent to $[b, a\,; c, a][b, a, a, c]$. The factor $[b, a, a, c]$ commutes obviously with c and commutes with a and b by Lemma 2.1, hence it commutes with A, B, and C. The factor $[b, a\,; c, a]$ is also $[c, a\,; b, a]^{-1}$. It commutes obviously with C, with $A = a$ by Lemma 2.4, and with $B = c$ by Corollary 3.52. □

Corollary 3.54. $[c, a, b, c, a] \equiv 1$ *and* $[c, a, b, c\,; c, a] \equiv 1$.

PROOF. We use part (iii) of Groves' Lemma with $C = [c, a, b]$, $B = c$, and $A = a$. Then $[B, A, C]^{-1}$ and $[C, B, B]$ are basic commutators of weight 5. By Lemma 2.5, $[C, A]$ is equivalent to $[c, a\,; b, a][c, a, a, b]$. The factor $[c, a, a, b]$ commutes obviously with b and c and commutes with a by Lemma 2.1, hence it commutes with A, B, and C. The factor $[c, a\,; b, a]$ commutes obviously with C, with $A = a$ by Lemma 2.4 and with $B = c$ by Corollary 3.52. □

Corollary 3.55. $[c, b\,; c, a\,; c, a] \equiv 1$.

PROOF. This follows from Theorem 3.50, Corollary 3.54, and Lemma 3.45. □

Corollary 3.56. $[c, b\,; c, a\,; a] \equiv 1$.

PROOF. We use part (v) of Groves' Lemma with $C = [c, b]$, $B = [c, a]$, and $A = a$ to show that $[C, B, A]$ is trivial. Observe that $[B, A, C]$ is a basic commutator of weight 5. Since $[C, B] = [c, b\,;\, c, a]$ and $[B, A] = [c, a, a]$, we also have that $[C, B\,;\, B, A]$ is trivial. We need yet to show that $[C, A, B]$, $[C, A\,;\, B, A]$, and $[C, A\,;\, C, B]^{-1}$ are trivial. For these, we use Lemmas 2.7 and 3.25 to write $[C, A] = [c, b, a]$ as $[b, a, c]^{-1}[c, a, b][c, b\,;\, c, a][c, b\,;\, b, a][c, a\,;\, b, a]$. Then $B = [c, a]$ commutes obviously with $[b, a, c]$ and $[c, a, b]$, commutes with $[c, b\,;\, c, a]$ by Corollary 3.55, commutes with $[c, b\,;\, b, a]$ by Lemma 3.24 and commutes with $[c, a\,;\, b, a]$ by Lemma 3.14. Similarly, $[B, A] = [c, a, a]$ commutes with $[b, a, c]$ and $[c, a, b]$ by Lemma 2.2 and commutes obviously with the other three factors of $[C, A]$. Finally, $[C, B] = [c, b\,;\, c, a]$ commutes obviously with $[b, a, c]$ and $[c, a, b]$ and is one of the three commuting factors of $w = [c, b\,;\, c, a][c, b\,;\, b, a][c, a\,;\, b, a]$ by Lemma 3.25. □

Lemma 3.57. $[w, c] \equiv [c, b, c\,;\, b, a, c][c, b, c\,;\, c, a, c]$.

PROOF. Since $w \equiv [c, a\,;\, b, a]([c, b\,;\, b, a][c, b\,;\, c, a])$ by Lemma 3.25 and $[c, a\,;\, b, a]$ commutes with c by Corollary 3.52, we have $[w, c] \equiv [[c, b\,;\, b, a][c, b\,;\, c, a], c]$ using the commutator identity $[uv, z] = [u, z]^v[v, z]$. Using this again and Lemma 2.3, we have $[w, c] \equiv [c, b, c\,;\, b, a, c]^{[c,b\,;\,c,a]}[c, b, c\,;\, c, a, c]$. □

Lemma 3.58. $[b, a, c, c, w] \equiv 1$ *and* $[[b, a, c, c], [w, c]] \equiv 1$.

PROOF. The second equivalence follows from the first. We use part (iv.b) of Groves' Lemma with $C = [b, a, c]$, $B = w$, and $A = c$. Since w is a product of weight 2 basic commutators and their inverses, $[C, B]$ is trivial. Here, $[C, A, A]$ is a basic commutator of weight 5. By Lemma 3.57, $[B, A] \equiv [c, b, c\,;\, b, a, c][c, b, c\,;\, c, a, c]$ where the two factors commute with $C = [b, a, c]$ by Lemmas 3.27 and 3.30. □

Lemma 3.59. $[c, b, c, a] \equiv [c, b, c\,;\, c, a, c][c, b\,;\, c, a][b, a, c, c]^{-1}[c, a, b, c]$.

PROOF. As a first step, we show that $[c, b, a, c] \equiv [b, a, c, c]^{-1}[c, a, b, c][w, c]$. By Lemma 2.7, we have $[c, b, a] \equiv [b, a, c]^{-1}[c, a, b]w$, so

$$\begin{aligned}[c, b, a, c] &\equiv [[b, a, c]^{-1}[c, a, b]w, c] = [[b, a, c]^{-1}[c, a, b], c]^w[w, c] \\ &= \left(([b, a, c, c]^{-1})^{[b,a,c]^{-1}[c,a,b]}[c, a, b, c]\right)^w [w, c].\end{aligned}$$

Then $[b, a, c, c]$ commutes with $[b, a, c]$ and $[c, a, b]$ by Lemma 3.40, so we may simplify the above to $[c, b, a, c] \equiv ([b, a, c, c]^{-1})^w[c, a, b, c]^w[w, c]$. Since w commutes with $[b, a, c, c]$ by Lemma 3.58 and with $[c, a, b, c]$ by Lemma 3.34, we obtain $[c, b, a, c] \equiv [b, a, c, c]^{-1}[c, a, b, c][w, c]$.

We use part (i) of Groves' Lemma with $C = [c, b]$, $B = c$, and $A = a$ to obtain

$$[c, b, c, a] = [c, b, c]^{-1}[c, b, a]^{-1}[c, b\,;\, c, a]([c, b, c][c, b, a][c, b, a, c])^{[c,a]}.$$

In this equation, we use the first paragraph to substitute $[b, a, c, c]^{-1}[c, a, b, c][w, c]$ for $[c, b, a, c]$ and we use Lemma 2.7 to substitute $[b, a, c]^{-1}[c, a, b]w$ for $[c, b, a]$. We obtain

$$\begin{aligned}[c, b, c, a] \equiv{}& [c, b, c]^{-1}w^{-1}[c, a, b]^{-1}[b, a, c][c, b\,;\, c, a] \\ & ([c, b, c][b, a, c]^{-1}[c, a, b]w[b, a, c, c]^{-1}[c, a, b, c][w, c])^{[c,a]}.\end{aligned}$$

Using the form given for $[w, c]$ in Lemma 3.57, $[c, a]$ commutes obviously with $[w, c]$ and with the weight 3 commutators. Since c commutes obviously with $[b, a, c, c]$ and $[c, a, b, c]$ and a commutes with these by Corollaries 3.53 and 3.54, $[c, a]$ commutes with these also. By Lemma 3.25, $w \equiv [c, b\,;\, c, a][c, b\,;\, b, a][c, a\,;\, b, a]$ and then $[c, a]$ commutes with these three factors of w by Corollary 3.55, Lemma 3.24, and Lemma 3.14. Next w commutes with its factor $[c, b\,;\, c, a]$ by Lemma 3.25 and obviously with the weight 3 basic commutators. Also $[c, a, b]$ commutes obviously with $[c, b\,;\, c, a]$ and commutes with $[c, b, c]$ and $[b, a, c]$ by Lemmas 2.11 and 3.24. Since $[c, b\,;\, c, a]$ commutes with $[c, b, c]$ and $[b, a, c]$, we may simplify the last displayed equivalence to obtain

$$[c, b, c, a] \equiv [c, b, c]^{-1}[b, a, c][c, b, c][b, a, c]^{-1}[c, b\,;\, c, a][b, a, c, c]^{-1}[c, a, b, c][w, c].$$

Since $[b, a, c]^{-1}$ commutes with $[c, b, c\,;\, b, a, c]$ by Lemma 3.27, the initial product of four commutators on the right above is equivalent to $[b, a, c\,;\, c, b, c]$. The factor $[w, c]$ above commutes with $[c, a, b, c]$ since $[c, a, b, c]$ commutes obviously with c, commutes with b by Lemma 2.10 and commutes with a by Corollary 3.54. $[w, c]$ commutes with $[b, a, c, c]$ by Lemma 3.58. Using Lemma 3.57 again, write $[w, c]$ as $[c, b, c\,;\, b, a, c][c, b, c\,;\, c, a, c]$ which commutes obviously with $[c, b\,;\, c, a]$ so we obtain

$$[c, b, c, a] \equiv [b, a, c\,;\, c, b, c][c, b, c\,;\, b, a, c][c, b, c\,;\, c, a, c][c, b\,;\, c, a][b, a, c, c]^{-1}[c, a, b, c].$$

The first two factors on the right are inverses, so the conclusion follows. □

Lemma 3.60. $[b, a, c, c\,;\, c, b] \equiv [b, a, c\,;\, c, b, c]$.

PROOF. As a preliminary step, we show that $[c, b, c\,;\, c, b, a] \equiv [b, a, c\,;\, c, b, c]$. Using Lemmas 2.7 and 3.24 we have $[c, b, a] \equiv (w[c, a, b])[b, a, c]^{-1}$. Since $[c, b, c]$ commutes obviously with w and commutes with $[c, a, b]$ by Lemma 2.11, we compute

$$[c, b, c\,;\, c, b, a] \equiv [[c, b, c], (w[c, a, b])[b, a, c]^{-1}] \equiv ([c, b, c\,;\, b, a, c]^{-1})^{[b,a,c]^{-1}}.$$

We then know that $[c, b, c\,;\, b, a, c]$ commutes with $[b, a, c]$ by Lemma 3.27.

We continue by using part (i) of Groves' Lemma with $C = [c, b, c]$, $B = [c, b]$, and $A = a$. Since $[C, B]$ is a basic commutator of weight 5, we obtain

$$[c, b, c, a] \equiv [c, b, c\,;\, c, b, a]([c, b, c, a]^{[c,b]})^{[c,b,a]}.$$

We use the first paragraph to substitute $[b, a, c\,;\, c, b, c]$ for $[c, b, c\,;\, c, b, a]$ above and we use Lemma 3.59 to substitute for $[c, b, c, a]$. We obtain

$$\begin{aligned}&[c, b, c; c, a, c][c, b\,;\, c, a][b, a, c, c]^{-1}[c, a, b, c]\\ &\qquad \equiv [b, a, c\,;\, c, b, c]([c, b, c\,;\, c, a, c][c, b\,;\, c, a][b, a, c, c]^{-1}[c, a, b, c])^{[c,b][c,b,a]}.\end{aligned}$$

Then $[c, a, b, c]$ commutes obviously with c, commutes with b by Lemma 2.10 and commutes with a by Corollary 3.54, so we may simplify the above to obtain

$$\begin{aligned}&[c, b, c\,;\, c, a, c][c, b\,;\, c, a][b, a, c, c]^{-1}\\ &\qquad \equiv [b, a, c\,;\, c, b, c]([c, b, c\,;\, c, a, c][c, b\,;\, c, a][b, a, c, c]^{-1})^{[c,b]^a}.\end{aligned}$$

The exponent $[c, b]^a$ above commutes with $[c, b\,;\, c, a]$ since $[c, b\,;\, c, a]$ commutes with a by Corollary 3.56 and commutes with $[c, b]$ by Lemma 3.13. To see that $[c, b]^a$ also commutes with $[c, b, c\,;\, c, a, c]$, write the exponent as $[c, b]w[b, a, c]^{-1}[c, a, b]$ using Lemma 2.7. $[c, b, c\,;\, c, a, c]$ commutes obviously with $[c, b]$ and w, commutes with $[b, a, c]$ by Lemma 3.30 and commutes with $[c, a, b]$ by Lemma 3.35. The factor $[c, b, c\,;\, c, a, c]$ commutes with $[c, b, c]$ and $[b, a, c]$ by Lemmas 3.29 and 3.30, so it commutes with $[b, a, c\,;\, c, b, c]$. Since $[b, a, c\,;\, c, b, c]$ commutes obviously with $[c, b\,;\, c, a]$, we may further simplify the last displayed equivalence to obtain

$$[b, a, c, c]^{-1} \equiv [b, a, c\,;\, c, b, c]([b, a, c, c]^{-1})^{[c,b]^a}$$

which we rewrite as

$$[b,a,c,c]^{-1}[b,a,c,c]^{[c,b][c,b,a]} \equiv [b,a,c\,;\,c,b,c].$$

Next, we show that $[c,b,a]$ commutes with both $[c,b]$ and $[b,a,c,c]$. For the former, we have $[c,b,a] \equiv [b,a,c]^{-1}[c,a,b]w$ by Lemma 2.7. Since $[c,b]$ commutes obviously with $[b,a,c]$ and $[c,a,b]$, we need to show that $[c,b]$ commutes with w. We use Lemma 3.25 to write $w \equiv [c,b\,;\,c,a][c,b\,;\,b,a][c,a\,;\,b,a]$ and note that $[c,b]$ commutes with the three factors by Lemmas 3.13, 3.12, and 3.24. Recall that $[b,a,c,c]$ commutes with $[c,a,b]$ and $[b,a,c]$ by Lemma 3.40 and commutes with w by Lemma 3.58. This shows that $[c,b,a]$ also commutes with $[b,a,c,c]$.

Now the exponent $[c,b][c,b,a]$ on the left side of the last display is equivalent to $[c,b,a][c,b]$, so we may recalculate the left side as

$$[b,a,c,c]^{-1}[b,a,c,c]^{[c,b,a][c,b]} \equiv [b,a,c,c]^{-1}[b,a,c,c]^{[c,b]} \equiv [b,a,c,c\,;\,c,b].$$

□

Lemma 3.61. $[c,b\,;\,b,a\,;\,b,a,c,c] \equiv 1$ *and* $[c,b,c\,;\,b,a,c\,;\,b,a,c,c] \equiv 1$.

PROOF. We use part (iv.a) of Groves' Lemma with $C = [b,a,c,c]$, $B = [b,a]$, and $A = [c,b]$ to show that $[b,a,c,c]$ commutes with the inverse of $[c,b\,;\,b,a]$. By Lemma 2.9, $[C,B]$ is trivial. By Lemma 3.60, $[C,A]$ is equivalent to $[b,a,c\,;\,c,b,c]$ from which it is obvious that $[C,A]$ commutes with both A and B.

Since $[b,a,c,c]$ commutes with $[c,b\,;\,b,a]$ by the paragraph above and commutes obviously with c, it commutes with $[c,b\,;\,b,a\,;\,c]$ which is equivalent to $[c,b,c\,;\,b,a,c]$ by Lemma 2.3. □

Lemma 3.62. $[b,a,c,c\,;\,c,b,c] \equiv [c,b,c\,;\,b,a,c]^2$.

PROOF. We use part (i) of Groves' Lemma with $C = [b,a,c]$, $B = [c,b]$, and $A = c$. Since $[C,B]$ is a basic commutator of weight 5, we obtain

$$1 \equiv [b,a,c,c]^{-1}[c,b,c\,;\,b,a,c]^{-1}([b,a,c,c]^{[c,b]})^{[c,b,c]}.$$

By Lemma 3.61, $[b,a,c,c]$ commutes with $[c,b,c\,;\,b,a,c]$ so we may rewrite the above equivalence as

$$[c,b,c\,;\,b,a,c] \equiv [[b,a,c,c],[c,b][c,b,c]] = [b,a,c,c\,;\,c,b,c][b,a,c,c\,;\,c,b]^{[c,b,c]}.$$

Since $[b,a,c,c\,;\,c,b]$ is equivalent to $[c,b,c\,;\,b,a,c]^{-1}$ by Lemma 3.60, it commutes with $[c,b,c]$ by Lemma 3.28 and the conclusion follows. □

Lemma 3.63. $[b,a,c,c\,;\,c,b,c] \equiv [b,a,c,c\,;\,c,b\,;\,c]$.

PROOF. We use part (i) of Groves' Lemma with $C = [b,a,c,c]$, $B = [c,b]$, and $A = c$. Since $[C,A]$ is a basic commutator of weight 5, we obtain

$$[b,a,c,c\,;\,c,b\,;\,c] \equiv [b,a,c,c;\,;\,b,c]^{-1}[b,a,c,c\,;\,c,b,c][b,a,c,c\,;\,c,b]^{[c,b,c]}.$$

Then $[b,a,c,c\,;\,c,b]$ is equivalent to $[c,b,c\,;\,b,a,c]^{-1}$ by Lemma 3.60 so it commutes with $[c,b,c]$ by Lemma 3.28. Since $[b,a,c,c\,;\,c,b,c]$ is equivalent to $[c,b,c\,;\,b,a,c]^2$ by Lemma 3.62, the conclusion follows. □

Lemma 3.64. $[b,a,c,c,b] \equiv [c,b,c\,;\,b,a,c]^{-2}$.

PROOF. As a preliminary step, we show that $[b,a,c,b,c] \equiv [c,b,c\,;\,b,a,c]^{-1}$. We may substitute $[b,a\,;\,c,b][b,a,b,c]$ for $[b,a,c,b]$ by Lemma 3.3. Observe that $[b,a,b,c]$ commutes obviously with c and commutes with a and b by Lemma 2.6. Using this, when we compute $[b,a,c,b,c]$ we find $[b,a,c,b,c] \equiv [[b,a\,;\,c,b],c] = [[c,b\,;\,b,a]^{-1},c] = ([c,b\,;\,b,a\,;\,c]^{-1})^{[c,b\,;\,b,a]^{-1}}$. By Lemma 2.3, $[c,b\,;\,b,a\,;\,c]$ is equivalent to $[c,b,c\,;\,b,a,c]$ which commutes obviously with the exponent $[c,b\,;\,b,a]$.

We now use part (i) of Groves' Lemma with $C = [b,a,c]$, $B = c$, and $A = b$. Then $[B,A,C]^{-1}$ is a basic commutator of weight 5. Substituting $[c,b,c\,;\,b,a,c]^{-1}$ for $[C,A,B]$, we have

$$[b,a,c,c,b] \equiv [b,a,c,c]^{-1}[b,a,c,b]^{-1}([b,a,c,c][b,a,c,b][c,b,c\,;\,b,a,c]^{-1})^{[c,b]}.$$

By Lemma 3.3, $[b,a,c,b] \equiv [c,b\,;\,b,a]^{-1}[b,a,b,c]$. Since $[b,a,b,c]$ commutes obviously with c and commutes with a and b by Lemma 2.6, we can substitute for $[b,a,c,b]$ and simplify to obtain

$$[b,a,c,c,b] \equiv [b,a,c,c]^{-1}[c,b\,;\,b,a]([b,a,c,c][c,b\,;\,b,a]^{-1}[c,b,c\,;\,b,a,c]^{-1})^{[c,b]}.$$

Then $[c,b\,;\,b,a]$ commutes with $[c,b]$ by Lemma 3.12 and commutes with $[b,a,c,c]$ by Lemma 3.61, while $[c,b,c\,;\,b,a,c]$ commutes obviously with $[c,b]$. We simplify the above to obtain $[b,a,c,c,b] \equiv [b,a,c,c]^{-1}[b,a,c,c]^{[c,b]}[c,b,c\,;\,b,a,c]^{-1}$. The conclusion then follows since $[b,a,c,c\,;\,c,b]$ is equivalent to $[c,b,c\,;\,b,a,c]^{-1}$ by Lemma 3.60. □

Lemma 3.65. $[c,b,b,c,a] \equiv [c,b,c\,;\,b,a,c]^2$.

PROOF. As a preliminary step, we show that $[c,b,b,a,c] \equiv [c,b,c\,;\,b,a,c]^2$. By Lemma 3.38, we have $[c,b,b,a] \equiv [b,a,b,c]^{-1}[c,a,b,b][c,b\,;\,b,a]^2$. Since c commutes obviously with $[b,a,b,c]$ and $[c,a,b,b]$, we have $[c,b,b,a,c] \equiv [[c,b\,;\,b,a]^2,c] = [c,b\,;\,b,a\,;\,c]^{[c,b\,;\,b,a]}[c,b\,;\,b,a\,;\,c]$. By Lemma 2.3, $[c,b\,;\,b,a\,;\,c] \equiv [c,b,c\,;\,b,a,c]$ which commutes obviously with $[c,b\,;\,b,a]$, so $[c,b,b,a,c] \equiv [c,b,c\,;\,b,a,c]^2$.

We now use part (i) of Groves' Lemma with $C = [c,b,b]$, $B = c$, and $A = a$. We use Lemma 3.38 to replace $[C,A] = [c,b,b,a]$ by $[b,a,b,c]^{-1}[c,a,b,b][c,b\,;\,b,a]^2$ and we use the above paragraph to replace $[C,A,B] = [c,b,b,a,c]$ by $[c,b,c\,;\,b,a,c]^2$. Since $[B,A,C]^{-1}$ is a basic commutator of weight 5, we obtain

$$\begin{aligned}[c,b,b,c,a] \equiv{}& [c,b,b,c]^{-1}[c,b\,;\,b,a]^{-2}[c,a,b,b]^{-1}[b,a,b,c]\\ &([c,b,b,c][b,a,b,c]^{-1}[c,a,b,b][c,b\,;\,b,a]^2[c,b,c\,;\,b,a,c]^2)^{[c,a]}.\end{aligned}$$

Since $[b,a,b,c]$ commutes obviously with c and with a and b by Lemma 2.6 while $[c,a,b,b]$ commutes obviously with b and c and with a by Lemma 3.37, we may simplify the above display to obtain

$$[c,b,b,c,a] \equiv [c,b,b,c]^{-1}[c,b\,;\,b,a]^{-2}([c,b,b,c][c,b\,;\,b,a]^2[c,b,c\,;\,b,a,c]^2)^{[c,a]}.$$

We now note that $[c,b,b,c]$ commutes with $[c,b]$ by Lemma 2.1, with $[b,a]$ by Lemma 3.4 and with $[c,a]$ by Lemma 3.7. Since $[c,a]$ commutes with $[c,b\,;\,b,a]$ by Lemma 3.24 and obviously with $[c,b,c\,;\,b,a,c]$, the conclusion follows. □

Lemma 3.66. *For every integer* ν, $([c,b,c\,;\,b,a,c]^{\nu})^c \equiv [c,b,c\,;\,b,a,c]^{-\nu}$.

PROOF. We prove that $[c,b,c\,;\,b,a,c\,;\,c] \equiv [c,b,c\,;\,b,a,c]^{-2}$. Since this may also be written as $[c,b,c\,;\,b,a,c]^{-1}[c,b,c\,;\,b,a,c]^c \equiv [c,b,c\,;\,b,a,c]^{-2}$, the conclusion follows.

We have $[c, b, c\,; b, a, c] \equiv [b, a, c, c\,; c, b]^{-1}$ by Lemma 3.60, so that

$$[c, b, c\,; b, a, c\,; c] \equiv [[b, a, c, c\,; c, b]^{-1}, c] = ([b, a, c, c\,; c, b\,; c]^{-1})^{[b,a,c,c\,;c,b]^{-1}}.$$

Since $[b, a, c, c\,; c, b\,; c]$ is equivalent to $[c, b, c\,; b, a, c]^2$ by Lemmas 3.62 and 3.63, while the exponent $[b, a, c, c\,; c, b]^{-1}$ is equivalent to $[c, b, c\,; b, a, c]$ by Lemma 3.60, we have $[c, b, c\,; b, a, c\,; c] \equiv [c, b, c\,; b, a, c]^{-2}$. □

Theorem 3.67. $[c, b, c\,; b, a, c] \equiv 1$.

PROOF. In the next two paragraphs we will prove that $[c, b, c\,; b, a, c]^3 \equiv 1$ and that $[c, b, c\,; b, a, c]^{-4} \equiv 1$.

To prove that $[c, b, c\,; b, a, c]^3 \equiv 1$, we use part (i) of Groves' Lemma with $C = [b, a, c, c]$, $B = c$, and $A = b$. Then $[C, B]$ is a basic commutator of weight 5 and $[C, A]$ is equivalent to $[c, b, c\,; b, a, c]^{-2}$ by Lemma 3.64, so we obtain

$$1 \equiv [c, b, c\,; b, a, c]^2 [b, a, c, c\,; c, b] \big(([c, b, c\,; b, a, c]^{-2})^c\big)^{[c,b]}.$$

We use Lemma 3.60 to substitute $[c, b, c\,; b, a, c]^{-1}$ for $[b, a, c, c\,; c, b]$ and we use Lemma 3.66 with $\nu = -2$ to substitute $[c, b, c\,; b, a, c]^2$ for $([c, b, c\,; b, a, c]^{-2})^c$. Since $[c, b]$ commutes obviously with $[c, b, c\,; b, a, c]$, we have $[c, b, c\,; b, a, c]^3 \equiv 1$.

To prove that $[c, b, c\,; b, a, c]^{-4} \equiv 1$, we use part (i) of Groves' Lemma with $C = [c, b, b, c]$, $B = c$, and $A = a$. Then $[C, B]$ is a basic commutator of weight 5 and $[B, A, C]^{-1}$ is trivial by Lemma 3.7. Since $[C, A]$ is equivalent to $[c, b, c\,; b, a, c]^2$ by Lemma 3.65, we obtain $1 \equiv [c, b, c\,; b, a, c]^{-2} \big(([c, b, c\,; b, a, c]^2)^c\big)^{[c,a]}$. We are then done since $([c, b, c\,; b, a, c]^2)^c \equiv [c, b, c\,; b, a, c]^{-2}$ by Lemma 3.66 with $\nu = 2$. □

Corollary 3.68. $[c, b\,; b, a\,; c] \equiv 1$.

PROOF. This follows from Lemma 2.3 and Theorem 3.67. □

Corollary 3.69. $[b, a, c, c, b] \equiv 1$.

PROOF. This follows from Lemma 3.64 and Theorem 3.67. □

Corollary 3.70. $[c, b, b, c, a] \equiv 1$.

PROOF. This follows from Lemma 3.65 and Theorem 3.67. □

Eventually, in Theorem 3.87, we will want to show that all basic commutators of weight 4 commute with a, b, and c. In the meantime, we will have several occasions where we will want to claim that $[c, a, b, c]$ and $[b, a, c, c]$ commute with a, b, and c. Rather than repeating more detailed justifications for these, we will state and prove the following corollary and cite this corollary in the proof of Theorem 3.87 and some intervening lemmas.

Corollary 3.71. *If y is a simple basic commutator of weight 4 other than $[c, a, c, c]$ or $[c, b, c, c]$ then y commutes with a, b, and c.*

PROOF. We will dismiss without further comment all of the cases where $[y, a]$, $[y, b]$, or $[y, c]$ is a basic commutator of weight 5.

When y is a simple left-normed basic commutator of weight 4 and all three letters occur in y, then y is one of $[b, a, a, c]$, $[c, a, a, b]$, $[b, a, b, c]$, $[c, a, b, b]$, $[b, a, c, c]$, or $[c, a, b, c]$. The commutators $[b, a, a, c, a]$, $[b, a, a, c, b]$, and $[c, a, a, b, a]$ are trivial by Lemma 2.1 while the commutators $[b, a, b, c, a]$ and $[b, a, b, c, b]$ are trivial by Lemma 2.6. The commutator $[c, a, b, c, b]$ is trivial by Lemma 2.10 and the commutator $[c,a,b,b,a]$ is trivial by Lemma 3.37. The commutators $[b,a,c,c,a]$, $[c,a,b,c,a]$, and $[b,a,c,c,b]$ are trivial by Corollaries 3.53, 3.54, and 3.69, respectively.

Suppose that y is a simple left-normed basic commutator of weight 4 in which only 2 letters occur and that y is not $[c,a,c,c]$ or $[c,b,c,c]$. Then y is one of $[b,a,a,a]$, $[b,a,a,b]$, $[b,a,b,b]$, $[c,a,a,a]$, $[c,a,a,c]$, $[c,b,b,b]$, or $[c,b,b,c]$. The commutators $[b,a,a,b,a]$, $[c,a,a,c,a]$, $[c,a,a,c,b]$, and $[c,b,b,c,b]$ are trivial by Lemma 2.1 while $[b,a,b,b,a]$ is trivial by Lemma 2.6. $[c,b,b,b,a]$ is trivial by Lemma 3.39 and $[c,b,b,c,a]$ is trivial by Corollary 3.70. □

Lemma 3.72. $[c,a,c,b] \equiv [c,a\,;c,b][c,a,b,c]$.

PROOF. This follows from Lemma 3.10 and Corollary 3.51. □

Lemma 3.73. $[c,b,c\,;c,a,c\,;c,a,c] \equiv 1$.

PROOF. This follows the same proof that is used in Lemma 3.26. We use part (i) of Groves' Lemma with $C = [c,b\,;c,a]$, $B = [c,a]$, and $A = c$. Here, $[C,B]$ is trivial by Corollary 3.55. □

Lemma 3.74. *$[c,a,c,c]$ and $[c,b,c,c]$ commute with $[c,b\,;c,a]$ and $[c,b,c\,;c,a,c]$.*

PROOF. We begin by proving that if A, B, and C are elements in any group G, then $[B,A,C]$ and $[C,A\,;B,A]$ are trivial in G if $[C,B]$, $[C,A,B]$, and $[B,A,A]$ are trivial in G. From part (i) of Groves' Lemma with $[C,B]$ and $[C,A,B]$ trivial, we have

$$1 = [C,A]^{-1}[B,A,C]^{-1}[C,A]^{[B,A]}.$$

We see from this that when we prove the triviality of $[B,A,C]$, we will also have the triviality of $[C,A\,;B,A]$. To show the triviality of $[B,A,C]$, we expand the equation above by writing $[C,A]^{[B,A]}$ as $[B,A]^{-1}[C,A][B,A]$, $[C,A]^{-1}$ as $A^{-1}C^{-1}AC$, $[B,A,C]^{-1}$ as $C^{-1}[B,A]^{-1}C[B,A]$, and $[C,A]$ as $C^{-1}A^{-1}CA$. We obtain

$$1 = (A^{-1}C^{-1}AC)(C^{-1}[B,A]^{-1}C[B,A])[B,A]^{-1}(C^{-1}A^{-1}CA)[B,A].$$

We reduce this to obtain $1 = A^{-1}C^{-1}A[B,A]^{-1}A^{-1}CA[B,A]$. Using the hypothesized triviality of $[B,A,A]$, we rewrite this as $1 = A^{-1}C^{-1}[B,A]^{-1}C[B,A]A = ([B,A,C]^{-1})^{A}$ from which the triviality of $[B,A,C]$ follows.

To show that $[c,a,c,c]$ commutes with $[c,b\,;c,a]$ and $[c,b,c\,;c,a,c]$, we use this calculation with $C = [c,b\,;c,a]$, $B = [c,a,c]$, and $A = c$. The triviality of $[C,B]$ and $[B,A,A]$ is obvious. After using Lemma 2.3 to write $[C,A]$ as $[c,b,c\,;c,a,c]$, we know that $[C,A,B]$ is trivial by Lemma 3.73. Then $[c,a,c,c]$ commutes with $[c,b\,;c,a]$ since $[B,A,C]$ is trivial and commutes with $[c,b,c\,;c,a,c]$ since $[C,A\,;B,A]$ is trivial. We again use Lemma 2.3 to write $[C,A]$ as $[c,b,c\,;c,a,c]$.

Similarly, to show that $[c,b,c,c]$ commutes with $[c,b\,;c,a]$ and $[c,b,c\,;c,a,c]$, we use the observation above with $C = [c,b\,;c,a]$, $B = [c,b,c]$, and $A = c$. Here, the triviality of $[C,A,B]$ follows from Lemma 3.29 after again using Lemma 2.3 to write $[C,A]$ as $[c,b,c\,;c,a,c]$. □

Lemma 3.75. $[c,a,c,b,c] \equiv [c,a,c\,;c,b,c]$.

PROOF. We have $[c,a,c,b] \equiv [c,a\,;c,b][c,a,b,c]$ by Lemma 3.72 and $[c,a,b,c]$ commutes with a,b and c by Corollary 3.71. Since $[c,a\,;c,b\,;c] \equiv [c,a,c\,;c,b,c]$ by Lemma 2.3, we obtain $[c,a,c,b,c] \equiv [[c,a\,;c,b][c,a,b,c],c] \equiv [c,a,c\,;c,b,c]$. □

Lemma 3.76. $[c, a, c, c, b] \equiv [c, a, c, c\,;\, c, b][c, a, c\,;\, c, b, c]$.

PROOF. We use part (i) of Groves' Lemma with $C = [c, a, c]$, $B = c$, and $A = b$ and observe that $[B, A, C]^{-1}$ is a basic commutator of weight 5. We obtain

$$[c, a, c, c, b] \equiv [c, a, c, c]^{-1}[c, a, c, b]^{-1}([c, a, c, c][c, a, c, b][c, a, c, b, c])^{[c,b]}.$$

In this, we use Lemma 3.75 to substitute for $[c, a, c, b, c]$. We use Lemma 3.72 to substitute $[c, b\,;\, c, a]^{-1}[c, a, b, c]$ for $[c, a, c, b]$, but we may cancel the occurrences of $[c, a, b, c]$ by Corollary 3.71. We obtain

$$[c, a, c, c, b] \equiv [c, a, c, c]^{-1}[c, b\,;\, c, a]([c, a, c, c][c, b\,;\, c, a]^{-1}[c, a, c\,;\, c, b, c])^{[c,b]}.$$

Since $[c, b]$ commutes obviously with $[c, a, c\,;\, c, b, c]$ and commutes with $[c, b\,;\, c, a]$ by Lemma 3.13 while $[c, a, c, c]$ commutes with $[c, b\,;\, c, a]$ by Lemma 3.74, we may simplify further to obtain the conclusion. □

Lemma 3.77. $[c, b, c, a, c] \equiv [c, b, c\,;\, c, a, c]^{c}$.

PROOF. It will suffice to show that $[c, b, c, a, c] \equiv [c, b, c\,;\, c, a, c][c, b, c\,;\, c, a, c\,;\, c]$. We have $[c, b, c, a] \equiv [c, b, c\,;\, c, a, c][c, b\,;\, c, a][b, a, c, c]^{-1}[c, a, b, c]$ by Lemma 3.59. Since $[c, a, b, c]$ and $[b, a, c, c]$ commute with a, b, and c by Corollary 3.71 and the commutator $[c, b, c\,;\, c, a, c]$ commutes obviously with $[c, b\,;\, c, a]$, we have

$$[c, b, c, a, c] \equiv [[c, b\,;\, c, a][c, b, c\,;\, c, a, c], c] = [c, b\,;\, c, a\,;\, c]^{[c,b,c\,;\,c,a,c]}[[c, b, c\,;\, c, a, c], c].$$

Since $[c, b\,;\, c, a\,;\, c]$ is equivalent to $[c, b, c\,;\, c, a, c]$ by Lemma 2.3, it commutes with this commutator as an exponent and the conclusion follows. □

Lemma 3.78. $[c, b, c, c, a] \equiv [c, b, c, c\,;\, c, a][c, b, c\,;\, c, a, c]^{c[c,a]}$.

PROOF. We begin by using part (i) of Groves' Lemma with $C = [c, b, c]$, $B = c$, and $A = a$. Since $[B, A, C]^{-1}$ is a basic commutator of weight 5, we obtain

$$[c, b, c, c, a] \equiv [c, b, c, c]^{-1}[c, b, c, a]^{-1}([c, b, c, c][c, b, c, a][c, b, c, a, c])^{[c,a]}.$$

We substitute $[c, b, c\,;\, c, a, c]^{c}$ for $[c, b, c, a, c]$ using Lemma 3.77 and substitute the product $[c, b, c\,;\, c, a, c][c, b\,;\, c, a][b, a, c, c]^{-1}[c, a, b, c]$ for $[c, b, c, a]$ using Lemma 3.59. Since $[b, a, c, c]$ and $[c, a, b, c]$ commute with a, b and c by Corollary 3.71, we obtain

$$\begin{aligned}[c, b, c, c, a] \equiv [c, b, c{,}c]^{-1}&[c, b\,;\, c, a]^{-1}[c, b, c\,;\, c, a, c]^{-1}\\ &([c, b, c, c][c, b, c\,;\, c, a, c][c, b\,;\, c, a][c, b, c\,;\, c, a, c]^{c})^{[c,a]}.\end{aligned}$$

Then both $[c, b\,;\, c, a]$ and $[c, b, c\,;\, c, a, c]$ commute with $[c, b, c, c]$ by Lemma 3.74 while $[c, a]$ commutes obviously with $[c, b, c\,;\, c, a, c]$ and commutes with $[c, b\,;\, c, a]$ by Corollary 3.55. We simplify the above display to obtain the conclusion. □

Lemma 3.79. $[c, b, c, c, a] \equiv [c, b, c\,;\, c, a, c]$.

PROOF. By Lemma 3.78, it will suffice to prove that $[c, b, c\,;\, c, a, c]$ is equivalent to $[c, b, c, c\,;\, c, a][c, b, c\,;\, c, a, c]^{c[c,a]}$. We use part (i) of Groves' Lemma with $C = [c, b, c, c]$, $B = c$, and $A = a$. Then $[C, B]$ is a basic commutator of weight 5 and we have $[C, A] \equiv [B, A, C]^{-1}[C, A]^{B[B,A]}$ or $[c, b, c, c, a] \equiv [c, b, c, c\,;\, c, a][c, b, c, c, a]^{c[c,a]}$. We use Lemma 3.78 to substitute for $[c, b, c, c, a]$ in this equivalence and obtain

$$\begin{aligned}&[c, b, c, c\,;c, a][c, b, c\,;\, c, a, c]^{c[c,a]}\\ &\qquad \equiv [c, b, c, c\,;\, c, a]([c, b, c, c\,;\, c, a][c, b, c\,;\, c, a, c]^{c[c,a]})^{c[c,a]}.\end{aligned}$$

We divide both sides of this display on the left by $[c, b, c, c\,;\, c, a]$ and then conjugate both sides of the resulting equivalence by $(c[c, a])^{-1}$ to obtain the conclusion. □

Lemma 3.80. $[c, a, c, c\,;\, c, b]^c \equiv [c, b, c\,;\, c, a, c]$.

PROOF. We use part (i) of Groves' Lemma with $C = [c, a, c]$, $B = [c, b]$, and $A = c$. Since $[C, B]$ is a basic commutator of weight 5, we obtain

$$1 \equiv [c, a, c, c]^{-1}[c, a, c\,;\, c, b, c][c, a, c, c]^{[c,b][c,b,c]}.$$

Since $[c, a, c, c]$ commutes with $[c, a, c\,;\, c, b, c]$ by Lemma 3.74, we have

$$\begin{aligned}[c, b, c\,;\, c, a, c] &\equiv [c, a, c, c]^{-1}[c, a, c, c]^{[c,b]^c} = [c, a, c, c]^{-1}c^{-1}[c, b]^{-1}c[c, a, c, c]c^{-1}[c, b]c \\ &\equiv c^{-1}[c, a, c, c]^{-1}[c, b]^{-1}[c, a, c, c][c, b]c = [c, a, c, c\,;\, c, b]^c.\end{aligned}$$

□

Lemma 3.81. $[c, b, c, c\,;\, c, a] \equiv [c, a, c, c\,;\, c, b]^{-1}$.

PROOF. We will prove that $[c, a, c\,;\, c, b, c] \equiv [c, b, c, c\,;\, c, a]^c$. With Lemma 3.80, we then have $[c, b, c, c\,;\, c, a]^c \equiv ([c, a, c, c\,;\, c, b]^{-1})^c$ and the conclusion follows.

We use part (i) of Groves' Lemma with $C = [c, b, c]$, $B = [c, a]$, and $A = c$. Since $[C, B]$ is a basic commutator of weight 5, we obtain

$$1 \equiv [c, b, c, c]^{-1}[c, b, c\,;\, c, a, c][c, b, c, c]^{[c,a][c,a,c]}.$$

Since $[c, b, c, c]$ commutes with $[c, b, c\,;\, c, a, c]$ by Lemma 3.74, we have

$$\begin{aligned}[c, a, c\,;\, c, b, c] &\equiv [c, b, c, c]^{-1}[c, b, c, c]^{[c,a]^c} = [c, b, c, c]^{-1}c^{-1}[c, a]^{-1}c[c, b, c, c]c^{-1}[c, a]c \\ &\equiv c^{-1}[c, b, c, c]^{-1}[c, a]^{-1}[c, b, c, c][c, a]c = [c, b, c, c\,;\, c, a]^c.\end{aligned}$$

□

Lemma 3.82. $[c, b, c\,;\, c, a, c\,;\, c] \equiv [c, a, c, c\,;\, c, b]^c$.

PROOF. We use part (i) of Groves' Lemma with $C = [c, a, c, c]$, $B = c$, and $A = b$. Then $[C, B]$ is a basic commutator of weight 5 and we obtain

$$[c, a, c, c, b] \equiv [c, a, c, c\,;\, c, b][c, a, c, c, b]^{c[c,b]}.$$

We use Lemma 3.76 to substitute $[c, a, c, c\,;\, c, b][c, a, c\,;\, c, b, c]$ for $[c, a, c, c, b]$ above and then divide on the left by $[c, a, c, c\,;\, c, b]$ to obtain

$$[c, a, c\,;\, c, b, c] \equiv ([c, a, c, c\,;\, c, b][c, a, c\,;\, c, b, c])^{c[c,b]}.$$

We conjugate both sides of this last equivalence using $[c, b]^{-1}$ which commutes obviously with $[c, a, c\,;\, c, b, c]$ and obtain $[c, a, c\,;\, c, b, c] \equiv [c, a, c, c\,;\, c, b]^c[c, a, c\,;\, c, b, c]^c$. The conclusion follows when we divide both sides of this equivalence on the right by $[c, a, c\,;\, c, b, c]^c$ and write $[c, a, c\,;\, c, b, c]$ as $[c, b, c\,;\, c, a, c]^{-1}$. □

Theorem 3.83. $[c, a, c, c, b] \equiv 1$.

PROOF. Using Lemmas 3.76 and 3.81 we have that $[c, a, c, c, b]$ is equivalent to $[c, a, c, c\,;\, c, b][c, a, c\,;\, c, b, c] \equiv [c, b, c, c\,;\, c, a]^{-1}[c, a, c\,;\, c, b, c]$, so it will suffice to prove $[c, b, c, c\,;\, c, a]^{-1} \equiv [c, b, c\,;\, c, a, c]$.

We use part (i) of Groves' Lemma with $C = [c, b, c, c]$, $B = c$, and $A = a$. Since $[C, B]$ is a basic commutator of weight 5, we obtain

$$1 \equiv [c, b, c, c, a]^{-1}[c, b, c, c\,;\, c, a][c, b, c, c, a]^{c[c,a]}.$$

We use Lemma 3.79 to substitute $[c, b, c\,; c, a, c]$ for $[c, b, c, c, a]$ above and obtain

$$1 \equiv [c, b, c\,; c, a, c]^{-1}[c, b, c, c\,; c, a][c, b, c\,; c, a, c]^{c[c,a]}.$$

Next, $[c, b, c\,; c, a, c]$ commutes with $[c, b, c, c]$ by Lemma 3.74 and commutes obviously with $[c, a]$ so we obtain

$$\begin{aligned}[c, b, c, c\,; c, a]^{-1} &\equiv [c, b, c\,; c, a, c]^{-1}[c, b, c\,; c, a, c]^{c[c,a]} = [[c, b, c\,; c, a, c], c[c, a]] \\ &= [[c, b, c\,; c, a, c], [c, a]][c, b, c\,; c, a, c\,; c]^{[c,a]} \equiv [c, b, c\,; c, a, c\,; c]^{[c,a]}\end{aligned}$$

By Lemmas 3.82 and 3.80, $[c, b, c\,; c, a, c\,; c] \equiv [c, a, c, c\,; c, b]^c \equiv [c, b, c\,; c, a, c]$ which commutes obviously with $[c, a]$. □

Corollary 3.84. $[c, b, c\,; c, a, c] \equiv 1$.

PROOF. We have $[c, a, c, c, b] \equiv [c, a, c, c\,; c, b][c, a, c\,; c, b, c]$ by Lemma 3.76. Then $[c, a, c, c, b]$ and $[c, a, c, c\,; c, b]$ are both trivial by Theorem 3.83. □

Corollary 3.85. $[c, b\,; c, a\,; c] \equiv 1$.

PROOF. This follows from Lemma 2.3 and Corollary 3.84. □

Corollary 3.86. $[c, b, c, c, a] \equiv 1$.

PROOF. This follows from Lemma 3.79 and Corollary 3.84. □

Theorem 3.87. *If $r = 3$, then $(\mathfrak{R}_5)^F = \gamma_5(F)$.*

PROOF. It is known that $(\mathfrak{R}_4)^F = \gamma_4(F)$. See [6] or [4]. By the Basic Lemma, we need to show that $[y, x]$ is trivial whenever $x \in \{a, b, c\}$ and y is a basic commutator of weight 4 on $\{a, b, c\}$. We will dismiss without further comment all of the cases where $[y, x]$ is a basic commutator of weight 5.

When y is a simple basic commutator of weight 4, we may assume that y is $[c, a, c, c]$ or $[c, b, c, c]$ by Corollary 3.71. The commutators $[c, a, c, c, a]$ and $[c, b, c, c, b]$ are trivial by Lemma 2.6 while $[c, a, c, c, b]$ and $[c, b, c, c, a]$ are trivial by Theorem 3.83 and Corollary 3.86, respectively.

If y is a basic commutator of weight 4, having pattern $[2, 2]$, then either y or else y^{-1}, depending upon the ordering of the weight 2 basic commutators, is one of $[c, a\,; b, a]$, $[c, b\,; b, a]$, or $[c, b\,; c, a]$. The commutators $[c, a\,; b, a\,; a]$, $[c, b\,; b, a\,; b]$, and $[c, b\,; c, a\,; b]$ are trivial by Lemma 2.4. The commutator $[c, a\,; b, a\,; b]$ is trivial by Lemma 3.2 and the commutator $[c, b\,; b, a\,; a]$ is trivial by Lemma 3.20. The commutators $[c, a\,; b, a\,; c]$, $[c, b\,; c, a\,; a]$, $[c, b\,; b, a\,; c]$, and $[c, b\,; c, a\,; c]$ are trivial by Corollaries 3.52, 3.56, 3.68, and 3.85, respectively. □

4. Rank 4

Lemma 4.1. $[d, a, b\,; c, a, b] \equiv 1$ *and* $[d, a\,; c, a\,; b] \equiv 1$.

PROOF. By Lemma 2.3, it will suffice to show that $[d, a, b\,; c, a, b]$ is trivial. We use part (iv.a) of Groves' Lemma with $C = [c, a, b]$, $B = [d, a]$, and $A = b$. Then $[C, B]$ and $[C, A, A]$ are basic commutators of weight 5. To show that $[C, A, B]$ is trivial, we need only show that $[C, A, d]$ and $[C, A, a]$ are trivial. The first is a basic commutator of weight 5 and the second is trivial by Theorem 3.87. □

Lemma 4.2. $[d, a, c\,; b, a, c] \equiv 1$ *and* $[d, a\,; b, a\,; c] \equiv 1$.

PROOF. We repeat the proof of Lemma 4.1, replacing $C = [c, a, b]$ with $C = [b, a, c]$ and replacing $A = b$ with $A = c$. □

Lemma 4.3. $[d, a, b, c, a] \equiv 1$.

PROOF. We use part (iii) of Groves' Lemma with $C = [d, a, b]$, $B = c$, and $A = a$. Then $[B, A, C]^{-1}$ and $[C, B, B]$ are basic commutators of weight 5. By Lemma 2.5, we have $[C, A] \equiv [d, a\,; b, a][d, a, a, b]$. The factor $[d, a, a, b]$ commutes with a and hence with A, B and C by Lemma 2.1. The factor $[d, a\,; b, a]$ commutes with $A = a$ by Lemma 2.4, with $B = c$ by Lemma 4.2 and commutes obviously with C. □

Lemma 4.4. $[c, a, b, d, c] \equiv 1$.

PROOF. We use part (iii) of Groves' Lemma with $C = [c, a, b]$, $B = d$, and $A = c$. Then $[B, A, C]^{-1}$, $[C, B, B]$, $[C, A, A]$, and $[C, A, B]$ are basic commutators of weight 5. By Theorem 3.87, $[C, A, C]$ is also trivial. □

Lemma 4.5. $[d, a, b, d, c] \equiv 1$.

PROOF. We use part (iii) of Groves' Lemma with $C = [d, a, b]$, $B = d$, and $A = c$. Then $[B, A, C]^{-1}$, $[C, B, B]$, $[C, A, A]$, and $[C, A, B]$ are basic commutators of weight 5. The commutator $[C, A]$ commutes obviously with d, commutes with a by Lemma 4.3 and commutes with b by Lemma 2.10, so $[C, A, C]$ is also trivial. □

Lemma 4.6. $[c, a\,; b, a\,; d] \equiv 1$.

PROOF. By Lemma 2.3, it will suffice to prove that $[c, a, d\,; b, a, d] \equiv 1$. Here, write w for $W(a, b, d)$. By Lemma 2.7, we have $[d, b, a] \equiv [b, a, d]^{-1}[d, a, b]w$ or $[b, a, d] \equiv [d, a, b]w[d, b, a]^{-1}$. Since w commutes obviously with $[c, a, d]$, it will suffice to show that both $[d, a, b]$ and $[d, b, a]$ commute with $[c, a, d]$.

To show that $[d, a, b\,; c, a, d] \equiv 1$, we use part (iv.a) of Groves' Lemma with $C = [d, a, b]$, $B = [c, a]$, and $A = d$. Then $[C, B]$ and $[C, A, A]$ are basic commutators of weight 5. Since $[C, A]$ commutes with c by Lemma 4.5 and commutes with a by Theorem 3.87, $[C, A, B]$ is also trivial.

To show that $[d, b, a\,; c, a, d] \equiv 1$, we use part (iv.a) of Groves' Lemma with $C = [c, a, d]$, $B = [d, b]$, and $A = a$. Then $[C, B]$ is a basic commutator of weight 5. By Lemma 2.5, $[C, A]$ is equivalent to $[c, a\,; d, a][c, a, a, d]$. The factor $[c, a, a, d]$ commutes with a, b, c and d by Lemma 2.1, so it commutes with A and B. The factor $[c, a\,; d, a]$ commutes with $A = a$ by Lemma 2.4. It commutes with b by Lemma 4.1 and commutes with d by Theorem 3.87, so it commutes with $B = [d, b]$. □

Lemma 4.7. $[b, a, c, d, a] \equiv 1$.

PROOF. We use part (iii) of Groves' Lemma with $C = [b, a, c]$, $B = d$, and $A = a$. Then $[B, A, C]^{-1}$ and $[C, B, B]$ are basic commutators of weight 5. By Lemma 2.5, $[C, A]$ is equivalent to $[b, a\,; c, a][b, a, a, c]$. The factor $[b, a, a, c]$ commutes with a, b, c, and d by Lemma 2.1 and hence with A, B, and C. The factor $[b, a\,; c, a]$ commutes obviously with C, commutes with $A = a$ by Theorem 3.87 and commutes with $B = d$ by Lemma 4.6. □

Lemma 4.8. $[c, a, b, d, a] \equiv 1$.

PROOF. Interchange b and c in the proof of Lemma 4.7. □

Lemma 4.9. $[d, b\,;\, b, a\,;\, c] \equiv 1$.

Proof. By Lemma 2.3, it will suffice to prove $[d, b, c\,;\, b, a, c] \equiv 1$. We use part (iv.a) of Groves' Lemma with $C = [b, a, c]$, $B = [d, b]$, and $A = c$. Then $[C, B]$ and $[C, A, A]$ are basic commutators of weight 5. Since $[C, A]$ commutes obviously with d and commutes with b by Theorem 3.87, $[C, A, B]$ is trivial also. □

Lemma 4.10. $[d, b, b, c, a] \equiv 1$.

Proof. We use part (iii) of Groves' Lemma with $C = [d, b, b]$, $B = c$, and $A = a$. Then $[B, A, C]^{-1}$ and $[C, B, B]$ are basic commutators of weight 5. With a change in notation, $[C, A]$ is equivalent to $[b, a, b, d]^{-1}[d, a, b, b][d, b\,;\, b, a]^2$ by Lemma 3.38. The factor $[b, a, b, d]$ commutes obviously with d and commutes with a, b, and c by Lemma 2.6, so it commutes with A, B, and C. The factor $[d, a, b, b]$ commutes obviously with b, c, and d and commutes with a by Theorem 3.87. The factor $[d, b\,;\, b, a]$ commutes obviously with C. It commutes with $A = a$ by Theorem 3.87 and commutes with $B = c$ by Lemma 4.9. □

Lemma 4.11. $[b, a, d\,;\, b, a, c] \equiv 1$.

Proof. Write w again for $W(a, b, c)$, so $[b, a, c] \equiv [c, a, b]w[c, b, a]^{-1}$ by Lemma 2.7. Since $[b, a, d]$ commutes obviously with w, it will suffice to show that $[b, a, d]$ commutes with both $[c, a, b]$ and $[c, b, a]$.

To see that $[b, a, d\,;\, c, a, b] \equiv 1$, we use part (iv.a) of Groves' Lemma with $C = [c, a, b]$, $B = [b, a]$, and $A = d$. Then $[C, B]$ and $[C, A, A]$ are basic commutators of weight 5, $[C, A, b]$ is trivial by Lemma 2.10 and $[C, A, a]$ is trivial by Lemma 4.8.

To see that $[b, a, d\,;\, c, b, a] \equiv 1$, we use part (iv.a) of Groves' Lemma with $C = [b, a, d]$, $B = [c, b]$, and $A = a$. Then $[C, B]$ is a basic commutator of weight 5 and $[C, A]$ is equivalent to $[b, a\,;\, d, a][b, a, a, d]$ by Lemma 2.5. The factor $[b, a, a, d]$ commutes with a, b and c by Lemma 2.1. The factor $[b, a\,;\, d, a]$ commutes with a and b by Theorem 3.87 and commutes with c by Lemma 4.2. □

Lemma 4.12. $[c, a, d\,;\, c, a, b] \equiv 1$.

Proof. We use part (iv.a) of Groves' Lemma with $C = [c, a, b]$, $B = [c, a]$, and $A = d$. Then $[C, B]$ and $[C, A, A]$ are basic commutators of weight 5, $[C, A, c]$ is trivial by Lemma 4.4 and $[C, A, a]$ is trivial by Lemma 4.8. □

Lemma 4.13. $[d, a, c\,;\, d, a, b] \equiv 1$.

Proof. We use part (iv.a) of Groves' Lemma with $C = [d, a, b]$, $B = [d, a]$, and $A = c$. Then $[C, B]$, $[C, A, A]$, and $[C, A, d]$ are basic commutators of weight 5 while $[C, A, a]$ is trivial by Lemma 4.3. □

Lemma 4.14. $[b, a, d, c] \equiv [b, a\,;\, d, c][b, a, c, d]$.

Proof. We have $[b, a, d, c] \equiv [b, a\,;\, d, c][b, a, d\,;\, b, a, c][b, a, c, d]^{[d,c]}$ by Lemma 2.8. Then $[b, a, c, d]$ commutes with $[d, c]$ using Lemma 2.10 and $[b, a, d\,;\, b, a, c]$ is trivial by Lemma 4.11. □

Lemma 4.15. $[c, a, d, b] \equiv [c, a\,;\, d, b][c, a, b, d]$.

Proof. We have $[c, a, d, b] \equiv [c, a\,;\, d, b][c, a, d\,;\, c, a, b][c, a, b, d]^{[d,b]}$ by Lemma 2.8. Then $[c, a, b, d]$ commutes with $[d, b]$ using Lemma 2.10 and $[c, a, d\,;\, c, a, b]$ is trivial by Lemma 4.12. □

Lemma 4.16. $[d, a, c, b] \equiv [d, a\,; c, b][d, a, b, c]$.

PROOF. We have $[d, a, c, b] \equiv [d, a\,; c, b][d, a, c\,; d, a, b][d, a, b, c]^{[c,b]}$ by Lemma 2.8. Then $[d, a, b, c]$ commutes with $[c, b]$ using Lemma 2.10 and $[d, a, c\,; d, a, b]$ is trivial by Lemma 4.13. □

There is a common generalization of Lemmas 2.5, 3.3, 3.72, 4.14, 4.15, and 4.16. If $i < m$ and $i \leq j < k$, then $[x_m, x_i, x_k, x_j] \equiv [x_m, x_i\,; x_k, x_j][x_m, x_i, x_j, x_k]$. The various cases of this seem to require individual proofs and our uses of this result are generally for some specific case.

Lemma 4.17. $[d, c\,; b, a\,; a] \equiv 1$.

PROOF. By Lemma 4.14, we have $[d, c\,; b, a] \equiv [b, a, c, d][b, a, d, c]^{-1}$, so it will suffice to show that both $[b, a, c, d]$ and $[b, a, d, c]$ commute with a. The former is known by Lemma 4.7. For the latter, we use part (iii) of Groves' Lemma with $C = [b, a, d]$, $B = c$, and $A = a$. Then $[B, A, C]^{-1}$ is a basic commutator of weight 5 and $[C, B]$ is equivalent to $[b, a\,; d, c][b, a, c, d]$ by Lemma 4.14. Since c commutes with $[b, a\,; d, c]$ by Lemma 2.4 and commutes with $[b, a, c, d]$ by Lemma 2.10, $[C, B, B]$ is trivial. By Lemma 2.5, $[C, A]$ is equivalent to $[b, a\,; d, a][b, a, a, d]$. The factor $[b, a, a, d]$ commutes with a, b, c, and d by Lemma 2.1. The factor $[b, a\,; d, a]$ commutes obviously with C, commutes with $B = c$ by Lemma 4.2 and commutes with $A = a$ by Lemma 2.4. □

Lemma 4.18. $[d, b\,; c, a\,; a] \equiv 1$.

PROOF. By Lemma 4.15, we have $[d, b\,; c, a] \equiv [c, a, b, d][c, a, d, b]^{-1}$, so it will suffice to show that both $[c, a, b, d]$ and $[c, a, d, b]$ commute with a. The former is known by Lemma 4.8. For the latter, we use part (iii) of Groves' Lemma with $C = [c, a, d]$, $B = b$, and $A = a$. Then $[B, A, C]^{-1}$ is a basic commutator of weight 5 and $[C, B]$ is equivalent to $[c, a\,; d, b][c, a, b, d]$ by Lemma 4.15. Since b commutes with $[c, a\,; d, b]$ by Lemma 2.4 and commutes with $[c, a, b, d]$ by Lemma 2.10, $[C, B, B]$ is trivial. By Lemma 2.5, $[C, A]$ is equivalent to $[c, a\,; d, a][c, a, a, d]$. The factor $[c, a, a, d]$ commutes with a, b, c, and d by Lemma 2.1. The factor $[c, a\,; d, a]$ commutes obviously with C, commutes with $B = b$ by Lemma 4.1 and commutes with $A = a$ by Lemma 2.4. □

Lemma 4.19. $[d, a\,; c, b\,; a] \equiv 1$.

PROOF. By Lemma 4.16, we have $[d, a\,; c, b] \equiv [d, a, c, b][d, a, b, c]^{-1}$, so it will suffice to show that both $[d, a, b, c]$ and $[d, a, c, b]$ commute with a. The former is known by Lemma 4.3. For the latter, we use part (iii) of Groves' Lemma with $C = [d, a, c]$, $B = b$, and $A = a$. Then $[B, A, C]^{-1}$ is a basic commutator of weight 5 and $[C, B]$ is equivalent to $[d, a\,; c, b][d, a, b, c]$ by Lemma 4.16. Since b commutes with $[d, a\,; c, b]$ by Lemma 2.4 and commutes with $[d, a, b, c]$ by Lemma 2.10, $[C, B, B]$ is trivial. By Lemma 2.5, $[C, A]$ is equivalent to $[d, a\,; c, a][d, a, a, c]$. The factor $[d, a, a, c]$ commutes with a, b, c, and d by Lemma 2.1. The factor $[d, a\,; c, a]$ commutes obviously with C, commutes with $B = b$ by Lemma 4.1 and commutes with $A = a$ by Lemma 2.4. □

Lemma 4.20. $[d, a\,; c, b\,; c] \equiv 1$.

PROOF. By Lemma 2.3, it will suffice to prove $[d, a, c\,; c, b, c] \equiv 1$. We use part (iv.a) of Groves' Lemma with $C = [c, b, c]$, $B = [d, a]$, and $A = c$. Then $[C, B]$ and $[C, A, A]$ are basic commutators of weight 5. Since $[C, A]$ commutes obviously with d and commutes with a by Theorem 3.87, $[C, A, B]$ is trivial also. □

Lemma 4.21. $[d, a\,; c, b\,; d] \equiv 1$.

PROOF. By Lemma 2.3, it will suffice to prove $[d, a, d\,; c, b, d] \equiv 1$. We use part (iv.a) of Groves' Lemma with $C = [d, a, d]$, $B = [c, b]$, and $A = d$. Then $[C, B]$ and $[C, A, A]$ are basic commutators of weight 5. Here, both $[C, A, c]$ and $[C, A, b]$ are trivial by Theorem 3.87. □

Lemma 4.22. $[d, b\,; c, a\,; c] \equiv 1$.

PROOF. By Lemma 2.3, it will suffice to prove $[d, b, c\,; c, a, c] \equiv 1$. We use part (iv.a) of Groves' Lemma with $C = [c, a, c]$, $B = [d, b]$, and $A = c$. Then $[C, B]$, $[C, A, A]$ and $[C, A, d]$ are basic commutators of weight 5 while $[C, A, b]$ is trivial by Theorem 3.87. □

Lemma 4.23. $[d, b\,; c, a\,; d] \equiv 1$.

PROOF. By Lemma 2.3, it will suffice to prove $[d, b, d\,; c, a, d] \equiv 1$. We use part (iv.a) of Groves' Lemma with $C = [d, b, d]$, $B = [c, a]$, and $A = d$. Then $[C, B]$ and $[C, A, A]$ are basic commutators of weight 5. Here, both $[C, A, c]$ and $[C, A, a]$ are trivial by Theorem 3.87. □

Lemma 4.24. $[d, c\,; b, a\,; d] \equiv 1$.

PROOF. By Lemma 2.3, it will suffice to prove $[d, c, d\,; b, a, d] \equiv 1$. We use part (iv.a) of Groves' Lemma with $C = [d, c, d]$, $B = [b, a]$, and $A = d$. Then $[C, B]$ and $[C, A, A]$ are basic commutators of weight 5. Here, both $[C, A, b]$ and $[C, A, a]$ are trivial by Theorem 3.87. □

Lemma 4.25. $[b, a, c, d\,; b, a] \equiv 1$.

PROOF. We use part (iv.b) of Groves' Lemma with $C = [b, a, c]$, $B = [b, a]$, and $A = d$. Then $[C, B]$ and $[C, A, A]$ are basic commutators of weight 5 while $[B, A, C]$ is trivial by Lemma 4.11. □

Lemma 4.26. $[d, a, c, c, b] \equiv 1$.

PROOF. We use part (iii) of Groves' Lemma with $C = [d, a, c]$, $B = c$, and $A = b$. Then $[B, A, C]^{-1}$ and $[C, B, B]$ are basic commutators of weight 5. By Lemma 4.16, $[C, A]$ is equivalent to $[d, a\,; c, b][d, a, b, c]$. The factor $[d, a, b, c]$ commutes obviously with c and d, commutes with b by Lemma 2.10, and commutes with a by Lemma 4.3. The factor $[d, a\,; c, b]$ commutes obviously with C, commutes with $B = c$ by Lemma 4.20 and commutes with $A = b$ by Lemma 2.4. □

Lemma 4.27. $[b, a, d, d, c] \equiv 1$.

PROOF. We use part (iii) of Groves' Lemma with $C = [b, a, d]$, $B = d$, and $A = c$. Then $[B, A, C]^{-1}$ and $[C, B, B]$ are basic commutators of weight 5. By Lemma 4.14, $[C, A]$ is equivalent to $[b, a\,; d, c][b, a, c, d]$. The factor $[b, a, c, d]$ commutes obviously with $B = d$, commutes with $A = c$ by Lemma 2.10. To see that

$[b, a, c, d]$ commutes with C, we note that it commutes obviously with d and commutes with $[b, a]$ by Lemma 4.25. The factor $[b, a\,; d, c]$ commutes obviously with C, commutes with $B = d$ by Lemma 4.24 and commutes with $A = c$ by Lemma 2.4. □

Lemma 4.28. $[c, a, d, d, b] \equiv 1$.

PROOF. We use part (iii) of Groves' Lemma with $C = [c, a, d]$, $B = d$, and $A = b$. Then $[B, A, C]^{-1}$ and $[C, B, B]$ are basic commutators of weight 5. By Lemma 4.15, $[C, A]$ is equivalent to $[c, a\,; d, b][c, a, b, d]$. The factor $[c, a, b, d]$ commutes obviously with d, commutes with c by Lemma 4.4, commutes with b by Lemma 2.10, and commutes with a by Lemma 4.8. The factor $[c, a\,; d, b]$ commutes obviously with C, commutes with $B = d$ by Lemma 4.23 and commutes with $A = b$ by Lemma 2.4. □

Lemma 4.29. $[d, a, c, d, b] \equiv 1$.

PROOF. We use part (iii) of Groves' Lemma with $C = [d, a, c]$, $B = d$, and $A = b$. Then $[B, A, C]^{-1}$ and $[C, B, B]$ are basic commutators of weight 5. By Lemma 4.16, $[C, A]$ is equivalent to $[d, a\,; c, b][d, a, b, c]$. The factor $[d, a, b, c]$ commutes obviously with c and d, commutes with b by Lemma 2.10, and commutes with a by Lemma 4.3. The factor $[d, a\,; c, b]$ commutes obviously with C, commutes with $B = d$ by Lemma 4.21 and commutes with $A = b$ by Lemma 2.4. □

Lemma 4.30. $[c, b\,; b, a\,; d] \equiv 1$.

PROOF. By Lemma 2.3, it will suffice to prove that $[c, b, d\,; b, a, d] \equiv 1$. For this, we use part (iv.a) of Groves' Lemma with $C = [b, a, d]$, $B = [c, b]$, and $A = d$. Then $[C, B]$ and $[C, A, A]$ are basic commutators of weight 5, $[C, A, b]$ is trivial by Theorem 3.87 and $[C, A, c]$ is trivial by Lemma 4.27. □

Lemma 4.31. $[c, b\,; c, a\,; d] \equiv 1$.

PROOF. By Lemma 2.3, it will suffice to prove that $[c, b, d\,; c, a, d] \equiv 1$. For this, we use part (iv.a) of Groves' Lemma with $C = [c, a, d]$, $B = [c, b]$, and $A = d$. Then $[C, B]$ and $[C, A, A]$ are basic commutators of weight 5, $[C, A, c]$ is trivial by Theorem 3.87 and $[C, A, b]$ is trivial by Lemma 4.28. □

Lemma 4.32. $[d, b\,; d, a\,; c] \equiv 1$.

PROOF. By Lemma 2.3, it will suffice to prove that $[d, b, c\,; d, a, c] \equiv 1$. For this, we use part (iv.a) of Groves' Lemma with $C = [d, a, c]$, $B = [d, b]$, and $A = c$. Then $[C, B]$ and $[C, A, A]$ are basic commutators of weight 5, $[C, A, d]$ is obviously trivial and $[C, A, b]$ is trivial by Lemma 4.26. □

Lemma 4.33. $[b, a, c, d, b] \equiv 1$.

PROOF. We use part (iii) of Groves' Lemma with $C = [b, a, c]$, $B = d$, and $A = b$. Then $[B, A, C]^{-1}$ and $[C, B, B]$ are basic commutators of weight 5. By Lemma 3.3, $[C, A]$ is equivalent to $[b, a\,; c, b][b, a, b, c]$. The factor $[b, a, b, c]$ commutes obviously with c and d and commutes with a and b by Theorem 3.87. The factor $[b, a\,; c, b]$ commutes obviously with C, commutes with $B = d$ by Lemma 4.30 and commutes with $A = b$ by Theorem 3.87. □

Lemma 4.34. $[c, a, c, d, b] \equiv 1$.

PROOF. We use part (iii) of Groves' Lemma with $C = [c, a, c]$, $B = d$, and $A = b$. Then $[B, A, C]^{-1}$ and $[C, B, B]$ are basic commutators of weight 5. By Lemma 3.72, $[C, A]$ is equivalent to $[c, a\,; c, b][c, a, b, c]$. The factor $[c, a, b, c]$ commutes obviously with c and d and commutes with a and b by Theorem 3.87. The factor $[c, a\,; c, b]$ commutes obviously with C, commutes with $B = d$ by Lemma 4.31 and commutes with $A = b$ by Theorem 3.87. □

Lemma 4.35. $[c, b, b, d, a] \equiv 1$.

PROOF. We use part (iii) of Groves' Lemma with $C = [c, b, b]$, $B = d$, and $A = a$. Then $[B, A, C]^{-1}$ and $[C, B, B]$ are basic commutators of weight 5. By Lemma 3.38, $[C, A]$ is equivalent to $[b, a, b, c]^{-1}[c, a, b, b][c, b\,; b, a]^2$. The first two factors commute obviously with d and commute with a, b and c by Theorem 3.87. The factors $[c, b\,; b, a]$ commute obviously with C, commute with $B = d$ by Lemma 4.30 and commute with $A = a$ by Theorem 3.87. □

Lemma 4.36. $[c, b, c, d, a] \equiv 1$.

PROOF. We use part (iii) of Groves' Lemma with $C = [c, b, c]$, $B = d$, and $A = a$. Then $[B, A, C]^{-1}$ and $[C, B, B]$ are basic commutators of weight 5. By Lemma 3.59 and Theorem 3.87, $[C, A]$ is equivalent to $[c, b\,; c, a][b, a, c, c]^{-1}[c, a, b, c]$. The factors $[b, a, c, c]$ and $[c, a, b, c]$ commute obviously with c and d and commute with a and b by Theorem 3.87. The factor $[c, b\,; c, a]$ commutes obviously with C, commutes with $B = d$ by Lemma 4.31 and commutes with $A = b$ by Theorem 3.87. □

Lemma 4.37. $[d, c\,; b, a\,; b] \equiv 1$.

PROOF. By Lemma 4.14, we have $[d, c\,; b, a] \equiv [b, a, c, d][b, a, d, c]^{-1}$. Since $[b, a, c, d]$ commutes with b by Lemma 4.33, it will suffice to prove that $[b, a, d, c]$ commutes with b. To prove that $[b, a, d, c, b] \equiv 1$, we use part (iii) of Groves' Lemma with $C = [b, a, d]$, $B = c$, and $A = b$. Then $[B, A, C]^{-1}$ is a basic commutator of weight 5. Since $[C, B] \equiv [d, c\,; b, a]^{-1}[b, a, c, d]$ by Lemma 4.14, $[d, c\,; b, a]$ commutes with $B = c$ by Lemma 2.4 and $[b, a, c, d]$ commutes with c by Lemma 2.10, we also have that $[C, B, B]$ is trivial. By Lemma 3.3, with a change of notation, $[C, A]$ is equivalent to $[b, a\,; d, b][b, a, b, d]$. The factor $[b, a, b, d]$ commutes obviously with d and commutes with a, b and c by Lemma 2.6. The factor $[b, a\,; d, b]$ commutes obviously with C, commutes with $B = c$ by Lemma 4.9 and commutes with $A = b$ by Theorem 3.87. □

If y is a basic commutator of weight 4 in which all four letters a, b, c, and d occur, then y commutes with a, b, c, and d. We need this in Theorem 4.53, below, but we also need the pieces of this in some intervening lemmas. We have already proved all of the pieces of this result. Rather than repeating the justifications in two or three later places, we will record the justifications here and cite this lemma.

Lemma 4.38. *The commutators* $[d, a, b, c]$, $[c, a, b, d]$, $[b, a, c, d]$, $[d, a\,; c, b]$, $[d, b\,; c, a]$, *and* $[d, c\,; b, a]$ *all commute with* a, b, c, *and* d.

PROOF. $[d, a, b, c]$ commutes with a by Lemma 4.3 and with b by Lemma 2.10.
$[c, a, b, d]$ commutes with a, b, and c by Lemmas 4.8, 2.10, and 4.4.
$[b, a, c, d]$ commutes with a, b, and c by Lemmas 4.7, 4.33, and 2.10.
$[d, a\,; c, b]$ commutes with a, b, c, and d by Lemmas 4.19, 2.4, 4.20, and 4.21.

$[d, b\,;\, c, a]$ commutes with a, b, c, and d by Lemmas 4.18, 2.4, 4.22, and 4.23.
$[d, c\,;\, b, a]$ commutes with a, b, c, and d by Lemmas 4.17, 4.37, 2.4, and 4.24. □

Lemma 4.39. $[c, b, d\,;\, c, a, b] \equiv 1$.

PROOF. We use part (iv.a) of Groves' Lemma with $C = [c, a, b], B = [c, b]$ and $A = d$. Then $[C, B]$ is a basic commutator of weight 5 and $[C, A]$ commutes with b, c and d by Lemma 4.38. □

Lemma 4.40. $[c, b, d\,;\, b, a, c] \equiv 1$.

PROOF. Replace $C = [c, a, b]$ by $C = [b, a, c]$ in the proof of Lemma 4.39. □

Lemma 4.41. $[d, b, c\,;\, b, a, d] \equiv 1$.

PROOF. We use part (iv.a) of Groves' Lemma with $C = [b, a, d]$, $B = [d, b]$, and $A = c$. Then $[C, B]$ is a basic commutator of weight 5. By Lemma 4.14, $[C, A]$ is equivalent to $[b, a\,;\, d, c][b, a, c, d]$. These factors commute with a, b, c, and d by Lemma 4.38. □

Lemma 4.42. *$W(a, b, c)$ commutes with d and $W(a, b, d)$ commutes with c.*

PROOF. By Lemma 3.25 $W(a, b, c)$ is equivalent to $[c, b\,;\, c, a][c, b\,;\, b, a][c, a\,;\, b, a]$. The three factors commute with d by Lemmas 4.31, 4.30 and 4.6. Similarly, with a change of notation, $W(a, b, d)$ is equivalent to $[d, b\,;\, d, a][d, b\,;\, b, a][d, a\,;\, b, a]$ and these three factors commute with c by Lemmas 4.32, 4.9, and 4.2. □

Lemma 4.43. $[c, b, d, a] \equiv [c, b\,;\, d, a][c, a, b, d][b, a, c, d]^{-1}$.

PROOF. Write w for $W(a, b, c)$ throughout this proof. As a preliminary step, we show that $[c, b, a, d] \equiv [c, a, b, d][b, a, c, d]^{-1}$. By Lemmas 2.7 and 3.24, we have $[c, b, a] \equiv w[c, a, b][b, a, c]^{-1}$. Since w commutes with d by Lemma 4.42, we have $[c, b, a, d] \equiv [[c, a, b][b, a, c]^{-1}, d] = ([c, a, b, d][b, a, c, d]^{-1})^{[b,a,c]^{-1}}$ where both factors commute with the exponent by Lemma 4.38.

To prove the conclusion of the lemma, we use part (i) of Groves' Lemma with $C = [c, b]$, $B = d$, and $A = a$ to obtain

$$[c, b, d, a] = [c, b, d]^{-1}[c, b, a]^{-1}[c, b\,;\, d, a]([c, b, d][c, b, a][c, b, a, d])^{[d,a]}.$$

We use Lemma 2.7 to substitute for $[c, b, a]$ above and we use the first paragraph to substitute $[c, a, b, d][b, a, c, d]^{-1}$ for $[c, b, a, d]$. The weight 3 factors commute obviously with the exponent $[d, a]$, the factors $[c, a, b, d]$ and $[b, a, c, d]$ commute with $[d, a]$ by Lemma 4.38 while w commutes with a by Theorem 3.87 and with d by Lemma 4.42 and hence with $[d, a]$. We substitute and remove the exponent to obtain

$$\begin{aligned}[c, b, d, a] = [c, b,d]^{-1}w^{-1}[c, a, b]^{-1}[b, a, c][c, b\,;\, d, a]\\ [c, b, d][b, a, c]^{-1}[c, a, b]w[c, a, b, d][b, a, c, d]^{-1}.\end{aligned}$$

We reduce this to obtain the conclusion. Obviously, w commutes with the weight 3 commutators and $[c, b\,;\, d, a]$ commutes with w by Lemma 4.38. The factor $[c, b, d]$ commutes obviously with $[c, b\,;\, d, a]$ and commutes with $[b, a, c]$ and $[c, a, b]^{-1}$ by Lemmas 4.40 and 4.39. The factors $[b, a, c]$ and $[c, a, b]$ commute obviously with $[c, b\,;\, d, a]$. □

Lemma 4.44. $[d, b, c, a] \equiv [d, b\,;\, c, a][d, c\,;\, b, a][b, a, c, d]^{-1}[d, a, b, c]$.

PROOF. Write w for $W(a,b,d)$ throughout this proof. As a preliminary step, we show that $[d,b,a,c] \equiv [d,c;b,a][b,a,c,d]^{-1}[d,a,b,c]$. By Lemmas 2.7 and 3.24, with a change of notation, we have $[d,b,a] \equiv w[d,a,b][b,a,d]^{-1}$. Since w commutes with c by Lemma 4.42 and $[b,a,d,c] \equiv [b,a;d,c][b,a,c,d]$ by Lemma 4.14, we have

$$\begin{aligned}[d,b,a,c] &\equiv [[d,a,b][b,a,d]^{-1},c] = ([d,a,b,c][b,a,d,c]^{-1})^{[b,a,d]^{-1}} \\ &\equiv ([d,a,b,c][b,a,c,d]^{-1}[d,c;b,a])^{[b,a,d]^{-1}}.\end{aligned}$$

The three factors commute with each other and with the exponent by Lemma 4.38.

To prove the conclusion of the lemma, we use part (i) of Groves' Lemma with $C = [d,b]$, $B = c$, and $A = a$ to obtain

$$[d,b,c,a] = [d,b,c]^{-1}[d,b,a]^{-1}[d,b;c,a]([d,b,c][d,b,a][d,b,a,c])^{[c,a]}.$$

We use Lemma 2.7 to substitute for $[d,b,a]$ in the equation above and we use the first paragraph to substitute $[d,c;b,a][b,a,c,d]^{-1}[d,a,b,c]$ for $[d,b,a,c]$. The weight 3 factors commute obviously with the exponent $[c,a]$, the factors $[d,c;b,a]$, $[b,a,c,d]$, and $[d,a,b,c]$ commute with $[c,a]$ by Lemma 4.38 while w commutes with a by Theorem 3.87 and with c by Lemma 4.42 and hence with $[c,a]$. We substitute and remove the exponent to obtain

$$\begin{aligned}[d,b,c,a] = [d,b,c]^{-1}w^{-1}[d,a,b]^{-1}[b,a,d][d,b;c,a] \\ [d,b,c][b,a,d]^{-1}[d,a,b]w[d,c;b,a][b,a,c,d]^{-1}[d,a,b,c].\end{aligned}$$

We reduce this to obtain the conclusion. Obviously, w commutes with the weight 3 commutators and $[d,b;c,a]$ commutes with w by Lemma 4.38. The factor $[d,b,c]$ commutes obviously with $[d,b;c,a]$ and commutes with $[b,a,d]$ and $[d,a,b]^{-1}$ by Lemmas 4.41 and 2.11. The factors $[b,a,d]$ and $[d,a,b]$ commute obviously with $[d,b;c,a]$. □

Lemma 4.45. $[c,b,d,d,a] \equiv 1$.

PROOF. We use part (iii) of Groves' Lemma with $C = [c,b,d]$, $B = d$, and $A = a$. Then $[B,A,C]^{-1}$ and $[C,B,B]$ are basic commutators of weight 5. By Lemma 4.43, $[C,A]$ is equivalent to $[c,b;d,a][c,a,b,d][b,a,c,d]^{-1}$ where the three factors commute with a, b, c, and d by Lemma 4.38. □

Lemma 4.46. $[d,b,c,c,a] \equiv 1$.

PROOF. We use part (iii) of Groves' Lemma with $C = [d,b,c]$, $B = c$, and $A = a$. Then $[B,A,C]^{-1}$ and $[C,B,B]$ are basic commutators of weight 5. By Lemma 4.44, $[C,A]$ is equivalent to $[d,b;c,a][d,c;b,a][b,a,c,d]^{-1}[d,a,b,c]$ where the four factors commute with a, b, c, and d by Lemma 4.38. □

Lemma 4.47. $[d,b,c,d,a] \equiv 1$.

PROOF. Replace $B = c$ by $B = d$ in the proof of Lemma 4.46. □

Lemma 4.48. $[d,c;c,a;b] \equiv 1$.

PROOF. By Lemma 3.3, with a change of notation, $[d,c;c,a]$ is equivalent to the product $[c,a,c,d][c,a,d,c]^{-1}$. Since $[c,a,c,d]$ commutes with b by Lemma 4.34, it will suffice to show that $[c,a,d,c]$ also commutes with b. For this, we use part (iii) of Groves' Lemma with $C = [c,a,d]$, $B = c$, and $A = b$. Then $[B,A,C]^{-1}$ is a basic commutator of weight 5 and $[C,B,B]$ is trivial by Theorem 3.87. By Lemma 4.15,

$[C, A]$ is equivalent to the product $[c, a\,;\, d, b][c, a, b, d]$ where both of the factors commute with a, b, c, and d by Lemma 4.38. □

Lemma 4.49. $[d, c\,;\, d, a\,;\, b] \equiv 1$.

PROOF. By Lemma 3.72, with a change of notation, $[d, c\,;\, d, a]$ is equivalent to the product $[d, a, c, d][d, a, d, c]^{-1}$. Since $[d, a, c, d]$ commutes with b by Lemma 4.29, it will suffice to show that $[d, a, d, c]$ also commutes with b. For this, we use part (iii) of Groves' Lemma with $C = [d, a, d]$, $B = c$, and $A = b$. Then $[B, A, C]^{-1}$ is a basic commutator of weight 5 and $[C, B, B]$ is trivial by Theorem 3.87. By Theorem 3.87, $[C, A]$ commutes with A and C. By Lemma 3.72, with another change in notation, $[C, A]$ is equivalent to the product $[d, a\,;\, d, b][d, a, b, d]$ where $B = c$ commutes with the first factor by Lemma 4.32 and with the second factor by Lemma 4.5. □

Lemma 4.50. $[d, b\,;\, c, b\,;\, a] \equiv 1$.

PROOF. By Lemma 2.5, $[d, b\,;\, c, b] \equiv [d, b, c, b][d, b, b, c]^{-1}$. Since $[d, b, b, c]$ commutes with a by Lemma 4.10, it will suffice to show that $[d, b, c, b]$ also commutes with a. For this, we use part (iii) of Groves' Lemma with $C = [d, b, c]$, $B = b$, and $A = a$. Then $[B, A, C]^{-1}$ is a basic commutator of weight 5 and $[C, B, B]$ is trivial by Theorem 3.87. $[C, A]$ is equivalent to $[d, b\,;\, c, a][d, c\,;\, b, a][b, a, c, d]^{-1}[d, a, b, c]$ by Lemma 4.44 and the four factors commute with a, b, c, and d by Lemma 4.38. □

Lemma 4.51. $[d, c\,;\, c, b\,;\, a] \equiv 1$.

PROOF. With a change of notation, $[d, c\,;\, c, b] \equiv [c, b, c, d][c, b, d, c]^{-1}$ by Lemma 3.3. Since $[c, b, c, d]$ commutes with a by Lemma 4.36, it will suffice to show that $[c, b, d, c]$ also commutes with a. For this, we use part (iii) of Groves' Lemma with $C = [c, b, d]$, $B = c$, and $A = a$. Then $[B, A, C]^{-1}$ is a basic commutator of weight 5 and $[C, B, B]$ is trivial by Theorem 3.87. $[C, A]$ is equivalent to $[c, b\,;\, d, a][c, a, b, d][b, a, c, d]^{-1}$ by Lemma 4.43 and the three factors commute with a, b, c, and d by Lemma 4.38. □

Lemma 4.52. $[d, c\,;\, d, b\,;\, a] \equiv 1$.

PROOF. By Lemma 3.72, with a change of notation, $[d, c\,;\, d, b]$ is equivalent to $[d, b, c, d][d, b, d, c]^{-1}$. Since $[d, b, c, d]$ commutes with a by Lemma 4.47, it will suffice to show that $[d, b, d, c]$ also commutes with a. For this, we use part (iii) of Groves' Lemma with $C = [d, b, d]$, $B = c$, and $A = a$. Then $[B, A, C]^{-1}$ is a basic commutator of weight 5 while $[C, B, B]$, $[C, A, A]$, and $[C, A, C]$ are trivial by Theorem 3.87. By Lemma 3.59 and Theorem 3.87, with a change of notation, $[C, A]$ is equivalent to $[d, b\,;\, d, a][b, a, d, d]^{-1}[d, a, b, d]$. The three factors commute with $B = c$ by Lemmas 4.32, 4.27 and 4.5, respectively. □

Theorem 4.53. *If* $r = 4$, *then* $(\mathfrak{N}_5)^F = \gamma_5(F)$.

PROOF. It is known that $(\mathfrak{N}_4)^F = \gamma_4(F)$. See [6] or [4]. By the Basic Lemma, we need to show that $[y, x]$ is trivial whenever $x \in \{a, b, c, d\}$ and y is a basic commutator of weight 4 on $\{a, b, c, d\}$. This is known by Theorem 3.87 if at most 3 letters of $\{a, b, c, d\}$ occur in $[y, x]$ and is known by Lemma 4.38 if all 4 letters of $\{a, b, c, d\}$ occur in y. We will also dismiss without further comment all of the cases where $[y, x]$ is a basic commutator of weight 5.

The commutators $[c, b\,;\, c, a]$, $[c, b\,;\, b, a]$, and $[c, a\,;\, b, a]$ commute with d by Lemmas 4.31, 4.30, and 4.6, respectively.

The commutators $[d,b\,;d,a]$, $[d,b\,;b,a]$, and $[d,a\,;b,a]$ commute with c by Lemmas 4.32, 4.9, and 4.2, respectively. The commutators $[b,a,a,d]$, $[b,a,b,d]$, $[b,a,d,d]$, and $[d,a,b,d]$ commute with c by Lemmas 2.1, 2.6, 4.27, and 4.5, respectively.

The commutators $[d,c\,;d,a]$, $[d,c\,;c,a]$, and $[d,a\,;c,a]$ commute with b by Lemmas 4.49, 4.48, and 4.1, respectively. The commutators $[c,a,a,d]$, $[d,a,a,c]$, $[c,a,c,d]$, $[d,a,c,c]$, $[c,a,d,d]$, and $[d,a,c,d]$ commute with b by Lemmas 2.1, 2.1, 4.34, 4.26, 4.28, and 4.29, respectively.

The commutators $[d,c\,;d,b]$, $[d,c\,;c,b]$, and $[d,b\,;c,b]$ commute with a by Lemmas 4.52, 4.51, and 4.50, respectively. The commutators $[c,b,b,d]$, $[d,b,b,c]$, $[c,b,c,d]$, $[d,b,c,c]$, $[c,b,d,d]$, and $[d,b,c,d]$ commute with a by Lemmas 4.35, 4.10, 4.36, 4.46, 4.45, and 4.47, respectively.

□

5. Rank 5 and greater

We now return to the subscripted alphabet $X = \{x_i\}_{1 \le i \le r}$ of Section 2. In Theorem 5.10, we will want to show that $[y, x_\ell]$ is trivial whenever y is a basic commutator of weight 4 in which four distinct letters of X occur and x_ℓ is another letter of X which is distinct from these four. Throughout this section, we will take the letters in such a commutator y to be x_i, x_j, x_k, and x_m. When y is a simple left-normed basic commutator, we can choose our indices so that y is $[x_m, x_i, x_j, x_k]$ where $i < m$ and $i < j < k$. If y is a commutator with pattern $[2,2]$, we can choose our indices so that either y or else y^{-1} is $[x_m, x_i\,; x_k, x_j]$ where $i < m$ and $j < k$, and $i < j$. The lemmas in this section will generally have multiple instances. The first lemma has four distinct instances on the ordered alphabet $\{a,b,c,d,e\}$.

Lemma 5.1. *If $i < j < \ell < k$ and $i < m$, then $[x_m, x_i, x_j, x_k, x_\ell] \equiv 1$.*

Proof. We use part (iii) of Groves' Lemma with $C = [x_m, x_i, x_j]$, $B = x_k$, and $A = x_\ell$. Then $[B,A,C]^{-1}$, $[C,B,B]$, $[C,A,A]$, and $[C,A,B]$ are basic commutators of weight 5 while $[C,A,C]$ is trivial by Theorem 4.53. □

Lemma 5.2. *If $i < m$, $i < j < k$, and $j < \ell$, then $[x_m, x_i\,; x_k, x_j\,; x_\ell] \equiv 1$.*

Proof. By Lemma 2.3, it will suffice to prove that $[x_m, x_i, x_\ell\,; x_k, x_j, x_\ell]$ is trivial. For this we use part (iv.a) of Groves' Lemma with $C = [x_k, x_j, x_\ell]$, $B = [x_m, x_i]$, and $A = x_\ell$. Then $[C,B]$ and $[C,A,A]$ are basic commutators of weight 5. $[C,A,B]$ is trivial since $[C,A,x_m]$ and $[C,A,x_i]$ are trivial by Theorem 4.53. □

Lemma 5.3. *If $i < \ell < j < k$ and $i < m$, then $[x_m, x_i, x_j, x_k, x_\ell] \equiv 1$.*

Proof. We use part (iii) of Groves' Lemma with $C = [x_m, x_i, x_j]$, $B = x_k$, and $A = x_\ell$. Then $[B,A,C]^{-1}$ and $[C,B,B]$ are basic commutators of weight 5 while $[C,A,A]$ and $[C,A,C]$ are trivial by Theorem 4.53. By Theorem 4.53, we may assume that m is not equal to ℓ, j, or k. Using Lemma 4.14, 4.15, or 4.16, according as $i < m < \ell < j$, $i < \ell < m < j$, or $i < \ell < j < m$, we have $[C,A] \equiv [x_m, x_i\,; x_j, x_\ell][x_m, x_i, x_\ell, x_j]$. The factor $[x_m, x_i, x_\ell, x_j]$ commutes obviously with $B = x_k$ and the factor $[x_m, x_i\,; x_j, x_\ell]$ commutes with $B = x_k$ by Lemma 5.2. □

Lemma 5.4. *If $i < m$, $i < \ell$, $\ell < j$, and $\ell < k$, then $[x_m, x_i, x_\ell]$ commutes with both $[x_j, x_\ell, x_k]$ and $[x_k, x_\ell, x_j]$.*

PROOF. To prove that $[x_m, x_i, x_\ell]$ commutes with $[x_j, x_\ell, x_k]$, we use part (iv.a) of Groves' Lemma with $C = [x_m, x_i, x_\ell]$, $B = [x_j, x_\ell]$, and $A = x_k$. Then $[C, B]$ and $[C, A, A]$ are basic commutators of weight 5. $[C, A, B]$ is trivial since $[C, A, x_\ell]$ by Theorem 4.53 while $[C, A, x_j]$ is either obviously trivial, if $j \geq k$ or is trivial by Lemma 5.1 if $j < k$.

To prove that $[x_m, x_i, x_\ell]$ commutes with $[x_k, x_\ell, x_j]$, interchange j with k in the first paragraph. □

Lemma 5.5. *If $i < j < k$, then $[x_k, x_j, x_i]$ commutes with all basic commutators of weight 2.*

PROOF. Write w for $W(x_i, x_j, x_k)$. Then $[x_k, x_j, x_i] \equiv [x_j, x_i, x_k]^{-1}[x_k, x_i, x_j]w$ by Lemma 2.7. The factors $[x_j, x_i, x_k]$ and $[x_k, x_i, x_j]$ commute obviously with basic commutators of weight 2. By Lemma 3.25, w is equivalent to the product $[x_k, x_j\,; x_k, x_i][x_k, x_j\,; x_j, x_i][x_k, x_i\,; x_j, x_i]$ where these weight 4 factors commute with all letters of X by Theorem 4.53. □

Lemma 5.6. *If $i < m$ and $j < k$, and $i < \ell < j$, then $[x_m, x_i\,; x_k, x_j\,; x_\ell] \equiv 1$.*

PROOF. We use part (v) of Groves' Lemma with $C = [x_m, x_i]$, $B = [x_k, x_j]$, and $A = x_\ell$ to show that $[C, B, A]$ is trivial. We see that $[C, A]$ commutes obviously with B and $[C, B]$. We see by Lemma 5.5 that $[B, A]$ commutes with C and with $[C, B]$. Write w for $W(x_\ell, x_j, x_k)$ so that $[B, A]$ is equivalent to $[x_j, x_\ell, x_k]^{-1}[x_k, x_\ell, x_j]w$ by Lemma 2.7. Since $[C, A]$ commutes obviously with w and commutes with $[x_j, x_\ell, x_k]$ and $[x_k, x_\ell, x_j]$ by Lemma 5.4, we also have that $[C, A; B, A]$ is trivial. □

Lemma 5.7. *If $i < \min\{j, k, \ell, m\}$, then $[x_\ell, x_i, x_m\,; x_k, x_i, x_j] \equiv 1$.*

PROOF. Since, if necessary, we may replace $[x_\ell, x_i, x_m\,; x_k, x_i, x_j]$ by its inverse and change notation, we may assume that $m \geq j$. If $m = j$, the conclusion follows from Theorem 4.53, so we may assume $m > j$. We use part (iv.a) of Groves' Lemma with $C = [x_k, x_i, x_j]$, $B = [x_\ell, x_i]$, and $A = x_m$. Then $[C, B]$ and $[C, A, A]$ are basic commutators of weight 5 while $[C, A, x_i]$ is trivial by Theorem 4.53. Since $[C, A, x_\ell]$ is a basic commutator of weight 5 if $\ell \geq m$, is trivial by Lemma 5.1 if $j < \ell < m$, is trivial by Theorem 4.53 if $\ell = j$ and is trivial by Lemma 5.3 if $i < \ell < j$, $[C, A, B]$ is trivial. □

Lemma 5.8. *If $\ell < i < j < k$ and $i < m$, then $[x_m, x_i, x_j, x_k, x_\ell] \equiv 1$.*

PROOF. We use part (iii) of Groves' Lemma with $C = [x_m, x_i, x_j]$, $B = x_k$, and $A = x_\ell$. Then $[B, A, C]^{-1}$ and $[C, B, B]$ are basic commutators of weight 5. If $m = j$, then $[x_m, x_i, x_j, x_k, x_\ell]$ is trivial by Theorem 4.53, so we may asume that $m \neq j$. If $m < j$, then $[C, A] \equiv [x_m, x_i\,; x_j, x_\ell][x_m, x_\ell, x_i, x_j][x_i, x_\ell, x_m, x_j]^{-1}$ by Lemma 4.43. If $m > j$, then $[C, A] \equiv [x_m, x_i\,; x_j, x_\ell][x_m, x_j\,; x_i, x_\ell][x_i, x_\ell, x_j, x_m]^{-1}[x_m, x_\ell, x_i, x_j]$ by Lemma 4.44. In either case, all of the factors commute with x_ℓ, x_i, x_j, and x_m by Theorem 4.53, so $[C, A, A]$ and $[C, A, C]$ are trivial. The factors $[x_m, x_i\,; x_j, x_\ell]$ and $[x_m, x_j\,; x_i, x_\ell]$ commute with $B = x_k$ by Lemma 5.2. The factor $[x_m, x_\ell, x_i, x_j]$ commutes obviously with $B = x_k$ as does the factor $[x_i, x_\ell, x_m, x_j]$ which occurs when $m < j$. When $m > j$, the factor $[x_i, x_\ell, x_j, x_m]$ commutes with x_k obviously when $m \leq k$ and commutes with x_k by Lemma 5.1 when $k < m$. □

Lemma 5.9. *If $\ell < i < m$ and $i < j < k$, then $[x_m, x_i\,; x_k, x_j\,; x_\ell] \equiv 1$.*

PROOF. We use part (v) of Groves' Lemma with $C = [x_m, x_i]$, $B = [x_k, x_j]$, and $A = x_\ell$ to show that $[C, B, A]$ is trivial. By Lemma 5.5, $[B, A, C]$ and $[C, A, B]$ are trivial. Since $[C, B]$ is a product of weight 2 basic commutators and their inverses, it also commutes with $[C, A]$ and $[B, A]$ by Lemma 5.5. Hence, $[C, B\,; C, A]$ and $[C, B\,; B, A]$ are trivial. Write w_1 for $W(x_\ell, x_i, x_m)$ and write w_2 for $W(x_\ell, x_j, x_k)$ so that $[C, A] \equiv [x_i, x_\ell, x_m]^{-1}[x_m, x_\ell, x_i]w_1$ and $[B, A] \equiv [x_j, x_\ell, x_k]^{-1}[x_k, x_\ell, x_j]w_2$ by Lemma 2.7. To see that $[C, A\,; B, A]$ is trivial, we note that w_1 and w_2 commute with all letters of X by Lemma 3.25 and Theorem 4.53 and that the factors $[x_i, x_\ell, x_m]$, $[x_m, x_\ell, x_i]$, $[x_j, x_\ell, x_k]$, and $[x_k, x_\ell, x_j]$ commute with each other by Lemma 5.7. □

Theorem 5.10. *If $r \geq 5$, then $(\mathfrak{R}_5)^F = \gamma_5(F)$.*

PROOF. It is known that $(\mathfrak{R}_4)^F = \gamma_4(F)$. See [6] or [4]. By the Basic Lemma, it will suffice to prove that $[y, x_\ell]$ is trivial whenever $x_\ell \in X$ and y is a basic commutator of weight 4 on X. By Theorem 4.53, we may assume that five distinct letters of X occur in $[y, x_\ell]$, so that letters x_i, x_j, x_k, and x_m, all different from x_ℓ, occur in y.

If y is a simple left-normed basic commutator of weight 4, we may choose notation so that y is $[x_m, x_i, x_j, x_k]$ where $i < m$ and $i < j < k$. Then $[y, x_\ell]$ is a basic commutator of weight 5 if $\ell > k$ and is trivial by Lemma 5.1 if $j < \ell < k$, by Lemma 5.3 if $i < \ell < j$ and by Lemma 5.8 if $\ell < i$.

If y is a basic commutator having pattern $[2, 2]$, then we may choose notation so that either y or else y^{-1} is $[x_m, x_i\,; x_k, x_j]$ where $i < m$ and $i < j < k$. Then $[x_m, x_i\,; x_k, x_j]$ commutes with x_ℓ by Lemma 5.2 if $\ell > j$, by Lemma 5.6 if $i < \ell < j$ and by Lemma 5.9 if $\ell < i$. □

Appendix A. A list of critical commutators in rank 3

Lemma 2.4	$[c,a;b,a;a]\equiv 1;$	Lemma 3.2	$[c,a;b,a;b]\equiv 1;$	Corol. 3.52	$[c,a;b,a;c$
Lemma 3.20	$[c,b;b,a;a]\equiv 1;$	Lemma 2.4	$[c,b;b,a;b]\equiv 1;$	Corol. 3.68	$[c,b;b,a;c$
Corol. 3.56	$[c,b;c,a;a]\equiv 1;$	Lemma 2.4	$[c,b;c,a;b]\equiv 1;$	Corol. 3.85	$[c,b;c,a;c$

Basic	$[b,a,a,a,a]\equiv 1;$	Basic	$[b,a,a,a,b]\equiv 1;$	Basic	$[b,a,a,a,c$
Lemma 2.1	$[b,a,a,b,a]\equiv 1;$	Basic	$[b,a,a,b,b]\equiv 1;$	Basic	$[b,a,a,b,c$
Lemma 2.1	$[b,a,a,c,a]\equiv 1;$	Lemma 2.1	$[b,a,a,c,b]\equiv 1;$	Basic	$[b,a,a,c,c$
Lemma 2.6	$[b,a,b,b,a]\equiv 1;$	Basic	$[b,a,b,b,b]\equiv 1;$	Basic	$[b,a,b,b,c$
Lemma 2.6	$[b,a,b,c,a]\equiv 1;$	Lemma 2.6	$[b,a,b,c,b]\equiv 1;$	Basic	$[b,a,b,c,c$
Corol. 3.53	$[b,a,c,c,a]\equiv 1;$	Corol. 3.69	$[b,a,c,c,b]\equiv 1;$	Basic	$[b,a,c,c,c$
Basic	$[c,a,a,a,a]\equiv 1;$	Basic	$[c,a,a,a,b]\equiv 1;$	Basic	$[c,a,a,a,c$
Lemma 2.1	$[c,a,a,b,a]\equiv 1;$	Basic	$[c,a,a,b,b]\equiv 1;$	Basic	$[c,a,a,b,c$
Lemma 2.1	$[c,a,a,c,a]\equiv 1;$	Lemma 2.1	$[c,a,a,c,b]\equiv 1;$	Basic	$[c,a,a,c,c$
Lemma 3.37	$[c,a,b,b,a]\equiv 1;$	Basic	$[c,a,b,b,b]\equiv 1;$	Basic	$[c,a,b,b,c$
Corol. 3.54	$[c,a,b,c,a]\equiv 1;$	Lemma 2.10	$[c,a,b,c,b]\equiv 1;$	Basic	$[c,a,b,c,c$
Lemma 2.6	$[c,a,c,c,a]\equiv 1;$	Theorem 3.83	$[c,a,c,c,b]\equiv 1;$	Basic	$[c,a,c,c,c$
Lemma 3.39	$[c,b,b,b,a]\equiv 1;$	Basic	$[c,b,b,b,b]\equiv 1;$	Basic	$[c,b,b,b,c$
Corol. 3.70	$[c,b,b,c,a]\equiv 1;$	Lemma 2.1	$[c,b,b,c,b]\equiv 1;$	Basic	$[c,b,b,c,c$
Corol. 3.86	$[c,b,c,c,a]\equiv 1;$	Lemma 2.6	$[c,b,c,c,b]\equiv 1;$	Basic	$[c,b,c,c,c$

Appendix B. Some important weight six basic commutators

Lemma 2.2	$[b,a,b\,;\,b,a,a] \equiv 1;$	Lemma 2.2	$[b,a,c\,;\,b,a,a] \equiv 1;$
Lemma 3.1	$[b,a,c\,;\,b,a,b] \equiv 1;$	Lemma 2.2	$[c,a,a\,;\,b,a,a] \equiv 1;$
Lemma 2.2	$[c,a,a\,;\,b,a,b] \equiv 1;$	Lemma 2.2	$[c,a,a\,;\,b,a,c] \equiv 1;$
Lemma 2.2	$[c,a,b\,;\,b,a,a] \equiv 1;$	Lemma 3.1	$[c,a,b\,;\,b,a,b] \equiv 1;$
Lemma 3.24	$[c,a,b\,;\,b,a,c] \equiv 1;$	Lemma 2.2	$[c,a,b\,;\,c,a,a] \equiv 1;$
Lemma 2.2	$[c,a,c\,;\,b,a,a] \equiv 1;$	Lemma 3.1	$[c,a,c\,;\,b,a,b] \equiv 1;$
Theorem 3.50	$[c,a,c\,;\,b,a,c] \equiv 1;$	Lemma 2.2	$[c,a,c\,;\,c,a,a] \equiv 1;$
Corol. 3.51	$[c,a,c\,;\,c,a,b] \equiv 1;$	Lemma 2.2	$[c,b,b\,;\,b,a,a] \equiv 1;$
Lemma 2.2	$[c,b,b\,;\,b,a,b] \equiv 1;$	Lemma 3.5	$[c,b,b\,;\,b,a,c] \equiv 1;$
Lemma 2.2	$[c,b,b\,;\,c,a,a] \equiv 1;$	Lemma 2.2	$[c,b,b\,;\,c,a,b] \equiv 1;$
Lemma 3.8	$[c,b,b\,;\,c,a,c] \equiv 1;$	Lemma 2.2	$[c,b,c\,;\,b,a,a] \equiv 1;$
Lemma 3.1	$[c,b,c\,;\,b,a,b] \equiv 1;$	Theorem 3.67	$[c,b,c\,;\,b,a,c] \equiv 1;$
Lemma 2.2	$[c,b,c\,;\,c,a,a] \equiv 1;$	Lemma 2.11	$[c,b,c\,;\,c,a,b] \equiv 1;$
Corol. 3.84	$[c,b,c\,;\,c,a,c] \equiv 1;$	Lemma 2.2	$[c,b,c\,;\,c,b,b] \equiv 1.$

Lemma 3.14	$[c,a\,;\,b,a\,;\,b,a] \equiv 1;$	Lemma 3.14	$[c,a\,;\,b,a\,;\,c,a] \equiv 1;$
Lemma 3.24	$[c,a\,;\,b,a\,;\,c,b] \equiv 1;$	Lemma 3.6	$[c,b\,;\,b,a\,;\,b,a] \equiv 1;$
Lemma 3.24	$[c,b\,;\,b,a\,;\,c,a] \equiv 1;$	Lemma 3.12	$[c,b\,;\,b,a\,;\,c,b] \equiv 1;$
Corol. 3.55	$[c,b\,;\,c,a\,;\,c,a] \equiv 1;$	Lemma 3.13	$[c,b\,;\,c,a\,;\,c,b] \equiv 1.$

Appendix C. Some important transformations of nonbasic commutators

Lemma 2.3: If $i < j$ and $k < m$, and $\ell \geq \max\{i, k\}$ then

$$[x_m, x_k\,;\, x_j, x_i\,;\, x_\ell] \equiv [x_m, x_k, x_\ell\,;\, x_j, x_i, x_\ell].$$

Instances of Lemma 2.3 on the alphabet $\{\, a, b, c \,\}$:

$[c, a\,;\, b, a\,;\, a] \equiv [c, a, a\,;\, b, a, a]$; $[c, a\,;\, b, a\,;\, b] \equiv [c, a, b\,;\, b, a, b]$;
$[c, a\,;\, b, a\,;\, c] \equiv [c, a, c\,;\, b, a, c]$; $[c, b\,;\, b, a\,;\, b] \equiv [c, b, b\,;\, b, a, b]$;
$[c, b\,;\, b, a\,;\, c] \equiv [c, b, c\,;\, b, a, c]$; $[c, b\,;\, c, a\,;\, b] \equiv [c, b, b\,;\, c, a, b]$;
$[c, b\,;\, c, a\,;\, c] \equiv [c, b, c\,;\, c, a, c]$.

Lemma 2.5: If $i < j$ and $i < k$ then $[x_j, x_i, x_k, x_i] \equiv [x_j, x_i\,;\, x_k, x_i][x_j, x_i, x_i, x_k]$.

Instances of Lemma 2.5 on the alphabet $\{\, a, b, c \,\}$:

$[b, a, b, a] \equiv [b, a, a, b]$; $[b, a, c, a] \equiv [c, a\,;\, b, a]^{-1}[b, a, a, c]$;
$[c, a, c, a] \equiv [c, a, a, c]$; $[c, a, b, a] \equiv [c, a\,;\, b, a][c, a, a, b]$;
$[c, b, c, b] \equiv [c, b, b, c]$.

Forms for $[b, a, c, b]$ and $[c, a, c, b]$ first occur as instances of Lemma 2.8, but these forms are soon simplified:

In Lemma 2.8, $[b, a, c, b] \equiv [b, a\,;\, c, b][b, a, c\,;\, b, a, b][b, a, b, c]^{[c,b]}$.
In Lemma 3.3, $[b, a, c, b] \equiv [b, a\,;\, c, b][b, a, b, c] \equiv [c, b\,;\, b, a]^{-1}[b, a, b, c]$.

In Lemma 2.8, $[c, a, c, b] \equiv [c, a\,;\, c, b][c, a, c\,;\, c, a, b][c, a, b, c]^{[c,b]}$.
In Lemma 3.10, $[c, a, c, b] \equiv [c, a\,;\, c, b][c, a, c\,;\, c, a, b][c, a, b, c]$.

After Corollary 3.51, this would simplify to $[c, a, c, b] \equiv [c, a\,;\, c, b][c, a, b, c]$, but this last form is not stated as a formal result.

Lemma 3.36: $[c, b, a, b] \equiv [b, a, b, c]^{-1}[c, b\,;\, b, a][c, a, b, b]$.

Lemma 3.38: $[c, b, b, a] \equiv [b, a, b, c]^{-1}[c, a, b, b][c, b\,;\, b, a]^2$.

Lemma 3.59: $[c, b, c, a] \equiv [c, b, c\,;\, c, a, c][c, b\,;\, c, a][b, a, c, c]^{-1}[c, a, b, c]$.

Lemma 4.14: $[b, a, d, c] \equiv [b, a\,;\, d, c][b, a, c, d]$.
Lemma 4.15: $[c, a, d, b] \equiv [c, a\,;\, d, b][c, a, b, c]$.
Lemma 4.16: $[d, a, c, b] \equiv [d, a\,;\, c, b][d, a, b, c]$.
Lemma 4.43: $[c, b, d, a] \equiv [c, b\,;\, d, a][c, a, b, d][b, a, c, d]^{-1}$.
Lemma 4.44: $[d, b, c, a] \equiv [d, b\,;\, c, a][d, c\,;\, b, a][b, a, c, d]^{-1}[d, a, b, c]$.

References

[1] Gilbert Baumslag, *Lecture notes on nilpotent groups*, Regional Conference Series in Mathematics, American Mathematical Society, Providence, 1971. MR44#315
[2] Marshall Hall Jr., *The Theory of Groups*, second edition, Chelsea Publishing Company, New York, 1976. MR21#1996, MR54#2765
[3] G. Havas and J. S. Richardson, *Groups of exponent five and class four*, Commun. Alg. **11** (1983), 287–304. MR84f:20035
[4] David A. Jackson, Anthony M. Gaglione, and Dennis Spellman, *Basic commutators as relators*, J. Group Theory **5** (2002), 351–363. MR2003f:20049
[5] Hanna Neumann, *Varieties of Groups*, Ergebnisse der Mathematik und ihrer Grenzgebiete, vol. 37, Springer-Verlag, New York, Berlin, 1967. MR37#6734
[6] Charles C. Sims, *Verifying nilpotence*, J. Symbolic Computation **3** (1987), 231–247. MR88k:20057

Department of Mathematics and Computer Science, Saint Louis University, St. Louis, Missouri 63103

E-mail address: jacksoda@slu.edu

Department of Mathematics, United States Naval Academy, Annapolis, Maryland 21402

E-mail address: amg@usna.edu

5147 Whitaker Avenue, Philadelphia, Pennsylvania 19124

Contemporary Mathematics
Volume **511**, 2010

Basic Commutators as relations: A computational perspective

Primož Moravec and Robert Fitzgerald Morse

Abstract. In this paper we provide a complete example of a nontrivial group computation. We detail the computations needed to show that the normal closure of the set of commutators from a basic sequence of commutators of weight five in a free group F of rank three is equal to the fifth term in the lower central series of F. These computations were completed using a standard desktop computer, and the easily accessible GAP software system and its packages. This result is not new: Jackson, Gaglione and Spellman provide a proof using commutator calculus. The intent of this paper is to provide an example for group theorists on how to complete a serious computation using GAP and to illustrate how using standard systems readily available today can be used to help solve real problems in group theory.

1. Introduction

In his seminal paper "Verifying Nilpotence" [**8**], Charles Sims provides a method for verifying the nilpotence of a finitely presented group. The two main ingredients of Sims's method are finding maximal nilpotent quotients of finitely presented groups and term rewriting processes. His method has recently been applied in showing all 4-Engel groups are locally nilpotent [**2**]. One can also find a discussion and illustration of Sims's method in [**4**].

In [**8**], to illustrate his method for verifying nilpotence, Sims considers the problem of determining whether, given a basic sequence of commutators in a free group, the normal closure of those basic commutators of weight exactly k is equal to the kth term of the lower central series. He reduces the problem to determining whether the finitely presented group with basic commutators of weight exactly k as relations is nilpotent or not. Sims then applies the computational processes he developed earlier in the paper to show that this is true for $k \leq 4$. Sims was also able to show that the result holds for the $k = 5$ and rank 2 case. A complete traditional "hand" proof for the $k = 5$ and arbitrary rank case is given by Jackson, Gaglione, and Spellman [**5**]. The bulk of their paper deals with the rank 3 case. They then use an inductive argument to show the result holds for all ranks.

2000 *Mathematics Subject Classification.* 20-04, 20F18.
Key words and phrases. Basic commutators, Verifying nilpotence.

In this note we revisit the analysis and computations of Sims in [**8**] in a modern setting utilizing computational tools, readily available to mathematicians, such as GAP [**10**] and its packages, particularly Holt's Knuth-Bendix package `KBMAG` [**3**] and Nickel's `nq` [**6**] package for computing nilpotent quotients of finitely presented groups. Following Sims's method for computationally showing a finitely presented group is nilpotent, we computationally complete the $k = 5$ and rank 3 case. While the results in this paper are not new and the computations discussed here almost certainly have been carried out in some level of detail before, our goal is to provide a complete exposition of how to carry out these nontrivial computations. Our exposition includes mapping the mathematics to the GAP computational environment, finding the right ordering of generators so that the Knuth-Bendix procedure terminates, checking output to see that it is correct, and putting the computations into a complete context for a proof of the general result for $k = 5$. To prove the general result we (1) use the analysis of Sims to deal only with one basic sequence, (2) use the computational results for the rank 2 and 3 cases to start our induction and (3) apply the induction arguments found in [**5**] to obtain the full result for arbitrary rank.

2. Preliminaries

Let $r > 1$ and let F be the free group on the ordered finite alphabet $X = \{x_1, x_2, \ldots, x_r\}$. Let $\mathfrak{C}$ be a basic sequence of commutators over the alphabet X that begins with the ordered alphabet. Set $\mathfrak{R}_{\mathfrak{C},k}$ to be the elements of the basic sequence $\mathfrak{C}$ of weight exactly k.

Proving the equality $(\mathfrak{R}_{\mathfrak{C},k})^F = \gamma_k(F)$, where $\gamma_k(F)$ is the kth term of the lower central series of F, is equivalent to showing that the finitely presented group

$$G_k(r) = \langle X \mid \mathfrak{R}_{\mathfrak{C},k} \rangle$$

is nilpotent. This is a consequence of the fact that $\mathfrak{R}_{\mathfrak{C},k}$ is a subset of $\gamma_k(F)$. Therefore the maximal nilpotent quotient of $G_k(r)$ is $F/\gamma_k(F)$. Hence if $G_k(r)$ is nilpotent then it has class $k-1$. The group $G_k(r)$ as defined is dependent on k, $\mathfrak{C}$, and the rank r. To apply computational methods in proving that $G_k(r)$ is nilpotent, we remove its dependence on $\mathfrak{C}$. Hence if $G_k(r)$ can be shown to be nilpotent for one given basic sequence of commutators, it is nilpotent for any basic sequence of commutators. Sims proves this independence for $k \leq 9$:

THEOREM 1 ([**8**]). *For $k \leq 9$, the isomorphism type of $G_k(r)$ depends only on k and r.*

To show for a fixed k that the group $G_k(r)$ is nilpotent, we need only to consider a fixed number of ranks. The following is again found in [**8**]:

PROPOSITION 2. *To show that $G_k(r)$ is nilpotent, it is enough to consider the case $|X| \leq k$.*

We are now able to state our main result.

THEOREM 3. *Let F be the free group on the finite set X, let $\mathfrak{C}$ be a basic sequence of commutators over X, and let $\mathfrak{R}_{\mathfrak{C},5}$ be the set of commutators of weight 5 from $\mathfrak{C}$. Then $(\mathfrak{R}_{\mathfrak{C},5})^F = \gamma_5(F)$.*

Following Theorem 1 and Proposition 2, we need to verify the nilpotency of the groups $G_5(r)$ for $2 \leq r \leq 5$ and a fixed basic sequence of commutators of weight r. In the next section we set up our computational environment to verify the nilpotence of $G_5(r)$ for $2 \leq r \leq 5$.

3. Computational infrastructure

In this section we provide the computational infrastructure needed to prove Theorem 3. This consists primarily of several GAP functions[1] and in fixing our mathematical objects and GAP objects.

Our fixed basic sequence of commutators begins with the ordered alphabet X. The ordering of commutators of equal weight is done lexicographically. We define a fully bracketed commutator over the symbol set X as follows: if $x \in X$ then $[x]$ is fully bracketed, and if c and d are fully bracketed commutators then $[c, d]$ is fully bracketed. This definition translates directly into GAP as a list of lists. We can then recursively compute the weight of such a commutator as given below.

```
##  Compute the weight of a fully bracketed commutator <comm>.
##
Weight :=
  function(comm)
    if Length(comm)=1 then return 1; fi;
    return Weight(comm[1])+Weight(comm[2]);
  end;
```

The actual elements of X can be any totally ordered data type in GAP. We choose integers so that we can use the natural lexicographical ordering of lists of lists of integers available in GAP. The following function creates fully bracketed commutators of a given weight from the lexicographical basic sequence.

```
##  Build the lexicographical basic sequence of commutators of
##  the given <weight> and rank <r>. The symbol set is the list
##  of integers [1..r].
##
BasicComSeq :=
  function(r, weight)
      if weight=1 then
          return List([1..r], x->[x]);
      fi;
      return
        Concatenation
          ( List
            ( Partitions(weight,2),
              p->Filtered
                  ( Cartesian
                    ( BasicComSeq(r, Maximum(p)),
                      BasicComSeq(r, Minimum(p))
                    ),
                    C->(p[1]=p[2]  and C[1]>C[2]) or
                       (p[1]<>p[2] and C[1][2]<=C[2])
                  )
            )
          );
  end;
```

[1]A file of the GAP functions and commands listed in this section and Section 4 can be found at the second author's website rfmorse.evansville.edu/basiccommutators.g

The following function uses BasicComSeq to give a full sequence of basic commutators up to the given weight.

```
## Build the lexicographical basic sequence of commutators
## of all weights up to <w> and rank <r>.
##
FullBasicComSeq :=
  function(r, w)
    return Concatenation(List([1..w], w->BasicComSeq(r,w)));
  end;
```

The basic commutators over the symbols [1,2,3] of weight at most 2 are

```
gap> PrintArray(FullBasicComSeq(3,2));
[ [       1 ],
  [       2 ],
  [       3 ],
  [  [ 2 ],  [ 1 ] ],
  [  [ 3 ],  [ 1 ] ],
  [  [ 3 ],  [ 2 ] ] ]
```

We use the integer entries of such fully bracketed commutators as indices for the generators of the free group to construct their associated words in the free group. The following function constructs this associated word.

```
##  The integer entries of the fully bracketed commutator
##  <comm> are associated with the corresponding generator of the free
##  group <F>. The commutator is then evaluated to a word in <F>.
##
Eval :=
  function(F,comm)
    if Length(comm)=1 then return F.(comm[1]); fi;
    return Comm(Eval(F,comm[1]), Eval(F,comm[2]));
  end;
```

Here is an example to compute the commutator $[[f_3, f_1], f_1]$.

```
gap> F   := FreeGroup(3);;
gap> Wrd := Eval(F,[[[3],[1]],[1]]);
f1^-1*f3^-1*f1*f3*f1^-1*f3^-1*f1^-1*f3*f1^2
gap> Wrd = Comm(Comm(F.3,F.1),F.1);
true
```

The function Eval is used to create a presentation for $G_5(2)$

```
gap> F := FreeGroup(2);;
gap> # Evaluate as words in the free group F the words in
gap> # the basic sequence of length 5.
gap> R := List(BasicComSeq(2,5),c->Eval(F,c));;
gap> G_5_2 := F/R;
<fp group on the generators [ f1, f2 ]>
```

The group G_5_2 has a maximal nilpotent quotient of class 4 and we give its simplified presentation for later comparison.

```
gap> NilpotencyClassOfGroup(NilpotentQuotient(G_5_2));
4
gap> PresentationFpGroup(SimplifiedFpGroup(G_5_2));
<presentation with 2 gens and 6 rels of total length 154>
```

To show that $G_5(2)$ is nilpotent we create a group isomorphic to $G_5(2)$ with generators associated with the commutators in a basic sequence of weight up to 4.

Here is the general situation. Let $\mathfrak{C}$ be a basic sequence of commutators over the set $X = \{x_1, \ldots, x_r\}$. Let $x_1, \ldots, x_r, c_{r+1}, \ldots, c_n$ be a basic sequence of commutators of weight less than k, with $x_1, \ldots, x_r$ being an ordering of the generating set X and $n = \sum_{w=1}^{k-1} W(r, w)$, where $W(r, w)$ is the number of basic commutators of weight w of rank r computed using Witt's formula (see [**1**]). Set

$$\mathfrak{G} = \langle x_1, \ldots, x_n \mid x_{r+1} = c_{r+1}, \ldots, x_n = c_n, \mathfrak{R}_{\mathfrak{C},k} \rangle.$$

Tietze transformations are then used to simplify our presentation of $\mathfrak{G}$ so that $\mathfrak{G}$ is recognized to be $G_k(r)$. Sims's procedure for verifying the nilpotency of $G_k(r)$ requires us to show that $G_k(r)$ has a maximal nilpotent quotient and that $\mathfrak{G}$ has a consistent polycyclic presentation. The group $G_5(2)$ has a maximal nilpotent quotient of class 4. What remains to be done is to set up $\mathfrak{G}$ for this case and apply the Knuth-Bendix rewriting procedure to show $\mathfrak{G}$ has a consistent polycyclic presentation.

We can build $\mathfrak{G}$ using the infrastructure in place. Moreover, we can express each basic commutator on $\mathfrak{G}$ of weight 2 or greater as a commutator in two generators. The following GAP function does this recursive substitution.

```
##  Express each basic commutator as a fully bracketed commutator
##  with indices in the lexicographical basic sequence.
##
Recursive :=
  function(r,w)
    if w=1 then return BasicComSeq(r,w); fi;
    return
      List
      ( BasicComSeq(r,w), c->
        [ [ Position
            ( FullBasicComSeq(r, Weight(c[1])),
              c[1]
            )
          ],
          [ Position
            ( FullBasicComSeq(r, Weight(c[2])),
              c[2]
            )
          ]
        ]
      );
end;
```

We now create $\mathfrak{G}$ as a GAP object.

```
gap> # Build the generators of the group.
gap> gencoms := Concatenation(
>                 List([1..4], w->Recursive(2,w)));;
gap> # Create the free group and a relation associated to each
gap> # generator
gap> F:= FreeGroup(Length(gencoms));;
gap> rels := List([1..Length(gencoms)],
>                   i->F.(i)/Eval(F,gencoms[i]));;
gap> # Create the relations of G_5(2) and append them
gap> # to the relations already created and form the FP group.
gap> Append(rels,List(Recursive(2,5),c->Eval(F,c)));;
```

```
gap> G := F/rels;
<fp group on the generators [ f1, f2, f3, f4, f5, f6, f7, f8 ]>
gap> # Give a similar simplified presentation as G_5_2
gap> PresentationFpGroup(SimplifiedFpGroup(G));
<presentation with 2 gens and 6 rels of total length 154>
```

From the GAP group G we can create a rewriting system and check to see if it is confluent. However, as the group is presented neither the rewriting algorithms in GAP nor the use of KBMAG package provide an answer. In the next section we discuss some strategies to show that G has a consistent polycyclic presentation.

4. Knuth Bendix rewriting procedure

Let $G = F/R$ be a finitely presented group. The Knuth-Bendix procedure is designed to construct a normal form for the elements of G in terms of the generators of F. The normal form of an element is the least word in the generators of F and their inverses that represents the element in G, with respect to a specific ordering on the set of all words in F. This procedure has its ring theoretic analogue, namely the Gröbner bases of ideals of polynomial rings.

KBMAG offers four types of orderings used in rewriting systems. These are shortlex, recursive, wtlex and wreathprod. Wreath product orderings are the ones on which reduction to normal form in polycyclic groups is based [**9**, page 395]. Thus they are particularly well suited for finding rewriting systems of polycyclic groups. A special case of wreath product orderings is the so called recursive ordering. This can be described as follows. Let u and v be strings in the generators of G. If one of u and v, say v, is empty, then $u \succeq v$. Otherwise, let $u = u'a$ and $v = v'b$, where a and b are generators. Then $u \succ v$ if and only if one of the following holds:

(1) $a = b$ and $u' \succ v'$,
(2) $a \succ b$ and $u \succ v'$,
(3) $b \succ a$ and $u' \succ v$.

It becomes clear that the recursive ordering is the one that follows the spirit of the concept of basic commutators. Thus we use this ordering as our primary choice.

We order the generators of the group in reverse order. This is motivated by a similar trick used commonly in computations of Gröbner bases.

Using the GAP infrastructure in Section 3, we create the same group $\mathfrak{G}$ as before, but reverse the ordering of the generators of the free group in creating it. We need only one more function to help us relabel the basic commutators in this reverse ordering.

```
##  Treat each numeric entry in each commutator as a difference
##  value from which a new number (label) can be created.
##
Relabel :=
  function(d,comm)
    if Weight(comm)=1 then return [d-comm[1]]; fi;
    return [Relabel(d,comm[1]),Relabel(d,comm[2])];
  end;
```

We now create the group $\mathfrak{G}$ in GAP whose recursive presentation is ordered in reverse for $r = 2$. We know that there are 8 commutators that will be generators of $\mathfrak{G}$ from our previous analysis.

```
gap> F := FreeGroup(8);;
gap> # Build the generating set of commutators
```

```
gap> gencoms := Concatenation(List(([1..4]), w->Recursive(2,w)));;
gap> # For r=2 we will have 1-8 generators -- reverse the labels
gap> gencoms := Reversed(List(gencoms,c->Relabel(9,c)));;
gap> rels := List([1..8], i->F.(i)/Eval(F,gencoms[i]));;
gap> # Build the relations of weight 5
gap> coms := List(Recursive(2,5),c->Relabel(9,c));;
gap> Append(rels, List(coms, c->Eval(F,c)));;
gap> G := F/rels;
<fp group on the generators [ f1, f2, f3, f4, f5, f6, f7, f8 ]>
```

We can do some simple checks to see that we have the right group. It has a maximal nilpotent quotient of class 4 and after applying Tietze transformations we arrive back to a presentation similar to that of $G_5(2)$.

```
gap> # The maximal nilpotent quotient is 4
gap> NilpotencyClassOfGroup(NilpotentQuotient(G));
4
gap> PresentationFpGroup(SimplifiedFpGroup(G));
<presentation with 2 gens and 6 rels of total length 154>
```

We now apply the Knuth-Bendix procedure to the GAP object `G` using the recursive ordering. It finds a confluent system of relations.

```
gap> R := KBMAGRewritingSystem(G);;
gap> SetOrderingOfKBMAGRewritingSystem(R,"recursive");;
gap> KnuthBendix(R);
true
```

The set up for the rank 3 case follows exactly the same pattern as the rank 2 case. There are 32 commutators that are the generators of $\mathfrak{G}$, with 48 relations, which are the basic commutators of weight 5 and rank 3.

```
gap> F := FreeGroup(32);;
gap> # Build the generating set of commutators
gap> gencoms := Concatenation(List(([1..4]), w->Recursive(3,w)));;
gap> # For r=3 we will have 1-32 generators -- reverse the labels
gap> gencoms := Reversed(List(gencoms,c->Relabel(33,c)));;
gap> rels := List([1..32], i->F.(i)/Eval(F,gencoms[i]));;
gap> # Build the relations of weight 5
gap> coms := List(Recursive(3,5),c->Relabel(33,c));;
gap> Append(rels, List(coms, c->Eval(F,c)));;
gap> G := F/rels;
<fp group with 32 generators>
```

Our check again works out. The group `G` has a maximal nilpotent quotient of class 4 and when we simplify the presentation we obtain a group with 3 generators and 48 relations.

```
gap> # The maximal nilpotent quotient is 4
gap> NilpotencyClassOfGroup(NilpotentQuotient(G));
4
gap> PresentationFpGroup(SimplifiedFpGroup(G));
<presentation with 3 gens and 48 rels of total length 1276>
```

We now have to set up some more parameters in `KBMAG`, in particular, a bound on the length of the equations, and the number of equations and states allowed before starting the rewriting process. We impose these constraints so that the rewriting process has a chance to terminate. As we see below, we do terminate but need to release one constraint and recheck to see if we truly have a confluent rewriting system.

```
gap> R := KBMAGRewritingSystem(G);;
gap> SetOrderingOfKBMAGRewritingSystem(R,"recursive");;
gap> O:=OptionsRecordOfKBMAGRewritingSystem(R);;
gap> O.maxstoredlen:=[50,50];;
gap> O.maxstates:=2^20;;
gap> O.maxeqns:=2^20;;
gap> KnuthBendix(R);
#WARNING: Because of the control parameters you set,
#         the system may
#         not be confluent. Unbind the parameters and
#         re-run KnuthBendix to check!
#I  System computed is NOT confluent.
false
gap> Unbind(O.maxstoredlen);
gap> KnuthBendix(R);
true
```

Determining whether the rewriting system R is confluent can take a significant amount of computational time depending on the speed of the computer used. These times range from one hour to eight minutes. The memory required is less than 50 kilobytes of RAM.

We now complete the proof of Theorem 3 using the analytical work from Jackson, Gaglione, and Spellman [**5**]. Our computations show the result is true for $r = 3$. Theorems 4.53 and 5.10 in [**5**] give the result for $r = 4$ and $r = 5$ respectively using the inductive result for $r = 3$ and $r = 4$ respectively. This completes the proof of the theorem.

5. Conclusion

Our computations in Section 4 show that computing a maximal nilpotent quotient is relatively automatic, at least in the finitely presented groups we are working with. However, the "naive" use of rewriting procedures provide no results even for the $r = 2$ case. With some finesse we are able to settle the question for $r = 3$. This case takes up nearly 20 pages in [**5**]. While even the $r = 2$ case took some work, Sims [**8**] was able to make this computation. For $r = 3$ we assume it was out of reach for the computational power available at the time.

We attempted to employ the computational methods outlined in Sections 3 and 4 to show that $G_6(2)$ is nilpotent and made no progress at all. It is still an open question whether $(\mathfrak{N}_{\mathfrak{C},k})^F = \gamma_k(F)$ for $k > 5$ and $r \geq 2$. Other possible approaches to the problem include showing that $G_k(r)$ is a homomorphic image of a nilpotent group or using Sievers's Free Group Algorithms package, FGA[**7**], to assist in making computations similar to those found in [**5**].

References

[1] Marshall Hall, Jr. *The theory of groups*. The Macmillan Co., New York, N.Y., 1959.

[2] George Havas and M. R. Vaughan-Lee. 4-Engel groups are locally nilpotent. *Internat. J. Algebra Comput.*, 15(4):649–682, 2005.

[3] D Holt. *KBMAG – Knuth-Bendix on Monoids and Automatic Groups*, 2009. A GAP Package, see [**10**].

[4] Derek F. Holt, Bettina Eick, and Eamonn A. O'Brien. *Handbook of computational group theory*. Discrete Mathematics and its Applications (Boca Raton). Chapman & Hall/CRC, Boca Raton, FL, 2005.
[5] D. A. Jackson, A. M. Gaglione, and D. Spellman. Weight five basic commutators as relators. In *Computational Group Theory and the Theory of Groups II*, Contemp. Math., Providence, RI. Amer. Math. Soc.
[6] W. Nickel. *nq – Nilpotent Quotients of Finitely Presented Groups*, 2003. A GAP Package, see **[10]**.
[7] C. Sievers. *FGA – Free Group Algorithms*, 2005. A GAP Package, see **[10]**.
[8] Charles C. Sims. Verifying nilpotence. *J. Symbolic Comput.*, 3(3):231–247, 1987.
[9] Charles C. Sims. *Computation with finitely presented groups*, volume 48 of *Encyclopedia of Mathematics and its Applications*. Cambridge University Press, Cambridge, 1994.
[10] The GAP Group. *GAP – Groups, Algorithms, and Programming, Version 4.4*, 2005. (`http://www.gap-system.org`).

DEPARTMENT OF MATHEMATICS, UNIVERSITY OF LJUBLJANA, JADRANSKA 21, 1000 LJUBLJANA, SLOVENIA
E-mail address: `primoz.moravec@fmf.uni-lj.si`
URL: `www.fmf.uni-lj.si/~moravec`

DEPARTMENT OF ELECTRICAL ENGINEERING AND COMPUTER SCIENCE, UNIVERSITY OF EVANSVILLE, EVANSVILLE IN 47722 USA
E-mail address: `rfmorse@evansville.edu`
URL: `rfmorse.evansville.edu`

Contemporary Mathematics
Volume **511**, 2010

Groups of Minimal Order which are not n-Power Closed

Luise-Charlotte Kappe and Gabriela Mendoza

Dedicated to Wolfgang P. Kappe on the occasion of his eightieth birthday.

ABSTRACT. A group in which the set of n-th powers of elements forms a subgroup is called n-power closed. It is well known that the squares of elements in a group do not necessarily form a subgroup and that the alternating group on four letters is minimal with this property. For given n, what are the groups of minimal order such that the n-th powers of elements do not form a subgroup? For odd n, it can be shown that the dihedral group of order $2p$ is minimal with this property, where p is the smallest prime dividing n. If n is even, the situation is more complex. The order of the minimal group with the property of being not n-power closed depends on the odd prime factors of n and the exact 2-power dividing n. With initial guidance from GAP, we determine the groups of minimal order which are not n-power closed for $n = 2k$, $4k$, and $8k$, where k is odd.

1. Introduction

It is well known that the squares of elements in a group do not necessarily form a subgroup and that the alternating group on four letters is minimal with this property. This raises the following question: For given n, what are the groups of minimal order such that the n-th powers of elements do not form a subgroup? Providing some answers to this question is the topic of this paper. To the best of our knowledge, this question is nowhere dealt with in the literature, except in a paper by Desmond McHale [6] for the case $n = 2$. He shows that there are two groups of order 12, namely A_4, the alternating group on four letters, and $T = \langle a, b \mid a^3 = b^4 = 1, a^b = a^{-1}\rangle$, in which the squares do not form a subgroup, and that these groups are of minimal order with this property.

To make our notions more precise, we define $G^{(n)} = \{g^n \mid g \in G\}$ to be the set of elements in a group G raised to the n-th power, $n \in \mathbb{N}$, and $G^n = \langle g^n \mid g \in G\rangle$ as the subgroup generated by such elements. According to [4], we say a group is n-power closed if $G^{(n)} = G^n$. A class of groups having the property of being n-power closed for all n is the class of regular p-groups, introduced by P. Hall in [2].

2000 *Mathematics Subject Classification.* Primary 20D25, 20D60, Secondary 20F99.

Key words and phrases. n-power closed, groups of minimal order which are not n-power closed.

However, regular p-groups are not the only nonabelian groups with this property. As can be easily verified, the nonabelian groups of order 8 have this property too.

Our investigations of this topic started with GAP [1]. We collected experimental material by determining the groups of smallest order which are not n-power closed for $n = 2, 3, \dots, 30$. The GAP output for these cases allowed us to correctly conjecture the groups of smallest order which are not n-power closed for n odd and for $n \equiv 2 \pmod 4$. To determine the groups of smallest order which are not n-power closed for $n \equiv 4 \pmod 8$, we used a combination of direct inspection and GAP analysis of the power structure of groups of relevant orders. For further details we refer to the exposition. This is the content of [7]. The investigations were continued in [8] providing the groups G of minimal order which are not n-power closed for $n \equiv 8 \pmod{16}$. In [7] and [8], the use of GAP in the proofs was eliminated with the exception of the determination of the power structure of 2-groups and a few other instances. To keep the length of this paper within acceptable bounds, we again refer in many cases to the help of GAP. The GAP programs used in this context can be found in the last section of the paper.

We can look at the above results from a different angle by defining for every $n \in \mathbb{N}$, $n \geq 2$, the minimal number $\mu(n)$, if for every group G of order less than $\mu(n)$ we have $G^{(n)} = G^n$ and there exists a group of order $\mu(n)$ with $G^{(n)} \neq G^n$. With this definition we have $\mu(2) = 12$ [6], and $\mu(n) = 2p$ for n odd and p the smallest prime dividing n (Corollary 3.2). For n even the situation is more complex. From our list of groups of minimal order with $|G| \leq 256$ which are not n-power closed for n even, we have determined $\mu(2^\alpha k)$ with k odd and $\alpha = 1, 2, 3$ (Corollaries 4.3, 5.4, and 6.8).

The case $n = 2^\alpha k$, $\alpha \geq 4$ and k odd, remains open. We will give further details about this at the end of Section 6 and also discuss what the difficulties are in arriving at an explicit solution.

2. Preparatory results

For the convenience of the reader, this section contains various preparatory results and some examples to be applied in the rest of the paper. For the most part they are given here without proof.

LEMMA 2.1. *Let $n \in \mathbb{N}$ and G a group. If G^n is abelian and $G^{(n)} = G^n$, then $G^{(nk)} = G^{nk}$ for all $k \in \mathbb{N}$.*

LEMMA 2.2. *Let G be a group, $n \in \mathbb{N}$, and $\phi : G \to G$ with $\phi(g) = g^n$. If $\gcd(|G|, n) = 1$, then ϕ is onto and $G^{(n)} = G^n = G$.*

LEMMA 2.3. *Let G be a group and $n, t \in \mathbb{N}$. If $\gcd(|G|, n) = t$, then $G^{(n)} = G^{(t)}$.*

Let H be a subgroup or a subset of a finite group G. Recall that $\exp(H)$ is the least common multiple of element orders in H. With this definition we obtain the following proposition which has many applications throughout the paper.

PROPOSITION 2.4. *Let $G = NH$ with N an abelian normal subgroup of G and $\gcd(|N|, |H|) = 1$. If n is an integer such that $\exp(H)$ divides n, then $G^{(n)} = G^n$ and $G^n \leq N$.*

PROOF. Let $q = \exp(H)$. Then $g^q \in N$ for all $g \in G$. Hence $G^{(q)} \subseteq N = N^{(q)} \subseteq G^{(q)}$. It follows that $G^{(q)} = N$ and thus $G^{(q)} = G^q$. Since G^q is abelian and q divides n, we obtain that $G^{(n)} = G^n$ by Lemma 2.1. $\square$

For the convenience of the reader, the next proposition lists some classes of groups which are n-power closed for all integers n. For a detailed definition of regular p-group we refer the reader to [3].

PROPOSITION 2.5. *Let G be abelian, a regular p-group, or a group of order 8. Then G is n-power closed for all integers n.*

The next lemma provides a sufficient condition such that $G^{(m)} \neq G^m$ implies $G^{(mt)} \neq G^{mt}$.

LEMMA 2.6. *Let G be a group and m, t integers. Then $G^{(m)} \neq G^m$ implies $G^{(mt)} \neq G^{mt}$ provided* $\gcd(t, \exp(G^{(m)})) = 1$.

In our final proposition we determine the power structure for a class of groups which are essential for our investigations.

PROPOSITION 2.7. *Let p be an odd prime and $\alpha \geqslant 2$. Then for*

$$T = T(p,\alpha) = \langle a, b \mid a^p = b^{2^\alpha} = 1, a^b = a^{-1}\rangle \tag{2.7.1}$$

we have $T^{(2^\beta)} \neq T^{2^\beta}$ for $1 \leq \beta < \alpha$ and $T^{(2^\beta)} = T^{2^\beta} = \langle a\rangle$ for $\beta \geq \alpha$.

PROOF. Every $x \in T$ can be written uniquely as $x = a^j b^k$, where $0 \leq j \leq p-1$ and $k = 0, 1, \dots, 2^\alpha - 1$. First let $1 \leq \beta < \alpha$. We observe that $[a, b^2] = 1$. It follows that $x^2 = (a^j b^k)^2 = a^{2j} b^{2k}$, when k is even, and $x^2 = (a^j b^k)^2 = b^{2k}$ if k is odd. Hence, $T^{(2)} = \langle b^2\rangle \cup (\cup_{j=1}^{p-1} a^j \langle b^4\rangle)$. Since $a, b^2 \in T^{(2)}$ but $ab^2 \notin T^{(2)}$, it follows that $T^{(2)}$ is not a subgroup, hence $T^{(2)} \neq T^2$. Raising $T^{(2)}$ to the power $2^{\beta-1}$, $1 \leq \beta < \alpha$, and observing again $[a, b^2] = 1$, we arrive at $T^{(2^\beta)} = \langle b^{2^\beta}\rangle \cup (\cup_{j=1}^{p-1} a^j \langle b^{2^{\beta+1}}\rangle)$. Since $a, b^{2^\beta} \in T^{(2^\beta)}$ but $ab^{2^\beta} \notin T^{(2^\beta)}$, we conclude that $T^{(2^\beta)}$ is not a subgroup, hence $T^{(2^\beta)} \neq T^{2^\beta}$. Now let $\beta \geq \alpha$, then $T^{(2^\alpha)} = \langle a\rangle = T^{2^\alpha}$ by Proposition 2.4, and $T^{(2^\beta)} = \langle a\rangle = T^{2^\beta}$ by Lemma 2.1 for $\beta > \alpha$. □

3. The case of odd powers

For an odd integer n we can completely determine $\mu(n)$, the smallest order for which there exists a group G with $|G| = \mu(n)$ such that $G^{(n)} \neq G^n$.

THEOREM 3.1. *Let n be an odd integer and p the smallest prime dividing n. Then $G^{(n)} = G^n$ for all G with $|G| \leq 2p$, except for*

$$G \cong D_p = \langle a, b \mid a^p = b^2 = 1,\ a^b = a^{-1}\rangle,$$

the dihedral group of order $2p$, for which $G^{(n)} \neq G^n$.

PROOF. Let $|G| < 2p$. If $\gcd(|G|, n) = 1$, then $G^{(n)} = G^n$ by Lemma 2.2. If $\gcd(|G|, n) \neq 1$, we can assume $|G| \geq p$, since p is the smallest prime dividing n. It follows that $\gcd(|G|, n) = q$, where q is a prime with $p \leq q < 2p$. Since $|G| < 2p$ and all other integers m with $p < m < 2p$ are products of primes less than p, this implies that $G \cong C_q$ and hence G is n-power closed as an abelian group.

Now, let G be a group of order $2p$. If G is abelian, then $G^{(n)} = G^n$. Otherwise, $G \cong D_p$. We have that $\gcd(|G|, n) = p$, since $|G| = 2p$ and p is the smallest prime dividing the odd integer n. It can be easily seen that $D_p^{(p)}$ is the set of elements of order dividing 2. Hence $|D_p^{(p)}| = p + 1$. Since $p + 1$ is not a divisor of $2p$, it follows that $D_p^{(p)}$ is not a subgroup of D_p. Hence $D_p^{(p)} \neq D_p^p$. Observing that

$\gcd(2p, n) = p$, it follows that $D_p^{(n)} = D_p^{(p)}$ by Lemma 2.3. Hence $G^{(n)} \neq G^n$ for $G \cong D_p$. □

As a simple corollary we obtain the minimal number $\mu(n)$ where n is an odd integer.

COROLLARY 3.2. *Let n be an odd integer, $n \geq 3$, and p the smallest prime dividing n. Then $\mu(n) = 2p$.*

4. The case $\mathbf{n} \equiv 2 \pmod 4$

The case of even n is more involved. The goal of this section is to determine the minimal number $\mu(n)$ in case $n \equiv 2 \pmod 4$, or equivalently $n = 2k$, k odd. In our first theorem we determine $\mu(n)$ for $n = 2^\alpha k$, $\alpha \geq 1$ and k not divisible by 3.

THEOREM 4.1. *Let n be an integer with $n = 2^\alpha k$, $\alpha \geq 1$, where k is odd. If $|G| \leq 12$, then $G^{(n)} = G^n$ with the exception when k is not divisible by 3 and then there exist two groups with $G^{(n)} \neq G^n$ if $\alpha = 1$, namely $G \cong A_4$, the alternating group on four letters, and $G \cong T(3,2)$, and if $\alpha > 1$ then there exists exactly one such group, namely A_4.*

PROOF. Let G be a group with $|G| < 12$, and $n = 2^\alpha k$ with $\alpha \geq 1$ and k odd. If G is abelian or has order 8, then $G^{(n)} = G^n$ by Proposition 2.5. The remaining nonabelian groups of order less than 12 are D_3, and D_5, the dihedral groups of order 6 and 10. It follows from Proposition 2.4 that D_3 and D_5 are n-power closed for n even.

Now let G be a nonabelian group of order 12. For $G \cong D_6 = D_3 \times C_2$ we have that $D_6^{(2)} \cong D_3^{(2)} \times C_2^{(2)} = \langle a \rangle \cong C_3$. So by Lemma 2.3 it follows that $D_6^{(n)} \subseteq D_6^{(2)} \cong C_3$. Thus $G^{(n)} = G^n$ by Lemma 2.1. Now let k be not divisible by 3. Consider $G \cong A_4$. Then $\gcd(|A_4|, n) = 2$ or 4. Hence $A_4^{(n)} = A_4^{(2^\beta)}$ for $\beta = 1, 2$ by Lemma 2.3. Since $A_4^{(2^\beta)}$ consists of the elements of order dividing 3, we have $|A_4^{(2^\beta)}| = 9$. Thus A_4 is not n-power closed.

Finally, let $G \cong T = T(3,2)$ of Proposition 2.7. If $\gcd(|T|, n) = 2$, then $T^{(n)} \neq T^n$ by Lemma 2.3 and Proposition 2.7. However, $T^{(4)} = T^4$, again by Proposition 2.7. Thus $T^{(n)} = T^n$ for $n = 2^\alpha k$, $\alpha \geq 2$. □

From Table 7.1 we see that there exist two groups of order 16 which are not 2-power closed. Thus we obtain $\mu(n) \leq 16$ for all n with $n \equiv 2 \pmod 4$. On the other hand, if we let n be divisible by 6, the groups A_4 and $T(3,2)$ will be n-power closed, since they have exponent 6. This leads to our next theorem.

THEOREM 4.2. *Let $n = 2k$, where k is odd and divisible by 3. If $|G| \leq 16$, then $G^{(n)} = G^n$ except when $|G| = 16$ and there exist two groups of order 16 with $G^{(n)} \neq G^n$.*

PROOF. First assume that $|G| < 16$. If G is abelian or has order 8, our claim follows by Proposition 2.5. Thus $G \cong D_p$, $p = 3, 5, 7$, or $|G| = 12$. If $G \cong D_p$, it follows by Proposition 2.4 that G is n-power closed since n is even. It remains to consider the case when $|G| = 12$. If $G \cong D_6$ or A_4, then $G^{(n)} = G^n = \{1\}$, since $\exp(G) = 6$ and 6 divides n. We have to consider $G \cong T = T(3,2)$. A straightforward calculation shows that $T^{(6)} = T^6 = \langle b^2 \rangle \cong C_2$, and hence $T^{(n)} = T^n$ by Lemma 2.1, since 6 divides n.

Now let $|G| = 16$. By Table 7.1 it follows that there exist two groups of order 16 with $G^{(2)} \neq G^2$. By Lemma 2.3 we obtain that $G^{(2)} = G^{(2k)}$, hence $G^{(2k)} \neq G^{2k}$. For all other groups of order 16 we have $G^{(n)} = G^n$ by Lemma 2.1 and the help of GAP. $\square$

As a corollary to Theorems 4.1 and 4.2, we obtain the minimal number $\mu(2k)$ for k odd.

COROLLARY 4.3. *Let $n = 2k$ be an integer with k odd. Then $\mu(n) = 16$, unless 3 does not divide n, in which case $\mu(n) = 12$.*

5. The case $\mathbf{n} \equiv 4 \pmod 8$

The goal of this section is to determine the minimal number $\mu(n)$ for $n = 4k$, k odd. As we can see from Table 7.1, there exist nine groups of order 64 which are not 4-power closed but all of order 16 and 32 are. It follows by Lemma 2.3 that for those nine groups of order 64 we have $G^{(4k)} \neq G^{4k}$ for all odd k. Thus $\mu(4k) \leq 64$. The question arises for which values of k do we have $\mu(4k) < 64$. By Theorem 4.1 it follows that $\mu(4k) = 12$ whenever k is not divisible by 3. Since the groups of order 16 are 4-power closed, the question arises what is the value of $\mu(12)$. Using the GAP function of Example 7.2 with $n = 12$, we obtain $\mu(12) = 40$. This is summarized in the following theorem. We note that GAP is only used for the groups of order 16 and 32.

THEOREM 5.1. *Let $n = 4k$ with k odd and divisible by 3. If $|G| \leq 40$, then $G^{(n)} = G^n$ except when k is not divisible by 5 and $G \cong T(5,3)$ or $G \cong K$, where*

$$K = \langle a, b \mid a^5 = b^8 = 1,\ a^b = a^2 \rangle.$$

PROOF. First let $|G| < 40$. We will show that G^{12} is abelian and $G^{(12)} = G^{12}$. Then it follows by Lemma 2.1 that $G^{(n)} = G^n$, since 12 divides n. In case G is abelian, of order 8 or a regular p-group, our claim follows by Proposition 2.5. If G has order 16 or 32, it follows from Table 7.1 that $G^{(4)} = G^4$. It can be easily shown by GAP that G^4 is abelian. Thus by Lemma 2.3 we obtain $G^{(4)} = G^{(12)}$ and hence $G^{(12)} = G^{12}$ by Lemma 2.1. Now let $G \cong D_q$ or let G have order 18, 21, 39, 20, or 28. Then G has a normal abelian Sylow q-subgroup P_q and $G = P_qH$, where $\gcd(|P_q|, |H|) = 1$ and $\exp(H)$ divides 12. Thus our claim follows from Proposition 2.4. We observe that any group of order 30 has a normal cyclic subgroup of order 15. Thus our claim follows again from Proposition 2.4. The remaining cases are those where $|G|$ is divisible by 12. Thus we can assume $|G| = 24$ or 36 and obviously $\exp(G) > 12$. If $|G| = 24$, then $\exp(G) = 24$ and G has a cyclic Sylow 2-subgroup and $P_3 \lhd G$. This yields $G \cong T(3,3)$, and hence $T^{(12)} = T^{12} = \langle b^4 \rangle$ by Proposition 2.7 and Lemma 2.1. If G has order 36, we can assume that $\exp(G) = 18$ or 36, and hence $P_3 \cong C_9 = \langle b \rangle$. Note that in a group of order 36 at least one of the Sylow subgroups is normal. Assume first that $P_2 \lhd G$. Since G is not abelian, we can assume $P_2 \cong C_2 \times C_2$ with $P_3 = \langle b \rangle$, where b induces an automorphism of order 3 on P_2. This implies that $\langle b^3 \rangle$ is a central normal subgroup of G. We obtain $G^{(12)} \subseteq \langle b^3 \rangle$. Since $b^{12} = b^3$ and $(b^{-1})^{12} = b^6$, it follows that $G^{(12)} = \langle b^3 \rangle$. Thus $G^{(12)} = G^{12}$ and G^{12} is abelian. Now assume that $P_3 \lhd G$. Then $G \cong C_9 \rtimes P_2$, hence $G^{(4)} = G^4$ by Proposition 2.4. Since G^4 is abelian, it follows that $G^{(12)} = G^{12}$ by Lemma 2.1 and G^{12} is abelian.

Finally let $|G| = 40$. We must have $P_5 \lhd G$ and $\exp(G) = 40$, since otherwise $G^{(4)} = G^4 = P_5$ and our claim follows from Lemma 2.1. Thus $P_2 \cong C_8$ and $G \cong T(5,3)$ or $G \cong K$. If $G \cong K$, it can be easily verified that $K^{(4)} = \langle a \rangle \cup \langle b^4 \rangle$ and hence $|K^{(4)}| = 6$. It follows by Lemma 2.6 that $K^{(4k)} \neq K^{4k}$ for any odd k not divisible by 5. □

It can be easily seen that the groups $T(5,3)$ and K of Theorem 5.1 are 24-power and 60-power closed. Since our goal is to determine the values of $\mu(n)$ for $n = 4k$, k odd, we have to choose $k = 15$ in our next step. Instead of running the GAP function of Example 7.2 with $n = 60$, we decided to do the calculations manually with the exception of $|G| = 48$. The main reason for this is to avoid missing any orders prime to 60 which might be values of $\mu(n)$. More about this in the next section.

THEOREM 5.2. *Let $n = 4k$ with k odd and divisible by 15. If $|G| \leq 56$, then $G^{(n)} = G^n$ except when k is not divisible by 7 and $G \cong T(7,3)$ or $G \cong L$, where*

$$L = \left\langle e_1, e_2, e_3, b \;\middle|\; \begin{array}{l} e_1^2 = e_2^2 = e_3^2 = b^7 = 1, \\ e_1^b = e_2, e_2^b = e_3, e_3^b = e_1e_3, [e_i, e_j] = 1, \quad 1 \leq i < j \leq 3 \end{array} \right\rangle.$$

PROOF. First let $|G| < 56$. We will show that G^{60} is abelian and $G^{(60)} = G^{60}$. Then it follows by Lemma 2.1 that $G^{(n)} = G^n$, since 60 divides n. If $|G| < 40$, our claim follows by Theorem 5.1, since 12 divides 60. Now let $|G| = 40$. Since 20 divides n, it suffices to consider the two nonabelian groups of exponent 40, namely K and $T(5,3)$, as given in Theorem 5.1. We obtain $G^{(20)} = G^{20} = \langle b^4 \rangle$ in both cases. Since G^{20} is abelian, it follows by Lemma 2.1 that $G^{(60)} = G^{60}$. Now let $40 < |G| < 56$. If $|G| = 42$, 44, 46, 50, 51, 52, or 55, the q-Sylow subgroup for the largest prime q dividing the order of G is abelian and normal in G. Thus $G \cong P_qH$, where $\exp(H)$ divides 60 and $\gcd(q, |H|) = 1$. Setting $t = \exp(H)$, it follows by Proposition 2.4 that $G^{(t)} = G^t$. Since G^t is abelian and t divides 60, Lemma 2.1 implies that $G^{(60)} = G^{60}$ and G^{60} is abelian.

The remaining orders with $40 < |G| < 56$ for which there exist nonabelian groups are 48 and 54. First let $|G| = 54$. We observe $P_3 \lhd G$ and $G = P_3H$ with $H \cong C_2$. If P_3 is abelian, then it follows $G^{(60)} = G^{60}$ as in the previous cases. If P_3 is nonabelian and has exponent 3, then $\exp(G) = 6$. Since 6 divides 60, our claim follows trivially. In case $\exp(P_3) = 9$ and $P_3' \neq 1$, we have

$$P_3 = \langle a, b \mid a^9 = b^3 = 1, a^b = a^4 \rangle$$

and $P_3^{(3)} = \langle a^3 \rangle = P_3^3$, a cyclic group of order 3. It follows that $G^{(6)} = \langle a^3 \rangle = G^6$. Observing that 6 divides 60, Lemma 2.1 implies that $G^{(60)} = G^{60}$ and G^{60} is abelian. Now let G have order 48. With the help of GAP it can be easily verified that $G^{(60)} = G^{60}$ and G^{60} is trivial or an abelian 2-group.

Finally, let G have order 56. First let $P_7 \lhd G$. Then $G^{(4)} = G^4 = P_7$, if $\exp(G) = 28$. Since G^4 is abelian and 4 divides n, Lemma 2.1 implies $G^{(60)} = G^{60}$ and G^{60} is abelian. So let $\exp(G) = 56$. Then $G \cong T(7,3) = T$. By Proposition 2.7 it follows that $T^{(4)} \neq T^4$. It follows by Lemma 2.6 that $T^{(4k)} \neq T^{4k}$ for any odd k not divisible by 7.

If $P_7 \ntriangleleft G$, a counting argument shows that $P_2 \lhd G$ and then $G \cong L$, the only case for which P_7 is not normal. It follows that $L^{(2)}$ and hence $L^{(4)}$ consist of all elements in L having order dividing 7. Thus $|L^{(4)}| = 49$ and $L^{(4)} \neq L^4$. It follows by Lemma 2.6 that $L^{(4k)} \neq L^{4k}$ for any odd k not divisible by 7. □

It can be easily verified that the groups L and $T(7,3)$ are n-power closed for $n = 420$. As can be seen from the next theorem, $\mu(420) = 64$, the maximum possible in the case $n = 4k$, k odd.

THEOREM 5.3. *Let $n = 4k$, where k is odd and divisible by 105. If $|G| \leq 64$, then $G^{(n)} = G^n$ except for nine groups of order 64 where $G^{(n)} \neq G^n$.*

PROOF. First let $|G| < 64$. We will show that G^{420} is abelian and that $G^{(420)} = G^{420}$. Then it follows by Lemma 2.1 that $G^{(n)} = G^n$, since 420 divides n. If $|G| < 56$, it follows from Theorem 5.2 that G^{420} is abelian and $G^{(420)} = G^{420}$, since 60 divides 420. If $|G| = 56$, it suffices to consider the groups $T(7,3)$ and L of Theorem 5.2. Since $\exp(L) = 14$ and 14 divides 420, our claim follows trivially. For $T(7,3) = T$, we have $T^{28} = \langle b^4 \rangle = T^{(28)}$. Since $\langle b^4 \rangle$ is abelian and 28 divides 420, it follows that T^{420} is abelian and $T^{(420)} = T^{420}$. Now let us assume that $56 < |G| < 64$. We only have to consider those orders where there exist nonabelian groups, namely 57, 58, 60, 62, and 63. If the order of G is 60, then $G^{(420)} = G^{420} = \{1\}$ trivially, since $\exp(G)$ divides 420. The groups of order 57, 58, and 62 have a normal cyclic Sylow q-subgroup for the largest prime dividing the order. It can be easily verified that the assumptions of Proposition 2.4 are satisfied for these groups. Hence it follows that G^{420} is abelian and $G^{(420)} = G^{420}$. If the order of G is 63, then $P_7 \lhd G$ and we can assume that $\exp(G) = 63$. Then $P_3 = \langle b \rangle \cong C_9$, $G = P_7 \rtimes \langle b \rangle$ and $\langle b^3 \rangle$ is a central subgroup of G. Since $P_3^{(3)} = \langle b^3 \rangle = P_3^3$, it follows that $G^{(21)} = \langle b^3 \rangle = G^{21}$. Now 420 divisible by 21 implies that G^{420} is abelian and $G^{(420)} = G^{420}$ by Lemma 2.1. Finally, let the order of G be 64. If G is one of the 9 groups given in Table 7.1, we have that $G^{(4)} \neq G^4$, and hence $G^{(4k)} \neq G^{4k}$ for all odd k by Lemma 2.6. Otherwise $G^{(4)} = G^4$ for the rest of the groups of order 64. It can be verified with the help of GAP that G^4 is abelian in each case. Thus we obtain by Lemma 2.1 that G^{420} is abelian and $G^{(420)} = G^{420}$. □

As a corollary to Theorems 4.1, 5.1, 5.2 and 5.3 we obtain now the minimal number $\mu(4k)$ in case k is odd.

COROLLARY 5.4. *Let $n = 4k$ be an integer with k odd, then $\mu(n) \leq 64$. If n satisfies one of the following conditions, then $\mu(n) < 64$:*

(i) $\mu(4k) = 12$, *if k is not divisible by 3;*
(ii) $\mu(4k) = 40$, *if k is divisible by 3 and not divisible by 5;*
(iii) $\mu(4k) = 56$, *if k is divisible by 15 and not divisible by 7.*

6. The case $\mathbf{n \equiv 8 \pmod{16}}$ and beyond

The goal of this section is to determine the minimal number $\mu(n)$ for $n = 8k$, k odd. As we see from Table 7.1, there exist 18 groups of order 256 which are not 8-power closed but all groups of order 2^α, $\alpha \leq 7$, are. It follows by Lemma 2.6 that for those 18 groups of order 256 we have $G^{(8k)} \neq G^{8k}$ for all odd k. Thus $\mu(8k) \leq 256$. The question arises for which values of k do we have $\mu(8k) < 256$. As already discussed in the previous case, $\mu(2^\alpha k) = 12$, $\alpha \geq 1$, whenever k is not divisible by 3. Following the preceeding case, this suggests that we should determine $\mu(24)$. Using the GAP function of Example 7.2 with $n = 24$ leads to the group L of Theorem 5.2. However, we can not conclude that $\mu(24k) = 56$ whenever k is not divisible by 7. Any group order relatively prime to 24, such as 55, would be missed in this procedure. Thus we checked the power structure for all groups with a given order either manually or with the help of GAP. The outgrowth of these calculations is the following theorem.

THEOREM 6.1. *Let $n = 2^\alpha k$, $\alpha \geq 3$ with k odd and divisible by 3. If $|G| \leq 55$, then $G^{(n)} = G^n$ except when k is divisible by 11 and not divisible by 5 and then*

$$G \cong \langle a, b \mid a^{11} = b^5 = 1, a^b = a^4 \rangle.$$

PROOF. First let $|G| < 55$. We will show that G^{24} is abelian and $G^{(24)} = G^{24}$. Then it follows by Lemma 2.1 that $G^{(n)} = G^n$, since 24 divides n. If $|G| < 40$, Theorem 5.1 implies that G^{12} is abelian and $G^{(12)} = G^{12}$. Since 12 divides 24, our claim follows by Lemma 2.1. Now let $|G| = 40$. It suffices to consider the two groups $G \cong T(5,3)$ and K of Theorem 5.1. By Proposition 2.4 our claim follows. The orders with $40 < |G| < 55$ to be considered are 42, 44, 46, 50, 52, and 54. If $|G| = 50$, our claim follows by Proposition 2.4. Since 5 does not divide the remaining orders, we obtain that G^{12} is abelian and $G^{(12)} = G^{12}$ by Theorem 5.2 and our claim follows by Lemma 2.1.

Now let $|G| = 55$ and $G' \neq 1$. Then $G = \langle a, b \mid a^{11} = b^5 = 1,\ a^b = a^4 \rangle$. We observe that $G^{(11)}$ consists of the 44 elements of order 5 and the identity and hence $|G^{(11)}| = 45$. Since 45 does not divide 55, this yields that $G^{(11)} \neq G^{11}$. Since 11 divides k and 5 does not, our claim follows by Lemma 2.6. □

The case $\mu(n) = 56$ is a corollary of Theorem 5.2 and the preceding theorem.

COROLLARY 6.2. *Let $n = 2^\alpha k$, $\alpha \geq 3$, with k odd and divisible by 15. If $|G| \leq 56$, then $G^{(n)} = G^n$ except when k is not divisible by 7 and $G \cong L$, the group of order 56 of Theorem 5.2.*

PROOF. First let $|G| < 56$. We will show that G^{120} is abelian and that $G^{(120)} = G^{120}$. Then Lemma 2.1 implies that $G^{(n)} = G^n$, since 120 divides n. If $|G| < 55$, our claim follows from the preceding theorem and Lemma 2.1, since 24 divides 120. In case $|G| = 55$, our claim follows from Proposition 2.4. Now let $|G| = 56$. Theorem 5.2 yields that $G \cong L$, since $\alpha \geq 3$, and thus $L^{(n)} \neq L^n$. □

Observing that $\exp(L) = 14$, the group L is $8k$-power closed whenever k is divisible by 7. In view of Corollary 6.2, we have to determine the value of $\mu(8k)$ when k is divisible by $3, 5$, and 7. By direct inspection or with the help of GAP, we arrive at $\mu(840) = 117$. These calculations are summarized in our next theorem.

THEOREM 6.3. *Let $n = 2^\alpha k$, $\alpha \geq 3$, with k odd and divisible by 105. If $|G| \leq 117$, then $G^{(n)} = G^n$ except when k is not divisible by 9 and 13 and $G \cong F$, where*

$$F = \langle a, b \mid a^{13} = b^9 = 1, a^b = a^4 \rangle.$$

PROOF. First let $|G| < 117$. We will show that G^{840} is abelian and that $G^{(840)} = G^{840}$. Then Lemma 2.1 implies that $G^{(n)} = G^n$, since 840 divides n. If $|G| \leq 64$, our claim follows from Theorem 5.3, since 420 divides 840. Now let $|G| > 64$. In case G is abelian or a regular p-group, the claim follows by Proposition 2.5. For G having order 72, 80, 81, 90, 96, 108, or 112, the claim can be verified using GAP (Example 7.1). In case of the remaining orders, the claim follows by Proposition 2.4.

Finally, let G be a group of order 117. Without loss of generality, we can assume that G is nonabelian and $\exp(G)$ does not divide n, hence $\exp(G) = 117$ and $P_3 \cong C_9$. Since $P_{13} \triangleleft G$, we have that $G \cong F$. Thus b induces an automorphism of order 3 on P_{13} and $\langle b^3 \rangle$ is a normal central subgroup of F. A simple calculation shows $F^{(3)} = \langle a \rangle \cup \langle b^3 \rangle$ and therefore $|F^{(3)}| = 15$ and $F^{(3)} \neq F^3$. By Lemma 2.6 it follows $F^{(n)} \neq F^n$. □

There are two modifications of the n-value of Theorem 6.3 to make the group F of order 117 n-power closed, namely having n divisible by 13 or by 9. Both lead to two different values for $\mu(8k)$ which are less than 256. We arrived at these values again by direct inspection of the group orders or by using GAP.

THEOREM 6.4. *Let $n = 8k$ with k divisible by $3, 5, 7,$ and 13. If $|G| \leq 144$, then $G^{(n)} = G^n$ except when the highest 3-power dividing n is 3 and $G \cong M$, where*

$$M = \langle a, b \mid a^9 = b^{16} = 1, a^b = a^{-1} \rangle.$$

PROOF. First let $|G| < 144$. We will show that for $n_0 = 10920 = 8 \cdot 3 \cdot 5 \cdot 7 \cdot 13$ the subgroup G^{n_0} is abelian and $G^{(n_0)} = G^{n_0}$. Then Lemma 2.1 implies that $G^{(n)} = G^n$, since n_0 divides n. If $|G| < 117$, our claim follows from the preceding theorem and Lemma 2.1, since 840 divides n_0. If $117 < |G| < 144$, it can be shown with the help of GAP or by using Proposition 2.4 that G^{840} is abelian and $G^{(840)} = G^{840}$. In other words, the condition that k is divisible by 13 is not needed for these orders. Once again, our claim follows by Lemma 2.1. If $|G| = 117$, then $G \cong F$ and $F^{(39)} = \langle b^3 \rangle = F^{39}$ and F^{39} is abelian. Since 39 divides n_0, Lemma 2.1 implies that F^{n_0} is abelian and $F^{(n_0)} = F^{n_0}$.

Finally let $|G| = 144$. With the help of GAP we see that G^{n_0} is abelian and $G^{(n_0)} = G^{n_0}$ unless $G \cong M$. A simple calculation shows that $\langle b^8 \rangle \cup \langle a^3 \rangle = M^{(24)} = \{1, a^3, a^6, b^8\}$. Since $a^3 b^8 \notin M^{(24)}$, it follows that $M^{(24)}$ is not a subgroup. Observing that 24 divides n, Lemma 2.6 implies $M^{(n)} \neq M^n$. □

THEOREM 6.5. *Let $n = 8k$ with k odd and divisible by 9, 5, and 7. If $|G| \leq 176$, then $G^{(n)} = G^n$ except when k is not divisible by 11 and then $G \cong T(11, 4)$.*

PROOF. First let $|G| < 176$. We will show that for $n_1 = 2520 = 8 \cdot 9 \cdot 5 \cdot 7$ the subgroup G^{n_1} is abelian and $G^{(n_1)} = G^{n_1}$. Then Lemma 2.1 implies that $G^{(n)} = G^n$, since n_1 divides n. If $|G| < 117$ or $117 < |G| < 144$, it follows by Theorem 6.3 or 6.4 that G^{840} is abelian and $G^{(840)} = G^{840}$. Since 840 divides n_1, our claim follows by Lemma 2.1. Now let $|G| = 117$. Then $G \cong F$ and Proposition 2.4 yields our claim. If $|G| = 144$, then $G \cong M$ and $M^{72} = \langle b^8 \rangle = M^{(72)}$. Since 72 divides n_1, our claim follows by Lemma 2.1. If $144 < |G| < 176$, our claim can be verified using Proposition 2.4 or with the help of GAP.

Finally, let $|G| = 176$. Further analysis of the group of order 176 given in Example 7.3 with the help of GAP, we obtain that $\exp(G) = 176$ and thus that $G \cong T(11, 4) = T$. By Proposition 2.7 we obtain $T^{(8)} \neq T^8$. Since k is not divisible by 11, it follows by Lemma 2.6 that $T^{(n)} \neq T^n$. □

We again ask what modifications of the k-value in Theorem 6.5 will make the group $T(11, 4)$ n-power closed. In this case, having k divisible by 11 leads to a new value of $\mu(n) < 256$. Using the same methods as before leads to the following theorem.

THEOREM 6.6. *Let $n = 8k$, with k odd, and k divisible by $9 \cdot 5 \cdot 7 \cdot 11$. If $|G| \leq 208$, then $G^{(n)} = G^n$ except when k is not divisible by 13 and $G \cong T(13,4)$ or $G \cong H = \langle a, b \mid a^{13} = b^{16} = 1, a^b = a^8 \rangle$.*

PROOF. First let $|G| < 208$. We will show that for $n_2 = 27720 = 8 \cdot 9 \cdot 5 \cdot 7 \cdot 11$ the subgroup G^{n_2} is abelian and $G^{(n_2)} = G^{n_2}$. By Lemma 2.1 we then conclude that $G^{(n)} = G^n$. If $|G| < 176$, it follows by Theorem 6.5 that $G^{(2520)}$ is abelian and $G^{(2520)} = G^{2520}$. Since 2520 divides n_2, our claim follows by Lemma 2.1. If $|G| = 176$, then $G \cong T(11,4) = T$ and $T^{(88)} = \langle b^8 \rangle = T^{88}$ and our claim again follows by Lemma 2.1 Now let $176 < |G| < 208$. If $\exp(G)$ divides n_2 or G is abelian, our claim follows trivially or by Proposition 2.5. In the remaining cases, the claim follows by Proposition 2.4 or with the help of GAP.

Finally let $|G| = 208$. Further analysis of the groups of order 208 given in Example 7.3 with the help of GAP yields $G \cong H$ and $G \cong T(13,4) = T$ as given in Proposition 2.7. For both groups we have $P_{13} \cong C_{13}$ and P_{13} is normal. Since the exponent of T and H is 208, we have that $\langle b \rangle \cong C_{16}$ and b induces an automorphism of order 2 or 4, respectively, on P_{13}. If b induces an automorphism of order 2 on P_{13}, then b^2 is central and $G \cong T(13,4)$. Proposition 2.7 implies $T^{(8)} \neq T^8$. We conclude $T^{(n)} \neq T^n$ by Lemma 2.6. If b induces an automorphism of order 4, then $G \cong H$ and $b^4 \in Z(H)$. It can be seen from direct computation that $H^{(8)} = \langle a \rangle \cup \langle b^8 \rangle$, which is not a subgroup of H since $|H^{(8)}| = 14$. Lemma 2.6 implies $H^{(n)} \neq H^n$. □

Our last modification of k in Theorem 6.6 leads to the maximum value of $\mu(8k)$.

THEOREM 6.7. *Let $n = 8k$ with k odd and divisible by $3^2 \cdot 5 \cdot 7 \cdot 11 \cdot 13$. If $|G| \leq 256$, then $G^{(n)} = G^n$ except for 18 groups of order 256 where $G^{(n)} \neq G^n$*

PROOF. First let $|G| < 256$. We will show that G^{n_3} is abelian and $G^{(n_3)} = G^{n_3}$ for $n_3 = 8 \cdot 9 \cdot 5 \cdot 7 \cdot 11 \cdot 13 = 360360$. Then Lemma 2.1 yields the desired result. Since n_2 divides n_3, our claim follows by Theorem 6.6 whenever $|G| < 208$. If $|G| = 208$, then $G \cong T(13,4)$ or $H^{(104)} = H^{104} = \langle b^8 \rangle$. The rest of our claim follows from Lemma 2.1. If $208 < |G| < 256$, the claim follows in a similar manner as in the previous cases. We omit the details.

Finally, let $|G| = 256$. Then our claim follows in a similar manner as in Theorem 5.3. Again, we omit the details. □

As a corollary to Theorem 4.1 and the preceding theorems of this section we finally obtain a characterization of the values of the minimal number $\mu(8k)$ in case k is odd. We omit the proof.

COROLLARY 6.8. *Let n be an integer with $n = 8k$ and k odd, then $\mu(n) \leq 256$. If n satisfies one of the following conditions, then $\mu(n) < 256$:*

(i) $\mu(8k) = 12$, *if k is not divisible by 3;*
(ii) $\mu(8k) = 55$, *if k is divisible by 33 and not divisible by 5;*
(iii) $\mu(8k) = 56$, *if k is divisible by 15 and not divisible by 7;*
(iv) $\mu(8k) = 117$, *if k is divisible by 105 and not divisible by 9 and 13;*
(v) $\mu(8k) = 144$, *if k is divisible by 3, 5, 7, and 13, and not divisible by 9;*
(vi) $\mu(8k) = 176$, *if k is divisible by 9, 5, and 7, and not divisible by 11;*
(vii) $\mu(8k) = 208$, *if k is divisible by 9, 5, 7, and 11, and not divisible by 13.*

In conclusion, we want to take a look at the situation when $\alpha > 3$, the cases beyond what has been discussed in this paper. The following example in [3] shows that $\mu(2^\alpha k) \leq 2^{2\alpha+2}$ for k odd and $\alpha \geq 1$.

EXAMPLE 6.9. *Let $\alpha \geq 1$ and*

$$G(\alpha) = \langle a, b \mid a^{2^{\alpha+1}} = b^{2^{\alpha+1}} = 1, a^b = a^{-1} \rangle.$$

We have $|G(\alpha)| = 2^{2\alpha+2}$ and $G^{(2^\alpha)} = \{a^{2^\alpha}, b^{2^\alpha}, 1\}$. Since $|G^{(2^\alpha)}| = 3$, it follows that $G^{(2^\alpha)}$ is not a subgroup, hence $G(\alpha)$ is not 2^α-power closed. However, $\exp(G(\alpha)) = 2^{\alpha+1}$. *Hence $G(\alpha)$ is 2^γ-power closed for all $\gamma \geq \alpha + 1$.*

In this paper we showed with the help of GAP [1] that $2^{2\alpha+2}$ is a sharp upper bound of $\mu(2^\alpha k)$ for $\alpha = 1, 2, 3$. In addition we determined all values of k where $\mu(2^\alpha k) < 2^{2\alpha+2}$ for the corresponding values of α. To prove that $2^{2\alpha+2}$ is also a sharp upper bound for $\alpha > 3$, it remains to be shown that all 2-groups of order $2^\rho, \rho < 2\alpha + 2$, are 2^α-power closed. Since the groups of order 512 are contained in the small groups library of GAP, it might be possible to attack the problem with the help of GAP. However the sheer size of the file, containing around a million groups of order 512, makes it impractical to do any explicit calculations with these groups. Attacking this problem for $\alpha = 4$ and then $\alpha > 4$ with group theoretic means seems to have a better chance of success. To solve this problem and determine the values of k for which $\mu(2^\alpha k)$ is less than the maximum possible at least in the case $\alpha = 4$ will be the topic of further investigations.

7. GAP Results

This section contains results determining and conjecturing the groups of smallest order which are not n-power closed that were obtained with the help of GAP. They are preceded by the files of various GAP functions used in the determination of these groups. For the documentation of GAP see [1].

EXAMPLE 7.1. GAP function that takes a group G and raises each element to the n-th power. If $G^{(n)} = G^n$, the function returns true. If $G^{(n)} \neq G^n$, the function returns false.

```
isAPowerOfASubgroup:=function(G,n)
local l,x;
 l:=[];
 for x in G do
    x:=x\^{}n;
    if not (x in l) then
       Add(l,x);
    fi;
 od;
 if Size(l)=Size(Subgroup(G,l)) then
    return true;
 fi;
 return false;
end;;
```

EXAMPLE 7.2. GAP function to determine the minimal nonabelian groups such that $G^{(n)} \neq G^n$. The function uses the GAP function `AllSmallGroups` and the function in Example 7.1. It returns a list of groups of minimal order which are not n-power closed for n as specified.

```
findAMinimalCounterexample:=function(n)
local i,J,L,K;
 i:=2;
 K:=[];
 while i<=1000 do
    L:=AllSmallGroups(Size,i,IsAbelian,false);
    for J in L do}
       if not isAPowerOfASubgroup(J,n) then
          Add(K,J);
       fi;
    od;
    if Length(K)>0 then
       Print("found ", Length(K), " groups of order ",i,"\n");
       return K;
    fi;
    i:=i+1;
 od;
 Print("didn't find a group of order <= ",i-1,"\n");
 return K;
end;;
```

EXAMPLE 7.3. The following is the GAP output for the determination of groups of minimal order which are not n-power closed for $n = 8k$, k odd and $k = 33$, 15, 105, 315, 1365, 4095, or 3465, respectively.

```
gap> findAMinimalCounterexample(264);
 found 1 groups of order 55
 [ <pc group of size 55 with 2 generators> ]
gap> findAMinimalCounterexample(120);
 found 1 groups of order 56
 [ <pc group of size 56 with 4 generators> ]
gap> findAMinimalCounterexample(840);
 found 1 groups of order 117
 [ <pc group of size 117 with 3 generators> ]
gap> findAMinimalCounterexample(2520);
 found 1 groups of order 176
 [ <pc group of size 176 with 5 generators> ]
gap> findAMinimalCounterexample(10920);
 found 1 groups of order 144
 [ <pc group of size 144 with 6 generators> ]
gap> findAMinimalCounterexample(32760);
 found 1 groups of order 176
 [ <pc group of size 176 with 5 generators> ]
gap> findAMinimalCounterexample(27720);
```

```
found 2 groups of order 208
[ <pc group of size 208 with 5 generators>,
  <pc group of size 208 with 5 generators> ]}
```

EXAMPLE 7.4. In this example GAP analyzes groups of order 16, 32, 64, 128, and 256, determining the n-th powers for $n = 2, 4, 8$ such that $G^{(n)} \neq G^n$. The returned lists were deleted for space purposes. The size of such lists is specified.

1) For $|G| = 16$ and $n = 2$, 4:

```
gap> L:=AllSmallGroups(Size,16,x->isAPowerOfASubgroup(x,2),false);
 [ <pc group of size 16 with 4 generators>,
   <pc group of size 16 with 4 generators> ]}
gap> L:=AllSmallGroups(Size,16,x->isAPowerOfASubgroup(x,4),false);
 [ ]
```

2) For $|G| = 32$, $n = 2$, 4:

```
gap> L:=AllSmallGroups(Size,32,x->isAPowerOfASubgroup(x,2),false);
 [ <pc group of size 32 with 5 generators>,
   <pc group of size 32 with 5 generators>,
   <pc group of size 32 with 5 generators>,... ]
gap> Size(L);
 16
gap> L:=AllSmallGroups(Size,32,x->isAPowerOfASubgroup(x,4),false);
 [ ]
```

3) For $|G| = 64$, $n = 2$, 4, 8:

```
gap> L:=AllSmallGroups(Size,64,x->isAPowerOfASubgroup(x,2),false);
 [ <pc group of size 64 with 6 generators>,
   <pc group of size 64 with 6 generators>,
   <pc group of size 64 with 6 generators>,...]
gap> Size(L);
 127
gap> L:=AllSmallGroups(Size,64,x->isAPowerOfASubgroup(x,4),false);
 [ <pc group of size 64 with 6 generators>,
   <pc group of size 64 with 6 generators>,
   <pc group of size 64 with 6 generators>,...]
gap> Size(L);
 9
gap> L:=AllSmallGroups(Size,64,x->isAPowerOfASubgroup(x,8),false);
 [ ]
```

4) For $|G| = 128$, $n = 2$, 4, 8:

```
gap> L:=AllSmallGroups(Size,128,
          x->isAPowerOfASubgroup(x,2),false);
 [ <pc group of size 128 with 7 generators>,
   <pc group of size 128 with 7 generators>,
   <pc group of size 128 with 7 generators>,...]
gap> Size(L);
 1391
```

```
gap> L:=AllSmallGroups(Size,128,
          x->isAPowerOfASubgroup(x,4),false);
 [ <pc group of size 128 with 7 generators>,
   <pc group of size 128 with 7 generators>,
   <pc group of size 128 with 7 generators>,...]
gap> Size(L);
 176
gap> L:=AllSmallGroups(Size,128,
          x->isAPowerOfASubgroup(x,8),false);
 [ ]
```

5) For $|G| = 256$, $n = 2, 4, 8, 16$:

```
gap> L:=AllSmallGroups(Size,256,
          x->isAPowerOfASubgroup(x,2),false);
 [ <pc group of size 256 with 8 generators>,
   <pc group of size 256 with 8 generators>,
   <pc group of size 256 with 8 generators>,...]
gap> Size(L);
 27290
gap> L:=AllSmallGroups(Size,256,
          x->isAPowerOfASubgroup(x,4),false);
 [ <pc group of size 256 with 8 generators>,
   <pc group of size 256 with 8 generators>,
   <pc group of size 256 with 8 generators>,...]
gap> Size(L);
 3052
gap> L:=AllSmallGroups(Size,256,
          x->isAPowerOfASubgroup(x,8),false);
 [ <pc group of size 256 with 8 generators>,
   <pc group of size 256 with 8 generators>,
   <pc group of size 256 with 8 generators>,...]
gap> Size(L);
 18
gap> L:=AllSmallGroups(Size,256,
          x->isAPowerOfASubgroup(x,16),false);
 [ ]
```

Table 7.1 on the next page provides a summary of the GAP calculations of Example 7.4. The first column refers to the order of the group. The second column refers to the number of groups of this order. The next three columns specify the power n to which these groups were raised and the data below refer to the number of groups with $G^{(n)} \neq G^n$. For example, of the 14 groups of order 16 there are two groups such that $G^{(2)} \neq G^2$ and there are none with $G^{(n)} \neq G^n$ for $n = 4$, 8, and 16.

Acknowledgments. The authors want to thank David Garrison for facilitating the GAP implementations and the careful reading of this paper and the anonymous

order	number of groups	$n = 2$	$n = 4$	$n = 8$	$n = 16$
16	14	2	0	0	0
32	51	16	0	0	0
64	267	127	9	0	0
128	2328	1391	176	0	0
256	56092	27290	3052	18	0

TABLE 7.1

referee for his constructive comments and helpful suggestions which led to a much better presentation of the material.

References

[1] The GAP Group, *GAP – Groups, Algorithms, and Programming, Version 4.4*, 2004. Website at http://www.gapsystems.org.

[2] P. Hall, *A contribution to the theory of groups of prime-power order*, Proc. London Math. Soc. **36** (1934), 29–95.

[3] B. Huppert, *Endliche Gruppen I*, Springer-Verlag, 1967.

[4] L.-C. Kappe and J.H. Ying, *On exact power margin groups*, Rend. Math. Univ. Padova **87** (1992), 245–265.

[5] F.L. Kluempen, *On the Power Structure of Finite p-Groups*, Dissertation, State University of New York at Binghamton, NY, 2001.

[6] D. MacHale, *Minimum counterexamples in Group Theory*, Mathematics Magazine **54** (1981), 23–27.

[7] G. Mendoza, *On the Power Structure of Finite Groups*, Master's thesis, State University of New York at Binghamton, NY, 2004.

[8] G. Mendoza, *On Some Minimality Conditions Involving Elements of Prime Order in a Group G*, Dissertation, State University of New York at Binghamton, NY, 2007.

DEPARTMENT OF MATHEMATICAL SCIENCES, STATE UNIVERSITY OF NEW YORK AT BINGHAMTON, BINGHAMTON, NY 13902-6000, USA

E-mail address: menger@math.binghamton.edu

DEPARTMENT OF MATHEMATICS, RIVERSIDE COMMUNITY COLLEGE, RIVERSIDE, CA 92506-12999, USA

E-mail address: gabriela.mendoza@rcc.edu

Contemporary Mathematics
Volume **511**, 2010

On the Covering Number of Small Alternating Groups

Luise-Charlotte Kappe and Joanne L. Redden

Dedicated to Bernhard H. Neumann on the hundredth anniversary of his birth.

ABSTRACT. According to Bernhard Neumann, every group with a noncyclic finite homomorphic image is the union of finitely many proper subgroups. The minimal number of subgroups needed to cover a group G is called the covering number of G, denoted by $\sigma(G)$. The goal of this paper is to determine the covering numbers for small alternating groups A_n. It is known that $\sigma(A_5) = 10$ and $\sigma(A_6) = 16$. We show that $\sigma(A_7) = 31$, $\sigma(A_8) = 71$, $127 \leq \sigma(A_9) \leq 157$, and $\sigma(A_{10}) = 256$.

1. Introduction

Let G be a group. A finite cover of G is a collection $\mathcal{A} = \{A_i \mid 1 \leq i \leq n\}$ of proper subgroups A_i of G whose union is G. A cover is called irredundant if after the removal of any subgroup, the remaining subgroups do not cover the group. A cover of size n is said to be minimal if no cover of G has fewer than n members. A minimal cover is always irredundant. The size of a minimal cover of a group G is denoted by $\sigma(G)$. We call $\sigma(G)$ the covering number of G. The idea of a minimal cover is due to J.H.E. Cohn [**5**]. According to a result of B.H. Neumann [**16**], a group is the union of finitely many proper subgroups if and only if it has a finite noncyclic homomorphic image. Thus we can restrict our attention to finite groups when determining covering numbers.

It is a well-known exercise that no group is the union of two proper subgroups. Thus $\sigma(G) > 2$ for all groups G. It is less well-known that Scorza [**17**] already proved in 1926 that $\sigma(G) = 3$ if and only if G has a homomorphic image isomorphic to the Klein-Four group. Scorza's paper was not well-known in the mathematical community for a long time and his result was rediscovered more than once. For further details we refer here to the interesting survey article by Serena [**18**]. In [**9**], Greco characterizes the groups with $\sigma(G) = 4$. It should be mentioned here that these results by Scorza and Greco are obtained without the benefit of Neumann's

2000 *Mathematics Subject Classification.* Primary 20D06, 20D60, Secondary 20F99.
Key words and phrases. alternating groups, finite union of proper subgroups, minimal number of subgroups.

result. In two subsequent papers, [**10**] and [**11**], Greco gives a partial characterization of groups with $\sigma(G) = 5$.

It can be easily seen that there are groups with $\sigma(G) = 6$, namely the elementary abelian 5-group of rank 2, the dihedral group of order 10, and the holomorph of the cyclic group of order 5. Tomkinson [**19**] showed that there are no groups with $\sigma(G) = 7$ and he conjectured that there are no groups with covering number 11, 13, and 15, respectively. This raises the question of which integers n can arise as covering numbers of groups. Tomkinson [**19**] showed the following result for finite solvable groups.

THEOREM 1.1. *Let G be a finite solvable group and let p^α be the order of the smallest chief factor having more than one complement. Then $\sigma(G) = p^\alpha + 1$.*

This result had already been conjectured by Cohn [**5**] who also provided examples of groups with covering number $p^\alpha + 1$ for every prime p and $\alpha \geq 1$. In [**19**], Tomkinson suggests that it might of interest to investigate minimal covers of nonsolvable and in particular simple groups. Cohn [**5**] had already shown $\sigma(A_5) = 10$ and $\sigma(S_5) = 16$. Bryce, Fedri, and Serena started this investigation in [**4**], where they calculated $\sigma(G)$ for some linear groups. The main result of [**4**] is that $\sigma(G) = \frac{1}{2}q(q+1)$ when q is even and $\sigma(G) = \frac{1}{2}q(q+1)+1$ when q is odd, where $G \cong \mathrm{PSL}(2,q)$, $\mathrm{PGL}(2,q)$, or $\mathrm{GL}(2,q)$ and $q \neq 2,5,7,9$. In [**13**], Lucido investigates the Suzuki groups and shows $\sigma(Sz(q)) = \frac{1}{2}q^2(q^2+1)$ where $q = 2^{2m+1}$. Holmes in [**12**] determines the covering numbers of some of the sporadic simple groups such as M_{11}, M_{22}, M_{23}, Ly, and $O'N$ and gives estimates for J_1 and M^cL. It should be mentioned here that some of the results in [**12**] have been obtained with the help of GAP [**7**], a first in this context.

A particular result in [**4**] is that the covering number of $\mathrm{PSL}(2,7)$ is 15, a number which is not of the form prime power plus one, and thus showing that Tomkinson's conjecture is not valid for 15. Recently, Abdollahi, Ashraf, and Shaker, in [**1**], showed that $\sigma(S_6) = 13$. Thus Tomkinson's conjecture is not valid for 13 either. On the other hand, Detomi and Lucchini [**6**] proved that there are no groups with covering number 11. Thus Tomkinson's conjecture holds in this case. Garonzi [**8**] in his dissertation investigates which integers n not of the form prime power plus one and $16 < n < 27$ are covering numbers of groups. He shows that there are no groups with covering number $n = 19$, 21, 22, or 25, whereas $\sigma(M_{11}) = 23$, a result already obtained in [**12**]. At this point in time, the smallest integer for which it is not known whether it is a covering number for a group or not is $n = 27$. Lastly, it should be mentioned here that the proof for the claim that $n = 11$ and 13 are not covering numbers for groups given by Zhang in [**20**] is not valid.

Maróti [**14**] looks at the covering numbers of symmetric and alternating groups. He proves that $\sigma(S_n) = 2^{n-1}$ if n is odd unless $n = 9$, and $\sigma(S_n) \leq 2^{n-2}$ if n is even. For $n \neq 7, 9$, Maróti shows for the alternating group that $\sigma(A_n) \geq 2^{n-2}$ with equality if and only if n is even but not divisible by 4. For $n = 7$ and 9 he establishes $\sigma(A_7) \leq 31$ and $\sigma(A_9) \geq 80$. The topic of this paper is the determination of $\sigma(A_n)$ for small values of n such as $n = 7$, 8, 9, and 10. In [**5**], Cohn already established $\sigma(A_5) = 10$ and from [**4**] it follows that $\sigma(A_6) = 16$, since $\mathrm{PSL}(2,9) \cong A_6$. It was already shown in [**14**] that $\sigma(A_{10}) = 256$. We establish it here with different methods. For $n = 7$ we show that the upper bound of 31 for $\sigma(A_7)$ given in [**14**] is also a lower bound, and for $n = 8$ we establish $\sigma(A_8) = 71$. Finally, for $n = 9$ we prove that $127 \leq \sigma(A_9) \leq 157$. It should be mentioned here that with the help of GAP, R.F. Morse [**15**] has found an improved lower bound of 141 for $\sigma(A_9)$. Further details are given in Section 4.

It is obvious that in determining a minimal covering of a group we can restrict ourselves to finding a minimal covering by maximal subgroups. The conjugacy classes of maximal subgroups for alternating groups can be found in GAP [**7**]. To determine a minimal covering by maximal subgroups, it suffies to find a minimal covering of the conjugacy classes of maximal cyclic subgroups by maximal subgroups of the group. This is the method used in [**12**] to determine the covering numbers for some of the sporadic simple groups. We adapt this method to the case of alternating groups where we can easily identify the generators of maximal cyclic subgroups by their cycle structure.

To make our notions more precise, we introduce the following notation for the disjoint cycle decomposition of a nontrivial permutation. Let $m_1, m_2, \ldots, m_t \in \mathbb{N}$ with $1 < m_1 < m_2 < \cdots < m_t$ and $k_1, \ldots, k_t \in \mathbb{N}$. If α is a permutation with disjoint cycle decomposition consisting of k_i cycles of length m_i, $i = 1, \ldots, t$, then we denote the class of α by $(m_1^{k_1}, \ldots, m_t^{k_t})$. If $k_i = 1$, we write just m_i instead of m_i^1. As customary, we suppress 1-cycles and the identity permutation is denoted by (1). For example, the permutation with disjoint cycle decomposition (12)(34)(567) belongs to the class $(2^2, 3)$.

Either the total of elements of a given cycle structure is contained in the subgroups of a conjugacy class of maximal subgroups or the elements are partitioned evenly into two isomorphic conjugacy classes of maximal subgroups. In either case, the elements with the respective cycle structure are either partitioned into the subgroups of a conjugacy class of maximal subgroups of A_n or there exists an intersection between some of the subgroups of the conjugacy class.

For each A_n under consideration, we provide two tables which have been obtained with the help of GAP [**7**]. The first table gives the information on the conjugacy classes of maximal subgroups of A_n, such as the isomorphism type and order of the class representative and the size of each class. The second table lists the order and the cycle structure of each element as well as the total of such elements in A_n together with the distribution of these elements over the various conjugacy classes. For each conjugacy class we list how many elements with a given cycle structure are contained in a representative. If the elements are partitioned over the representatives we indicate this with a P.

The goal is to obtain upper and lower bounds for $\sigma(A_n)$ from the tables and the inventory. For a lower bound we determine minimal coverings by maximal subgroups for certain elements with a particular cycle structure. If the elements are partitioned into the subgroups of a particular conjugacy class, these bounds are not hard to come by, especially if there exists only one conjugacy class containing the elements in question. In such cases, the union of several distinct conjugacy classes provides a lower bound for a minimal covering of the whole group, and hopefully, the lower bound obtained is also an upper bound. This is the situation for A_8 and A_{10}. When the elements with a certain cycle structure are not partitioned into the subgroups of a conjugacy class, finding a minimal covering for these elements is much harder and very much depends on the arithmetic nature of n. This is the case when $n = 7$ and when $n = 9$. For $n = 7$, the critical case is finding a minimal covering for the elements of order 6 and for $n = 9$ this role is played by the elements of order 9. In the case of A_7, the question reduces to a purely combinatorial optimization problem which can be easily solved (see Lemmas 2.1 and 2.2). On the other hand, in the case of A_9 we have not yet found such a reduction to a combinatorial setting but there are indications that such a thing exists. This would allow us to solve the optimization problem with some algorithms of integer linear programming available for machine computation. For further details we refer to Section 4.

2. The Alternating Group A_7

In [**14**] it was already shown that $\sigma(A_7) \leq 31$. Here we will show that this upper bound is also a lower bound. We start with a purely combinatorial lemma which enables us to establish the minimal number of subgroups needed to cover the elements of order 6 in A_7.

LEMMA 2.1. *Consider the bipartite graph B with vertex sets T_7 and D, where T_7 consists of the 21 transpositions in S_7 and D of the 105 products of two disjoint 2-cycles in S_7. Connect (ij) in T_7 with any vertex in D having (ij) as a factor. Then the maximum number of vertices which can be removed from T_7 such that all vertices of D are still connected to T_7 is 6. Furthermore, only the removal of six vertices of the form (kl) where $i \in \{k, l\}$, $i = 1, \ldots, 7$, from T_7 results in a bipartite graph in which all vertices of D are still connected to T_7. The removal of any other set of six vertices from T_7 results in a graph where not all vertices of D are connected to T_7.*

PROOF. The vertices of T_7 can be represented as the edges of K_7, the complete graph on the vertex set $V = \{1, 2, \ldots, 7\}$. Note that there is a one-to-one correspondence between the transpositions (ij) of T_7 and the edges ij of K_7. Note that no two disjoint edges of K_7 can be removed, because that would separate a vertex of D from T_7. Thus, only a set of edges such that no two are disjoint can be removed. We find all maximal sets of this kind. Observe that there are two types of maximal sets of this form. The first type consists of edges removed that do *not* have a common vertex. If this is the case, they must form a triangle, $\{ij, ik, jk\} \subseteq T_7$. In the second case, the removed edges do have a common vertex. In this case, the set consists of the six edges that contain a fixed number i, that is, all edges ij for $j \in V \setminus \{i\}$.

Furthermore, in K_7, only the removal of six edges having a common vertex results in a graph where T_7 and D are still connected. Thus removing from K_7 six edges which do not share a common vertex results in at least one pair of disjoint edges being removed. As a result, D and T_7 will no longer be connected. □

In this context it should be mentioned that Lemma 2.1 can be derived from results in [**2**] and [**3**] dealing with related problems in a more general setting. The same is true for the results of the next lemma. In [**3**], a general formula is given for a related quantity from which a formula for the values we are interested in can be derived. Here instead we obtain these quantities with the help of GAP.

LEMMA 2.2. *Let L be a graph with vertices $T_7 = \{(ij); i \leq i < j \leq 7\}$, the 21 transpositions of S_7. Connect (ij) and (kl) with an edge if and only if (ij) and (kl) are disjoint. Let H be a subgraph of L with $v(H)$ vertices and $e(H)$ edges. Consider $e^*(m) = \min(e(H); v(H) = m)$. Then for $6 < m < 15$ we have the values of $e^*(m)$ given in Table 2.1.*

m	7	8	9	10	11	12	13	14
$e^*(m)$	4	8	12	15	20	27	33	39

TABLE 2.1

PROOF. Counting the edges in a subgraph H of L is equivalent to counting the pairs of disjoint transpositions in a subset H of T_7. This is done in the following GAP code.

```
gap> ## Representation of all tranpositions in S7 as lists
gap> ## of two elements from the set 1..7.
gap> ##
gap> T7 := Combinations([1..7],2);;
gap> ## Given a set H of transpositions for S7 return the
gap> ## number of disjoint transpositions
gap> ##
gap>
gap> e := function(H);
> return Size(Filtered(Combinations(H,2),
                x->IsEmpty(Intersection(x))));
> end;
function( H ) ... end
gap>
gap> ## Iterate through all subsets of T7 of fixed size m
gap>
gap> ## and find e*(m) for m = 7..14 and print the results
gap>
gap> ##
gap>
gap> for i in [7..14] do
>     Print("e*(",i,"): ", Minimum(List(Combinations(T7,i),
               H->e(H))),"\n");
> od;
e*(7): 4
e*(8): 8
e*(9): 12
e*(10): 15
e*(11): 20
e*(12): 27
e*(13): 33
e*(14): 39
```

The output of this code is what is shown in Table 2.1. □

We can now determine the minimal coverings of the elements of order 6 in A_7. To that end, we set $MS2 = \{S(i,j); 1 \leq i < j \leq 7\}$, where $\{i,j\}$ and its complement in $\{1,\ldots,7\}$ are the domains of transitivity of $S(i,j)$ under the action of A_7, and consider the subsets $\mathfrak{Q}(i)$ and $\mathfrak{R}(i)$, $i = 1,\ldots,7$, of $MS2$, where $\mathfrak{Q}(i) = \{S(k,l) \in MS2; i \in \{k,l\}\}$ and $\mathfrak{R}(i) = MS2 \setminus \mathfrak{Q}(i)$.

PROPOSITION 2.3. *The minimal covering number for elements of order* 6 *in* A_7 *is* 15. *A minimal covering is obtained if and only if the* 15 *subgroups are the subgroups of an* $\mathfrak{R}(i)$, $i = 1, \ldots, 7$. *In particular, any covering containing* 14 *or fewer subgroups from MS2 contains more than* 15 *subgroups.*

PROOF. By Table 2.4, located below, the 210 elements with cycle structure $(2^2, 3)$ generate 105 cyclic subgroups of order 6 with pairwise trivial intersections. Thus, there is a one-to-one correspondence between the 105 cyclic subgroups of order 6 and the 105 elements with cycle structure (2^2). The cyclic subgroups of order 6 are partitioned into the 35 subgroups of $MS1$ with three cyclic subgroups in each. In addition, the cyclic subgroups of order 6 are contained in the subgroups of $MS2$ with 10 in each. However, since $10 \cdot 21 > 105$, these cyclic subgroups are not partitioned into the subgroups of $MS2$. As mentioned earlier, there is a one-to-one correspondence between the 21 transpositions of S_7 and the 21 subgroups of $MS2$. Furthermore, a cyclic subgroup of order 6 is contained in $S(i,j)$ if and only if (ij) is a factor in the cycle decomposition of the element of order 2 in the cyclic subgroup. Thus there is a one-to-one correspondence between the bipartite graph B of Lemma 2.1 and the subgroups of $MS2$ and cyclic subgroups of order 6 contained in them. It follows by the first part of Lemma 2.1 that the maximum number of subgroups which can be removed from $MS2$ is six and the 15 subgroups in an $\mathfrak{R}(i)$, $i = 1, \ldots, 7$, cover all cyclic subgroups of order 6 in A_7. On the other hand, it follows from the second part of Lemma 2.1 that removal of six subgroups from $MS2$ other than those contained in a $\mathfrak{Q}(i)$, $i = 1, \ldots, 7$, does not result in a covering of the cyclic subgroups of order 6 by the 15 remaining subgroups of $MS2$.

It remains to be shown that any covering of the cyclic subgroups of order 6 using subgroups from $MS1$ and 14 or fewer subgroups from $MS2$ needs more than 15 subgroups total. To that end, we want to know how many distinct cyclic subgroups of order 6 are contained in the union of m arbitrary representatives from $MS2$, $0 < m < 15$. This number is $10m - d$, where d is the number of cyclic subgroups of order 6 which were counted twice. We observe that d depends on the choice of the m subgroups. To determine d, we translate the question into a graph theoretic problem as follows.

Consider the graph L with vertices $S(1,2), \ldots, S(6,7)$, the 21 subgroups of $MS2$. Connect $S(i,j)$ and $S(k,l)$ with an edge if and only if the intersection of $\{i,j\}$ and $\{k,l\}$ is empty. Consider a subgraph H of L with m vertices and denote the number of edges of H by $e(H)$. We claim now that $e(H) = d$. This can be seen as follows: Each representative of $MS2$ contains 10 copies of C_6. The representatives $S(i,j)$ and $S(k,l)$ share exactly one cyclic subgroup of order 6 if the intersection of $\{i,j\}$ and $\{k,l\}$ is empty. Otherwise they share no cyclic subgroups of order 6. Thus it follows $e(H) = d$. Consider a set of m subgroups from $MS2$ with an associated subgraph H, then the union of the m subgroups contains $N(m,H)$ cyclic subgroups of order 6, where $N(m,H) = 10m - e(H)$. Let $e^*(m) = \min(e(H))$, where H is any subgraph of L with m vertices. It follows that $N(m,H) \leq 10m - e^*(m)$.

If $N(m,H)$ cyclic subgroups of order 6 are covered by representatives from $MS2$, we have $105 - N(m,H)$ copies of C_6 to be covered by subgroups from $MS1$. Since each representative in $MS1$ contains 3 copies of C_6, we need at least $t(m,H)$ subgroups from $MS1$ to cover the remaining ones. Denoting with $\lceil x \rceil$ the ceiling of x, we have

$$t(m,H) = \lceil (105 - 10m + e(H))/3 \rceil \geq t^*(m),$$

where $t^*(m) = \lceil (105 - 10m + e^*(m))/3 \rceil$. We have to show

$$m + t^*(m) > 15. \tag{2.3.1}$$

If $m = 1, 2, \ldots, 6$, then obviously $e^*(m) = 0$ and (2.3.1) holds. For m satisfying $6 < m < 15$, the values $e^*(m)$ have been determined in Lemma 2.2. This results in the values for $m + t^*(m)$ as given in Table 2.2. Thus (2.3.1) holds for $6 < m < 15$, proving our claim. □

m	7	8	9	10	11	12	13	14
$m + t^*(m)$	20	19	18	17	16	16	16	16

TABLE 2.2

Now we are ready to establish the main theorem of this section.

THEOREM 2.4. *The covering number of A_7 is* 31.

PROOF. We first show that $\sigma(A_7) \leq 31$. Proposition 2.3 implies that the union of the 15 subgroups of $\mathfrak{R}(1)$ contains all elements of order 6. A closer inspection of the subgroups in $\mathfrak{R}(1)$ shows that the elements of order 5 not contained in the union are exactly the 5-cycles with $\{1\}$ as a fixed point. By Table 2.4, it follows that the missing elements of order 5 coincide with the elements of order 5 in $A(1) \cong A_6$, the subgroup in $MS3 = \{A(i); 1 \leq i \leq 7\}$ having $\{1\}$ as a fixed point. We conclude that

$$A(1) \cup \bigcup_{H \in \mathfrak{R}(1)} H \cup \bigcup_{H \in MS4} H = A_7,$$

hence $\sigma(A_7) \leq 31$. Next we will show that for covering the elements of order 6 and 5, more than 15 subgroups are needed. If we choose any set of subgroups different from an $\mathfrak{R}(i)$ to cover the elements of order 6, we need more than 15 subgroups by Proposition 2.3. If we choose an $\mathfrak{R}(i)$ to cover the elements of order 6, only 15 subgroups are needed again by Proposition 2.3. Since the elements of order 5 are partitioned into the 21 subgroups of $MS2$ and $\mathfrak{R}(i)$ is a proper subset of $MS2$, the subgroups of $\mathfrak{R}(i)$ do not cover all elements of order 5. Thus more than 15 subgroups are needed to cover the elements of order 6 and 5.

Finally we claim $\sigma(A_7) > 30$. By Table 2.4, at least 15 subgroups are needed to cover the elements of order 7. These subgroups come from $MS4$ or $MS5$. Hence they are distinct from the more than 15 subgroups needed to cover the elements of order 5 and 6. We conclude $\sigma(A_7) > 15 + 15 = 30$. This together with the above implies $\sigma(A_7) = 31$. □

The tables in this section provide information on the conjugacy classes of maximal subgroups of A_7 and an inventory of the orders and cycle structure of the elements of the group. This information was obtained with the help of GAP. The first table, Table 2.3, gives the isomorphism type, the order of each class representative and the the size of each class. The second table, Table 2.4, shows the distribution of the elements of A_7 across the conjugacy classes of maximal subgroups. In addition to the cycle structure and the order, we list the total of the elements with this cycle structure in the whole group as well as in each subgroup of a conjugacy class. If the elements are partitioned into the subgroups of a class,

Label	**Isomorphism Type**	**Order**	**Size**
$MS1$	$(C_3 \times A_4) : C_2$	72	35
$MS2$	S_5	120	21
$MS3$	A_6	360	7
$MS4$	$\mathrm{PSL}(3,2)$	168	15
$MS5$	$\mathrm{PSL}(3,2)$	168	15

TABLE 2.3. Conjugacy classes of maximal subgroups of A_7.

we denote this with P to the right of the quantity. For the split classes, $MS5$ and $MS6$, we also checked with the help of GAP if the total or only half of the elements with a given cycle structure were contained in the union of subgroups of that class. It turns out the union contains the total in each case.

Order	**C.S.**	**Size**	$MS1$ (35)	$MS2$ (21)	$MS3$ (7)	$MS4$ (15)	$MS5$ (15)
1	(1)	1	1	1	1	1	1
2	(2^2)	105	21	25	45	21	21
3	(3)	70	10	20	40		
3	(3^2)	280	16		40,P	56	56
4	(2,4)	630	18,P	30,P	90,P	42,P	42,P
5	(5)	504		24,P	144		
6	$(2^2,3)$	210	6,P	20			
7	(7)	720				48,P	48,P

TABLE 2.4. Inventory of elements and cycle structure of A_7 across conjugacy classes of maximal subgroups.

3. The Alternating Group A_8

In this section we determine the covering number for A_8. Maróti in [**14**] already estimated that $69 < \sigma(A_8) \leq 99$. Here we will establish the exact covering number of A_8.

THEOREM 3.1. *The covering number of* A_8 *is* 71.

PROOF. By Table 3.2 below, we see that there are 2688 elements of order 15 with cycle structure $(3,5)$ in A_8. These only occur in the conjugacy class $MS1$ and are partitioned into the 56 subgroups of the class with 48 such elements in each subgroup. Thus any covering of A_8 needs to contain these 56 subgroups. The class $MS1$ does not contain any elements of order 6 and cycle structure $(2,6)$. A minimal covering of these elements by subgroups of A_8 consists of the 15 subgroups in the class $MS5$. This can be seen as follows. The elements with cycle structure $(2,6)$ occur in the conjugacy classes $MS2$, $MS4$, $MS5$, and $MS6$. By Table 3.2, the total of 3360 elements with this cycle structure is partitioned into the 15 subgroups of $MS5$, with 224 elements in each subgroup and similarly for the isomorphic class

$MS6$. Thus there exists a covering of the elements with cycle structure $(2,6)$ by 15 subgroups. We have to show that using subgroups from other classes than $MS5$ does not decrease the number of sugroups needed for the covering. The elements with cycle structure $(2,6)$ are also partitioned into the 28 subgroups of $MS2$ with 120 such elements in each subgroup. However, these elements are not partitioned into the 35 subgroups of $MS4$ with 192 elements in each subgroup. Replacing some or all of the 15 subgroups of $MS5$ in the covering by subgroups from $MS2$, $MS4$, or $MS6$ does not decrease the number of subgroups needed for the overing, since the replacements contain fewer or the same amount of elements. Since the subgroups needed to cover the elements with cycle structure $(2,6)$ are distinct from the subgroups in $MS1$, we need at least $56+15$ subgroups for a covering of A_8. Hence $\sigma(A_8) \geq 71$.

Conversely, from Table 3.2 it follows that

$$\bigcup_{H \in MS1} H \cup \bigcup_{K \in MS5} K = A_8.$$

We conclude $\sigma(A_8) \leq 71$. This together with the above yields $\sigma(A_8) = 71$. □

Label	**Isomorphism Type**	**Order**	**Size**
$MS1$	$\mathrm{GL}(2,4) : C_2$	360	56
$MS2$	S_6	720	28
$MS3$	A_7	2520	8
$MS4$	$((A_4 \times A_4) : C_2) : C_2$	576	35
$MS5$	$(C_2 \times C_2 \times C_2) : PSL(3,2)$	1344	15
$MS6$	$(C_2 \times C_2 \times C_2) : PSL(3,2)$	1344	15

TABLE 3.1. Conjugacy classes of maximal subgroups of A_8.

Order	**C.S.**	**Size**	$MS1$ (56)	$MS2$ (28)	$MS3$ (8)	$MS4$ (35)	$MS5$ (15)	$MS6$ (15)
1	(1)	1	1	1	1	1	1	1
2	(2^2)	210	45	60	105	42	42	42
2	(2^4)	105		15		33	49	49
3	(3)	112	22	40	70	16		
3	(3^2)	1120	40	40,P	280	64	224	224
4	(4^2)	1260				108	252	252
4	$(2,4)$	2520	90	180	630	72,P	168,P	168,P
5	(5)	1344	24,P	144	504			
6	$(2,6)$	3360		120,P		192	224,P	224,P
6	$(2^2,3)$	1680	90	120	210,P	48,P		
7	(7)	5760			720,P		384,P	384,P
15	$(3,5)$	2688	48,P					

TABLE 3.2. Inventory of elements and cycle structure of A_8 across conjugacy classes of maximal subgroups.

The two tables in this section provide information on the conjugacy classes of the maximal subgroups of A_8 and an inventory of the orders and cycle structure of the elements of the group. The first table, Table 3.1, lists the conjugacy classes and isomorphism types of the maximal subgroups.

The second table, Table 3.2, shows the distribution of the elements of A_8 across the conjugacy classes of maximal subgroups. In addition to the cycle structure and the order, we list the total of the elements with this cycle structure in the whole group as well as in each subgroup of a conjugacy class. If the elements are partitioned into the subgroups of a class, we denote this with P to the right of the quantity. For the split classes, $MS5$ and $MS6$, we also checked with the help of GAP if all or only half of the elements with a given cycle structure were contained in the union of subgroups of that class. It turns out the union contains the total in each case.

4. The Alternating Group A_9

In this section we give an estimate for the covering number of A_9. Maróti in [**14**] only gives a very rough lower bound for $\sigma(A_9)$. He shows $\sigma(A_9) \geq 80$. We start with a proposition giving us a covering of the elements of order 9 in A_9. This result was obtained with the help of GAP.

PROPOSITION 4.1. *There exists an irredundant covering of the elements of order 9 in A_9 by 112 subgroups.*

PROOF. From Table 4.2 below, it follows that the 40320 elements of order 9 occur in the conjugacy classes $MS5$, $MS7$, and $MS8$. They are partitioned into the 280 subgroups of $MS5$ with 144 elements in each subgroup. The split conjugacy classes $MS7$ and $MS8$ each contain half of the elements of order 9. These 20160 elements are not partitioned into the 120 subgroups of the class $MS7$ or $MS8$, respectively. Each subgroup in a class contains 504 elements of order 9. We build an irredundant cover from the bottom up using a greedy algorithm with the help of GAP. We take in one subgroup after another from the class such that the number of elements of order 9 in the union is maximized at each step. We first create the maximal subgroups of A_9 and check if the elements of order 9 in the classes $MS7$ and $MS8$ are mutually exclusive.

```
## Create A9 and its set of maximal subgroups
##
gap> A9 := AlternatingGroup(9);;
gap> MS := MaximalSubgroups(A9);;

gap> ## Create the sets of conjugate maximal subgroups MS7 and MS8
gap> ## of size 120 each
gap> ##
gap> MS7 := ConjugateSubgroups(A9, First(MS, x->Size(x)=1512));;
gap> Length(MS7);
120
gap> MS8 := ConjugateSubgroups(A9,
              First(MS, x->Size(x)=1512 and not x in MS7));;
gap> Length(MS8);
120
```

```
gap> ## Check to see MS7 and MS8 are mutually exclusive, the sets
gap> ## of elements order 9 in MS7 and MS8 are mutually exclusive,
gap> ## and all elements of order 9 are in either MS7 or MS8.
gap> ##
gap> IsEmpty(Intersection(MS7,MS8));
true
gap> U1 := Filtered(Union(MS7), x->Order(x)=9);;
gap> U2 := Filtered(Union(MS8), x->Order(x)=9);;
gap> IsEmpty(Intersection(U1,U2));
true
gap> ForAll(Filtered(A9,x->Order(x)=9), y->y in U1 or y in U2);
true
```

The following is the GAP code for our greedy algorithm.

```
gap> ## Greedily choose subgroups based in the number of elements of
gap> ## order 9 they contain from MS7 to find a cover
gap> ##
gap> ## Start the greedy algorithm with the first subgroup in MS7
gap> ## since all subgroups in MS7 have 504 elements of order 9.
gap> ##
gap> CV := [MS7[1]];;

gap> ## Add subgroups from CV\M7 which maximizes the number
gap> ## elements of order 9.
gap> ##
gap> while Filtered(Union(CV), x->Order(x)=9) <> U1 do
>     diff   := Filtered(MS7, x->not x in CV);
>     maxlst := List(diff,
                    s->Size(Filtered(Union(Concatenation(CV,[s])),
                    x->Order(x)=9)));
      index  := Position(maxlst,Maximum(maxlst));
>     Add(CV, diff[index]);
>    od;

gap> Size(CV);
56
gap> Size(Filtered(Union(CV), x->Order(x)=9));
20160
```

As seen in the GAP code for the greedy algorithm, the process continues until the union of the test cover contains all 20160 elements of order 9 found in the union of the conjugacy class. This yields a covering of 56 conjugates from each class $MS7$ and $MS8$, for a total of 112 subgroups needed to cover all elements of order 9. □

In the next lemma we determine minimal coverings for the elements of order 15 and 12 in A_9.

LEMMA 4.2. *The minimal number of subgroups needed to cover the elements with cycle structure* $(3,5)$ *and* $(2,3,4)$ *in* A_9 *is* 45.

PROOF. By Table 4.2 below, the elements with cycle structure $(3,5)$ occur in the subgroups of the conjugacy classes $MS1$, $MS2$, and $MS4$. Those with cycle structure $(2,3,4)$ occur in the subgroups of the conjugacy classes $MS1$, $MS2$, $MS3$ and $MS5$. In all cases, with the exception of the elements with cycle structure $(2,3,4)$ in $MS5$, the elements are partitioned into the subgroups of the respective classes. In particular, the elements with cycle structure $(3,5)$ and $(2,3,4)$ are contained in the union of the 9 subgroups of $MS4$ and the 36 subgroups of $MS3$. Thus there exists a covering of these elements by 45 subgroups. It remains to be shown that this covering is minimal.

By removing subgroups from $MS4$ in the above covering, we are losing elements with cycle structure $(3,5)$. These elements have to be replaced by taking in subgroups from other classes. Without loss of generality, we can assume that these replacements come from $MS2$, the class which, next to $MS4$, contains the most such elements. We observe that $MS2$ also contains elements with cycle structure $(2,3,4)$, allowing us to remove subgroups from $MS3$ in the covering.

As can be seen from Table 4.2 below, the quantitative breakdown of this process is as follows. Removing k subgroups, $1 \leq k \leq 9$, from $MS4$ in the covering requires the replacement of $2688k$ elements with cycle structure $(3,5)$ by at least s subgroups from $MS2$, where $s = \lceil \frac{2688k}{288} \rceil$. This exchange brings 188 elements with cycle structure $(2,3,4)$ into the covering. This allows us to remove not more than $t = \lfloor \frac{180s}{420} \rfloor$ subgroups from $MS3$, where $\lfloor a \rfloor$ denotes the floor of the real number a. Therefore, the number of subgroups needed in such a covering contains at least $45 - k - t + s$ subgroups. It can be easily verified that $45 - k - t + s > 45$ for $1 \leq k \leq 9$. Thus it follows that the covering by the subgroups of $MS3$ and $MS4$ is a minimal covering of the elements with cycle structure $(3,5)$ and $(2,3,4)$. □

Now we are ready to give an estimate for the covering number of A_9.

THEOREM 4.3. *For the covering number of* A_9 *we have* $127 \leq \sigma(A_9) \leq 157$.

PROOF. By Table 4.2 below, the union of the 36 subgroups in $MS3$ and the 9 subgroups in $MS4$ contains all elements of A_9 except the elements of order 9 with cycle structure (9) and the elements of order 3 with cycle structure (3^3). By Proposition 4.1, these elements are all contained in the union of 112 subgroups from $MS7$ and $MS8$. Hence there exists a covering of A_9 by $112 + 45$ subgroups. It follows that $\sigma(A_9) \leq 157$.

To establish a lower bound for $\sigma(A_9)$, we consider the elements of order 12 with cycle structure $(2,3,4)$, of order 15 with cycle structure $(3,5)$, and of order 9 with cycle structure (9). By Lemma 4.2, a minimal covering of the elements with cycle structure $(3,5)$ and $(2,3,4)$ consists of the 45 subgroups of $MS4$ and $MS3$. For the elements of order 9, we only can give a lower bound for the number of subgroups needed to cover these elements. There are $8!$ elements of order 9 in A_9. They are partitioned into the two conjugacy classes $MS7$ and $MS8$, with half of the elements in each class. However, the 20160 elements of order 9 in a class are not partitioned into the 120 subgroups of each of the classes. There are 504 such elements in each subgroup with overlap between some of the subgroups. Since $\frac{20160}{504} = 40$, the number of subgroups needed from each class is at least 40, disregarding the fact that there is overlap between the subgroups. Thus we need at least 41 subgroups from each class to cover the elements of order 9. This together with the above yields $\sigma(A_9) \geq 127$, the desired result. □

The information necessary for the work on A_9 was obtained from GAP and is contained in the two tables in this section. The first table, Table 4.1, contains

Label	**Isomorphism Type**	**Order**	**Size**
$MS1$	$(A_4 \times A_5) : C_2$	1440	126
$MS2$	$(C_3 \times A_6) : C_2$	2160	84
$MS3$	S_7	5040	36
$MS4$	A_8	20160	9
$MS5$	$(((C_3 \times ((C_3 \times C_3) : C_2) : C_2) : C_3) : C_2$	648	280
$MS6$	$((C_3 \times C_3) : Q_8) : C_3$	216	840
$MS7$	$\mathrm{PSL}(2,8) : C_3$	1512	120
$MS8$	$\mathrm{PSL}(2,8) : C_3$	1512	120

TABLE 4.1. Conjugacy classes of maximal subgroups for A_9.

the information on the conjugacy classes of maximal subgroups of A_9. Table 4.2 displays the inventory of elements and cycle structures as they appear in a class representative for each of the conjugacy classes of maximal subgroups. As with A_7 and A_8, the group A_9 has a pair of isomorphic conjugacy classes. Using GAP, we determined that the union of one of these conjugacy classes contains the total of a certain cycle structure or half of the total. The notation is the same as in the previous cases and an asterisk denotes that only half of the elements of a certain cycle structure are contained in the union of the conjugacy class. The questions arise

Order	**C.S.**	**Size**	$MS1$ (126)	$MS2$ (84)	$MS3$ (36)	$MS4$ (9)	$MS5$ (280)	$MS6$ (840)	$MS7$ (120)	$MS8$ (120)
1	(1)	1	1	1	1	1	1	1	1	1
2	(2^2)	378	78	90	126	210	27			
2	(2^4)	945	45	45	105	105,P	54	9	63	63
3	(3)	168	28	42	70	112	6			
3	(3^2)	3360	160	120	280	1120	12,P	24	168	168
3	(3^3)	2240		80			80	56	56	56
4	(4^2)	11340	180			1260,P		54		
4	$(2,4)$	7560	240	360	840	2520	54			
5	(5)	3024	24,P	144	504	1344				
6	$(2,6)$	30240		360,P	840,P	3360,P	108,P	72	504	504
6	$(2^2,3)$	7560	300	450	630	1680	54			
7	(7)	25920			720,P	5760			216,P	216,P
9	(9)	40320					144,P		504*	504*
10	$(2^2,5)$	9072	72,P		504					
12	$(2,3,4)$	15120	120,P	180,P	420,P		108			
15	$(3,5)$	24192	192,P	288,P		2688,P				

TABLE 4.2. Inventory of elements and cycle structure of A_9 across conjugacy classes of maximal subgroups.

of what is the exact covering number for A_9 and what are the methods to obtain it. At the moment we do not know if the covering given in Proposition 4.1 is a minimal covering for the elements of order 9. It is conceivable that by choosing other subgroups which provide maximal contributions of elements of order 9 we can make subgroups chosen earlier redundant and thus lower the number of subgroups needed. One potential approach to finding the minimal number of subgroups needed to cover the elements of order 9 is to reduce the question to a combinatorial optimization problem as has been done in the case of A_7. The 20160 elements of order 9 in $MS7$ or $MS8$ are grouped into 3360 cyclic subgroups of order 9. These cyclic subgroups are distributed over the 120 subgroups of a class with 84 subgroups of order 9 in each representative. As can be shown with the help of GAP, each C_9 is contained in exactly three representatives of the class. This setting reflects a bipartite graph as used in Lemma 2.1 with the 120 maximal subgroups of the class as the vertices of S and the 3360 cyclic subgroups of order 9 as the vertices of T, where each vertex in T is connected to exactly 3 vertices in S and each vertex in S is connected to 84 vertices of T. The question is what is the maximal number of vertices which can be removed from S such that all vertices of T still remain connected to S. Denoting the maximum number by $\max(S)$, our current results tell us that $64 \leq \max(S) \leq 79$. What is lacking at the moment to obtain an exact answer for $\max(S)$ is an appropriate labeling system for the vertices of S and T which led to a solution of a similar problem in the case of A_7.

Robert Morse, [**15**], has used the fact that each C_9 is contained in exactly three representatives of $MS7$ (or $MS8$ respectively), to show that $\max(S) \leq 72$, resulting in $\sigma(A_9) \geq 2 \cdot 48 + 45 = 141$. These results were obtained with the help of GAP. The following is an outline of the procedure. Of the 120 subgroups in $MS7$, 111 subgroups can be chosen such that they can be partitioned into 37 distinct subsets of 3 subgroups each so the union of each subset contains cyclic subgroups of order 9 not contained in the union of any of the other subsets. Hence at least one of the subgroups in each triplet has to appear in the covering, providing a lower bound of 37 for the cover. However some of these unique cyclic subgroups do not occur in all of the subgroups of a triplet, requiring the inclusion of these subgroups in the covering. There are 11 occurrences of this phenomenon among the 37 triplets. Thus 11 more subgroups in addition to the 37 have to be included in any covering of the cyclic subgroups of order 9 in $MS7$, yielding a lower bound of at least 48 subgroups needed to cover the elements of order 9. Repeated trials of randomly selecting 48 subgroups from this search space gives coverings of up to 16500 elements of order 9 from the 20160 required. This indicates that the actual value may be close to the upper bound of 56 obtained in Proposition 4.1.

5. The Alternating Group A_{10}

In [**14**], Maróti showed that $\sigma(A_n) = 2^{n-2}$ whenever $n \equiv 2 \bmod 4$. In this section we will show that $\sigma(A_{10}) = 256$, using a method different from the one given in [**14**].

THEOREM 5.1. *The covering number of* A_{10} *is* 256.

PROOF. From Table 5.2 below, we see that there are 172800 elements of order 21 with cycle structure $(3, 7)$ in A_{10}. They are only contained in the conjugacy class $MS2$ and they are partitioned across the 120 subgroups of the class with 1440 such elements in each subgroup. Thus any covering of A_{10} needs to contain these 120 subgroups. Similarly, as can be seen from Table 5.2, the 403200 elements of order 9 in A_{10} are only contained in the conjugacy class $MS4$. In fact, they are partitioned into the 10 subgroups of this class with 40320 such elements in each subgroup. Thus any covering of A_{10} needs to contain these 10 subgroups. The union of the 130 subgroups in the classes $MS2$ and $MS4$ however does not contain the elements with cycle structure (5^2), $(2^3, 4)$, $(2, 8)$, and $(4, 6)$. They are all contained in the 126 subgroups of the conjugacy class $MS6$. Thus A_{10} is the union of the subgroups in the three conjugacy classes $MS2$, $MS4$, and $MS6$. We conclude $\sigma(A_{10}) \leq 256$.

On the other hand, none of the 126 subgroups in the class $MS6$ can be dispensed with because the subgroups form a minimal covering of the elements with cycle structure (5^2). This can be seen as follows. By Table 5.2, the 72576 elements with cycle structure (5^2) occur in the conjugacy classes $MS5$, $MS6$, and $MS7$. In $MS6$, these elements are partitioned across the 126 subgroups of the class with 576 elements each. Thus there exists a covering of the elements with cycle structure (5^2) by 126 subgroups. It can be easily seen that replacing subgroups from $MS6$ in the covering by subgroups from $MS5$ or $MS7$ does not decrease the number of subgroups needed for the covering since, by Table 5.2, each of the subgroups of $MS5$ and $MS7$ contains fewer elements of cycle structure (5^2) than a subgroup in $MS6$. In addition, the 120 subgroups of the class $MS2$ have to be contained in any covering of A_{10}, since they are the only class containing the elements with cycle structure $(3, 7)$. Since these elements are partitioned into the subgroups of the class, all subgroups are needed. Similarly, the elements of order 9 are partitioned into the 10 subgroups of $MS4$ and this is the only class containing these elements. Hence at least $120 + 126 + 10$ subgroups are required to cover A_{10}. It follows that $\sigma(A_{10}) > 255$. We conclude $\sigma(A_{10}) = 256$, the desired result. □

Label	Isomorphism Class	Order	Size
$MS1$	$(A_4 \times A_6) : C_2$	8640	210
$MS2$	$(C_3 \times A_7) : C_2$	15120	120
$MS3$	S_8	40320	45
$MS4$	A_9	181440	10
$MS5$	$((C_2 \times C_2 \times C_2 \times C_2) : A_5) : C_2$	1920	945
$MS6$	$(A_5 \times A_5) : C_4$	14400	126
$MS7$	$M(10)$	720	2520

TABLE 5.1. Conjugacy classes of maximal subgroups of A_{10}.

The final two tables contain the information for A_{10} obtained from GAP. The first of these lists the classes of maximal subgroups, giving the label we use, their isomorphism class, the order of a class representative and the size of the conjugacy class.

The second table presents an inventory of the order and cycle structure for each of the elements in a class representative for each of the conjugacy classes of maximal subgroups of A_{10}. Table 5.2 shows the distribution of the cycle structures across the classes of maximal subgroups of A_{10}. We observe that A_{10} does not contain any split classes of maximal subgroups. For the notation used, we refer to Section 2.

Order	C.S.	Size	$MS1$ (210)	$MS2$ (120)	$MS3$ (45)	$MS4$ (10)	$MS5$ (945)	$MS6$ (126)	$MS7$ (2520)
1	(1)	1	1	1	1	1	1	1	1
2	(2^2)	630	138	168	238	378	30	130	
2	(2^4)	4725	225	315	525	945	125	225	45
3	(3)	240	48	72	112	168		40	
3	(3^2)	8400	360	420	1120	3360	80	400	
3	(3^3)	22400	320	560		2240,P			80
4	(4^2)	56700	540		1260,P	11340	300	900	270
4	$(2,4)$	18900	720	1260	2940	7560	60	600	
4	$(2^3,4)$	18900	360		1260		140	1200	
5	(5)	6048	144	504	1344	3024		48,P	
5	(5^2)	72576					384	576,P	144
6	$(2,6)$	151200	720,P	2520	6720	30240	160,P		
6	$(2^2,3)$	25200	1200	1680	2800	7560	240	1000	
6	$(2^2,3^2)$	25200	120,P	420	1120			400	
7	(7)	86400		720,P	5760	25920			
8	$(2,8)$	226800			5040,P		240,P	3600	180
9	(9)	403200				40320,P			
10	$(2^2,5)$	90720	432,P	1512	4032	9072,P		720,P	
12	$(4,6)$	151200	720,P				160,P	2400	
12	$(2,3,4)$	151200	1440	2520	3360,P	15120,P		1200,P	
15	$(3,5)$	120960	1152	1008,P	2688,P	24192		960,P	
21	$(3,7)$	172800		1440,P					

TABLE 5.2. Inventory of elements and cycle structure of A_{10} across conjugacy classes of maximal subgroups.

Acknowledgements. The authors would like to thank Lucas Sabalka and Tom Zaslavsky for facilitating the proof of Lemma 2.1 and Robert F. Morse for sharing with us his results for the lower bound of the covering in the case of the elements of order 9 in A_9. The authors would also like to thank the referee for his comments making this a better paper. Lastly, we want to thank Marge Pratt for her expert help in formatting the manuscript.

References

[1] A. Abdollahi, F. Ashraf and S.M. Shaker, *The symmetric group of degree six can be covered by* 13 *and no fewer proper subgroups*, Bull. Malays. Math. Sci. Soc., **30** (2007), 57-58.

[2] B. Ábrego, S. Fernández-Merchant, M.G. Neubauer, and W. Watkins, *Sum of squares of degrees in a graph*, Journal of Inequalities in Pure and applied Mathematics, (to appear).

[3] R. Ahlswede and G.H.O. Katona, *Graphs with maximal number of pairs of edges*, Acta Math. Acad. Sci. Hungar. 32 (1978), 97-120.

[4] R.A. Bryce, V. Fedri and L. Serena, *Subgroup Coverings of Some Linear Groups*, Bull. Austral. Math. Soc., **60** (1999), 239-244.

[5] J.H.E. Cohn, *On n-sum Groups*, Math. Scand., **75** (1994), 44-58.

[6] E. Detomi and A. Lucchini, *On the structure of primitive n-sum groups*, CUBO, A Mathematical Journal, **10** (2008), 195-210.

[7] The GAP Group, *GAP - Groups, Algorithms, and Programming*, Version 4.4.7, 2006, (http://www.gap-system.org).

[8] M. Garonzi, *Finite groups that are the union of at most 25 proper subgroups*, Journal of Algebra, to appear.

[9] D. Greco, *I gruppi che sono somma di quattro sottogruppi*, Rend. Accad. delle Scienze di Napoli, **18** (1951), 74-85.

[10] D. Greco, *Su alcuni gruppi finiti che sono somma di cinque sottogruppi*, Rend. Sem. Mat. Univ. Padova, **22** (1953), 313-333.

[11] D. Greco, *Sui gruppi che sono somma di quattro o cinque sottogruppi*, Rend. Accad. delle Scienze di Napoli, **23**, (1956), 49-56.

[12] P.E. Holmes, *Subgroup coverings of some sporadic groups*, J. Combin. Theory, Ser. A, **113** (2006), 1204-1213.

[13] M.S. Lucido, *On the covers of finite groups*, Proceedings Groups-St. Andrews 2001 at Oxford, Lecture Notes LMS 305 (2003), 395-399.

[14] Attila Maróti, *Covering Symmetric Groups with Proper Subgroups*, Journal of Combinatorial Theory, **110** (2005), 97-111.

[15] R.F. Morse, Personal communication.

[16] B.H. Neumann, *Groups covered by Permutable Subsets*, J. London Math. Soc., **29** (1954), 236-248.

[17] G. Scorza, *I gruppi che possone pensarsi come somma di tre lori sottogruppi*, Boll. Un. Mat. Ital., **5** (1926), 216-218.

[18] L. Serena, *On finite covers of groups by subgroups*, Adv. Group Theory (2002), Aracne, Rome 2003, 173-190.

[19] M.J. Tomkinson, *Groups as the Union of Proper Subgroups*, Math.Scand., **81** (1997), 189-198.

[20] J. Zhang, *Finite groups as the union of proper subgroups*, Serdica Math. J., **32** (2006), 259-268.

Department of Mathematical Sciences, State University of New York at Binghamton, Binghamton, NY 13902-6000, USA

E-mail address: menger@math.binghamton.edu

Department of Mathematics, Elmira College, Elmira, NY 14901, USA

E-mail address: mathprofessor@mac.com

Contemporary Mathematics
Volume **511**, 2010

Certain homological functors of 2-generator p-groups of class 2

Arturo Magidin and Robert Fitzgerald Morse

ABSTRACT. Using a new classification of 2-generator p-groups of class 2, we compute various homological functors for these groups. These functors include the nonabelian tensor square, nonabelian exterior square and the Schur multiplier. We also determine which of these groups are capable and which are unicentral.

1. Introduction

In [1] a new classification for the 2-generator p-groups of nilpotency class two is given that corrects and simplifies previous classifications for these groups published in [4, 21, 29]. Initial attempts to compute the nonabelian tensor squares for odd prime can be found in [4], with some corrections as well as the computation of other homological functors in [3]. For the $p = 2$ case, computations of the nonabelian tensor square were started in [21].

Because these previous computations were based on a classification that we now know to have been incomplete, we present these computations afresh in light of the new classification. The goal of this paper is to give a complete description of the groups that appear in the diagram in Figure 1 below for each 2-generator p-group of class two. The group $H_2(G)$ in the diagram is the second integral homology group of G, also known as the Schur multiplier. The groups $G \otimes G$ and $G \wedge G$ are the nonabelian tensor square of G and the exterior square of G, respectively, and have become the focus of increasing interest of late in part because of their application in determining certain group theoretic properties of G. The group $\Gamma(G/G')$ is the Whitehead universal quadratic functor, and $J_2(G)$ is known to be isomorphic to $\pi_3 SK(G)$, the third homotopy group of the suspension of an Eilenberg-Mac Lane space of G. Although it does not appear in the diagram (because we will not be computing it) there is also a natural map from $H_3(G)$, the third integral homology group of G, into $\Gamma(G/G')$ whose image is the kernel of the map ψ.

The nonabelian tensor product was explicitly introduced by Brown and Loday [12], extending certain classical ideas of Whitehead and generalizing the tensor product of abelian groups when two groups act on each other and satisfying certain

2000 *Mathematics Subject Classification.* 20F18, 20J99.
Key words and phrases. p-groups, nonabelian tensor square, capability.

$$\begin{array}{ccccccccc}
 & & & & 0 & & 0 & & \\
 & & & & \downarrow & & \downarrow & & \\
 & & \Gamma(G/G') & \xrightarrow{\psi} & J_2(G) & \longrightarrow & H_2(G) & \longrightarrow & 0 \\
 & & \downarrow & & \downarrow & & \downarrow & & \\
0 & \longrightarrow & \nabla(G) & \longrightarrow & G\otimes G & \xrightarrow{\theta} & G\wedge G & \longrightarrow & 1\\
 & & \downarrow & & {\scriptstyle\kappa}\downarrow & & {\scriptstyle\kappa'}\downarrow & & \\
 & & 1 & & G' & = & G' & & \\
 & & & & \downarrow & & \downarrow & & \\
 & & & & 1 & & 1 & &
\end{array}$$

FIGURE 1. All rows and columns in the diagram are exact.

compatibility conditions. These conditions are always satisfied when a group acts on itself by conjugation, leading to the nonabelian tensor square of a group: given a group G, its nonabelian tensor square $G \otimes G$ is generated by the symbols $g \otimes h$ with $g, h \in G$, subject to the relations

$$gg' \otimes h = ({}^g g' \otimes {}^g h)(g \otimes h) \quad \text{and} \quad g \otimes hh' = (g \otimes h)({}^h h \otimes {}^h h'),$$

for all $g, g', h, h' \in G$, with ${}^x y = xyx^{-1}$. The nonabelian exterior square $G \wedge G$ is the quotient of $G \otimes G$ modulo the central subgroup generated by the elements $g \otimes g$ for all $g \in G$; the image of the generator $g \otimes h$ in $G \wedge G$ is denoted by $g \wedge h$.

The nonabelian exterior square of G is implicit in [25], where Miller interprets $H_2(G)$ as the group that contains all commutator relations satisfied by the commutators in G modulo those that are universally satisfied. In an unpublished note [14] R.K. Dennis made this explicit, and extended the nonabelian exterior square to the nonabelian tensor square.

The study of $G \otimes G$ from a purely group theoretic view, in particular computing the nonabelian tensor squares of certain groups explicitly and describing them in more standard terms, was begun by Brown, Johnson, and Robertson in [11]. As can be expected, the definition by generators and relations quickly becomes too large and unwieldy, but with the help of computers they were able to use Tietze transformations to simplify the generators and relations given by the definition to compute the nonabelian tensor squares of all nonabelian groups up to order 30. For larger finite groups and infinite groups, crossed pairings have been used as a computational method. If G and L are groups, then a function $\phi\colon G \times G \to L$ is called at *crossed pairing* if

$$\phi(gg', h) = \phi({}^g g', {}^g h)\phi(g, h) \quad \text{and} \quad \phi(g, hh') = \phi(g, h)\phi({}^h g, {}^h h'),$$

for all $g, g', h, h' \in G$. From the definition, it is clear that a crossed pairing determines a unique homomorphism $\phi^*\colon G \otimes G \to L$ such that $\phi^*(g \otimes g') = \phi(g, g')$. The basic technique is to make an educated "guess" on the structure of L, usually based on some explicit computations made with the help of a computer algebra package, and then construct a crossed pairing from $G \otimes G$ into the conjectured group L and try to prove that the induced map on the nonabelian tensor square is an isomorphism. This last step can be difficult, and can sometimes lead to the incorrect identification of the nonabelian tensor square with a group that is in fact a proper quotient of it (e.g., see the comments after the proof of Theorem 4.4 in [21]). The crossed pairing method was used in [2] to determine the nonabelian tensor square of the free nilpotent groups of class two and finite rank, and in [5, 9] for the free 2-Engel groups of finite rank; the latter is particularly computationally intensive. They were also behind the attempts for the 2-generator p-groups of class two [3, 4, 21].

More recent work is based on ideas of Rocco [27], also discovered independently by Ellis and Leonard [18], which identify the nonabelian tensor square with a subgroup of the commutator subgroup of a suitable group $\nu(G)$. Then we can use commutator calculus to study the structure of the nonabelian tensor square. This has also helped define effective algorithms for some classes of groups, such as polycyclic groups [8], which allow the use of computer tools such as GAP [19] to compute the nonabelian tensor square of a given polycyclic group, even if it is infinite. Using this approach, the nonabelian tensor square of the free nilpotent groups of finite rank and arbitrary class was obtained in [7].

The new ideas, such as Rocco's construction, proved sufficient to fully describe the nonabelian tensor squares in the case of the free nilpotent groups of finite rank in large part because the tensor squares had no torsion. In our case, with finite groups, the techniques did not seem to be *quite* enough to get us all the way home precisely because of the issue of the orders of elements. We often found ourselves in the following situation: looking at a generator, say for example $a \otimes [a, b]$, the identities of G and commutator calculus in Rocco's construction (described below) allowed us to prove $(a \otimes [a, b])^{p^k} = 1$ held for a certain k; computations with GAP seemed to indicate that this was in fact the order of the element; but we did not see how to show that p^k was *exactly* the order using only the relations we had on hand. This is, of course, the usual problem one faces when dealing with a group that is given by generators and relations.

The standard way to address this problem is to use homomorphic images of the group to give lower bounds to the order of the elements, and if one is lucky these lower bounds will agree with established upper bounds and yield the order of the element. Crossed pairings allow us to find such homomorphic images of $G \otimes G$. We use them exclusively to find mappings from $G \otimes G$ to finite cyclic groups that provide lower bounds for the orders of specific generators of $G \otimes G$. As a corollary we sometimes deduce independence conditions for a generator with the other generators of the nonabelian tensor square. Our use of crossed pairings is much more restricted than in prior work in this area and we turned to them only when we could find no other way to determine the order of a particular generator. We believe this judicious and limited use of crossed pairings helps us to avoid the difficulties and pitfalls that seem to have beset their use in the past.

Hence our method of computation is a hybrid approach that uses the commutator calculus initiated by Rocco [27] and expanded on by Blyth and Morse [8], and the cross pairing method used by Kappe and others. Moreover, we also take advantage of new structural results of the nonabelian tensor square established by Blyth, Fumagalli, and Morigi [10].

Capable and unicentral groups. A group is called *capable* if it is a central factor of some group. Capability plays an important role in P. Hall's classification scheme for p-groups up to isoclinism [20]. A *unicentral* group G is a group whose largest capable image is $G/Z(G)$. These two group theoretic properties can be determined by computing a central subgroup of G defined by the nonabelian exterior square of G called the *nonabelian exterior center* (see Section 5).

The determination of the capable groups among the 2-generator p-groups of class 2 was begun in [3, 24] for odd prime and the 2-group case was continued in [23] (using different techniques); these efforts relied on the incorrect descriptions contained in [4, 21]. This led to a mischaracterization of the capable groups among these groups when $p > 2$ in [3] (e.g., compare Corollary 4.3 there with Theorem 6.2 of [24]).

In Section 5 we identify all 2-generator p-groups of class 2 that are capable and those that are unicentral. Moreover, within this class of groups, we provide formulas that count of number of capable and unicentral groups of order p^n.

Notation and conventions. Our groups will be written multiplicatively unless we explicitly state otherwise, with 1 denoting the identity element. When the groups are explicitly stated to use additive notation, the identity element will be denoted by 0.

There are two standard actions of a group on itself by conjugation: the left action and the right action. Most of the work in this area uses the left action convention; Rocco's paper, however, uses the right action. The relevant formulas were "translated" into left action in [8], and we use the translated formulas, so our work inside the group $\nu(G)$ defined below uses left actions. However, our definitions inside the groups G use right actions, since those are more common in dealing with commutator collection and nilpotent groups; in any case, in a nilpotent group of class two we have that the two definitions of the commutator, $[x,y] = xyx^{-1}y^{-1}$ and $[x,y] = x^{-1}y^{-1}xy$, actually agree, so the two conventions lead to the same formulas in that restricted setting.

Because we will work mostly with left actions, we write our commutators left normed, so that $[x,y,z] = [[x,y],z]$.

The classification of 2-generated p-groups of class 2. Our computations are based on the following classification of the 2-generator p-groups of class 2.

THEOREM 1 ([1]). *Let p be a prime, $n > 2$ a positive integer. Every 2-generated p-group of class exactly 2 and order p^n corresponds to an ordered 5-tuple of integers, $(\alpha,\beta,\gamma;\rho,\sigma)$, such that:*

(i) $\alpha \geq \beta \geq \gamma \geq 1$;
(ii) $\alpha + \beta + \gamma = n$;
(iii) $0 \leq \rho, \sigma \leq \gamma$;

where $(\alpha,\beta,\gamma;\rho,\sigma)$ corresponds to the group presented by

$$G = \left\langle a, b \;\middle|\; [a,b]^{p^\gamma} = [a,b,a] = [a,b,b] = 1, a^{p^\alpha} = [a,b]^{p^\rho}, b^{p^\beta} = [a,b]^{p^\sigma} \right\rangle. \tag{2}$$

Moreover:

FAMILY 1. *If $\alpha > \beta$, then G is isomorphic to:*

(a) $(\alpha,\beta,\gamma;\rho,\gamma)$ *when* $\rho \leq \sigma$.
(b) $(\alpha,\beta,\gamma;\gamma,\sigma)$ *when* $0 \leq \sigma < \sigma + \alpha - \beta \leq \rho$ *or* $\sigma < \rho = \gamma$.

(c) $(\alpha, \beta, \gamma; \rho, \sigma)$ *when* $0 \leq \sigma < \rho < \min(\gamma, \sigma + \alpha - \beta)$.

FAMILY 2. *If* $\alpha = \beta > \gamma$ *or* $\alpha = \beta = \gamma$ *and* $p > 2$, *then* G *is isomorphic to* $(\alpha, \beta, \gamma; \min(\rho, \sigma), \gamma)$.

FAMILY 3. *If* $\alpha = \beta = \gamma$ *and* $p = 2$, *then* G *is isomorphic to:*

(a) $(\alpha, \beta, \gamma; \min(\rho, \sigma), \gamma)$ *when* $0 \leq \min(\rho, \sigma) < \gamma - 1$.
(b) $(\alpha, \beta, \gamma; \gamma - 1, \gamma - 1)$ *when* $\rho = \sigma = \gamma - 1$.
(c) $(\alpha, \beta, \gamma; \gamma, \gamma)$ *when* $\min(\rho, \sigma) \geq \gamma - 1$ *and* $\max(\rho, \sigma) = \gamma$.

The groups listed in Family 1(a) through Family 3(c) are pairwise non-isomorphic.

The parameters α, β, and γ are fixed for each family. Below, we will often refer to a group $G = G_p(\alpha, \beta, \gamma; \rho, \sigma)$ *"with presentation as in Theorem 1."* When we do this, we mean that G is the p-group with presentation (2), and that ρ and σ are the canonical values given in the descriptions of each Family in the statement of the theorem.

Main results. Our main results are a description of the invariant factors of the nonabelian tensor square, the nonabelian exterior square, $\Gamma(G/G')$, $\nabla(G)$, $H_2(G)$, and $J_2(G)$, for each 2-generator p-group of class 2 in terms of its canonical parameters α, β, γ, ρ, and σ. This detailed description is a homomorphic image of the presentation below.

Theorem. *Let* $G = G_p(\alpha, \beta, \gamma; \rho, \sigma)$ *be a 2-generator p-group of class* 2 *with presentation as in Theorem 1. Then* $G \otimes G$ *is a homomorphic image of the abelian group generated by the six elements*

$$a \otimes a, \quad b \otimes b, \quad (a \otimes b)(b \otimes a), \quad a \otimes b, \quad a \otimes [a, b], \quad b \otimes [a, b],$$

subject to the relations:

$$\begin{aligned}
&(a \otimes a)^{2^{\alpha+1}} = (b \otimes b)^{2^{\beta+1}} = \big((a \otimes b)(b \otimes a)\big)^{2^{\beta}} = 1,\\
&(a \otimes b)^{2^{\alpha+1}} = (a \otimes [a, b])^{2^{\min(\rho+1,\gamma)}} = (b \otimes [a, b])^{2^{\min(\sigma+1,\rho+1,\gamma)}} = 1,\\
&(a \otimes a)^{2^{\alpha}} = (a \otimes [a, b])^{2^{\rho}}, \qquad (b \otimes b)^{2^{\beta}} = (b \otimes [a, b])^{2^{\sigma}},\\
&(a \otimes b)^{2^{\beta}} = (a \otimes [a, b])^{2^{\sigma}} (b \otimes [a, b])^{2^{\beta-1}}
\end{aligned}$$

if $p = 2$; *and if* $p > 2$ *then subject to the relations:*

$$\begin{aligned}
&(a \otimes a)^{p^{\alpha}} = (b \otimes b)^{p^{\beta}} = \big((a \otimes b)(b \otimes a)\big)^{p^{\beta}} = 1,\\
&(a \otimes b)^{p^{\alpha}} = (a \otimes [a, b])^{p^{\rho}} = (b \otimes [a, b])^{p^{\min(\rho,\sigma)}} = 1,\\
&(a \otimes b)^{p^{\beta}} = (a \otimes [a, b])^{p^{\sigma}}.
\end{aligned}$$

The exterior squares are generated by the (images of the) latter three generators listed above. To obtain a basis for $G \otimes G$, we replace the generator $a \otimes [a, b]$ with $(a \otimes b)^{p^{\beta-\sigma}}(a \otimes [a, b])$ when $\sigma < \rho$. When $p = 2$, the last three generators may need sundry modifications, depending on the family from Theorem 1 to which G belongs. Some straightforward order considerations then yield the invariants and bases for $\nabla(G)$, $H_2(G)$, and $J_2(G)$.

Previous attempts. As we mentioned above, there were prior attempts at computing the nonabelian tensor square of 2-generator p-groups of class 2 using a different description (which was missing family 1(c) from Theorem 1).

In addition to the missing groups, there were a few problems. In Theorem 4.4 of [4], the authors mistakenly give a proper homomorphic image of the tensor square of the groups that correspond to our Family 1(b); their attempted correction in Proposition 3.5 of [3] has either one invariant too large, or is again a proper homomorphic image, depending on exactly how their parameters compare to one another (we will not specify the details, since their notation is incompatible with ours; see the discussion in [1] on how to translate into our parameters, and compare their formula (3.5.1) with Theorem 36). For $p=2$, [21] incorrectly computes the tensor square for the groups in our Family 3(b) when $\gamma > 1$, giving instead a proper homomorphic image by replacing one invariant of order $2^{\gamma+1}$ with one of order 2^{γ}.

As far as we are aware, we are presenting the first complete and correct (we hope!) computations of the nonabelian tensor squares for this class of groups. Moreover, we compute for the first time the other homological functors for these groups when $p=2$.

Organization of the paper. In Section 2, we describe the group $\nu(H)$ introduced by Rocco in which most of our computations will take place, as well as summarizing some of the relevant commutator identities we will use. We then present some recent structural results that will play a key part in our analysis. With these in hand, we perform our initial tensor calculations for any group G in our classification. We close Section 2 by setting up our use of crossed pairings.

In Section 3 we finish the computations of the nonabelian tensor squares and exterior squares, first dealing with the odd p case and then with $p=2$; these are the main results from which the rest will follow. In Section 4 we use these descriptions to compute $\Gamma(G/G')$, $\nabla(G)$, $H_2(G)$, and $J_2(G)$. Section 5 describes and computes the nonabelian tensor and exterior centers, and determines which groups in our class are capable and which are unicentral. In the final section we provide some details on how GAP was used to perform the explicit computation of these functors for specific groups in our class, as these computations were often key in motivating and directing the theorems in the paper.

2. Preliminaries

In this section we review the setting used in computing the homological functors found in Figure 1, list some applicable structural results, make some initial tensor calculations for the groups in Theorem 1, and set up our use of crossed pairings.

2.1. The group $\nu(H)$.

The computational setting for this paper is based on the following group construction.

DEFINITION 3. Let be H be a group with presentation $\langle \mathcal{H}|\mathcal{R}\rangle$ and let H^{φ} be an isomorphic copy of H via the mapping $\varphi : h \mapsto h^{\varphi}$ for all $h \in H$. We define the group $\nu(H)$ with the presentation

$$(4)\qquad \big\langle \mathcal{H}, \mathcal{H}^{\varphi} \mid \mathcal{R}, \mathcal{R}^{\varphi}, {}^{x}[g,h^{\varphi}] = [{}^{x}g, ({}^{x}h)^{\varphi}] = {}^{x^{\varphi}}[g,h^{\varphi}], \forall x,g,h \in H\big\rangle.$$

The following theorem gives some structural characteristics of $\nu(H)$. Most importantly it shows that the nonabelian tensor square of H embeds naturally as a subgroup of $[\nu(H), \nu(H)]$.

THEOREM 5 ([18,27]). *Let H be a group.*

(i) *The groups H and H^φ isomorphically embed into $\nu(H)$.*
(ii) *If H is nilpotent of class c then $\nu(H)$ is nilpotent of class at most $c+1$.*
(iii) *If H is a p-group then $\nu(H)$ is a p-group.*
(iv) *The map $\phi : H \otimes H \to [H, H^\varphi] \lhd \nu(H)$ defined by $\phi(g \otimes h) = [g, h^\varphi]$ for all g and h in H is an isomorphism.*

By Theorem 5(iv), any tensor calculation in $H \otimes H$ can be performed by making computations within the subgroup $[H, H^\varphi]$ in $\nu(H)$. These computations use a specialized commutator calculus that takes into account the interactions between elements of H and H^φ. These rules mimic known identities between elements of $H \otimes H$ and record new identities as well. Much of this commutator calculus can be found in [8, 10, 27]. We provide a list of the identities we use in this paper for the convenience of the reader.

Since all our computations will take place in $\nu(G)$, we will use $[G, G^\varphi]$ and $G \otimes G$ interchangeably, and likewise use $[g, h^\varphi]$ and $g \otimes h$ for $g, h \in G$ synonymously, identifying the two descriptions via the corresponding isomorphism.

LEMMA 6. *Let H be a group. The following relations hold in $\nu(H)$:*

(i) $[h_1^\varphi, h_2, h_3] = [h_1, h_2^\varphi, h_3] = [h_1, h_2, h_3^\varphi] = [h_1^\varphi, h_2^\varphi, h_3] = [h_1^\varphi, h_2, h_3^\varphi] = [h_1, h_2^\varphi, h_3^\varphi]$ *for all h_1, h_2, h_3 in H;*
(ii) $[h_1, [h_2, h_3]^\varphi] = [h_2, h_3, h_1^\varphi]^{-1}$ *for all h_1, h_2, h_3 in H;*
(iii) $[h, h^\varphi]$ *is central in $\nu(H)$ for all h in H;*
(iv) $[h_1, h_2^\varphi][h_2, h_1^\varphi]$ *is central in $\nu(H)$ for all h_1, h_2 in H;*
(v) $[h, h^\varphi] = 1$ *for all h in H'.*

LEMMA 7. *Let H be a group with commuting elements x, y. Then in $\nu(H)$:*

(i) $[x^n, y^\varphi] = [x, y^\varphi]^n = [x, (y^\varphi)^n]$ *for all integers n;*
(ii) $[x, y^\varphi]$ *is central in $\nu(H)$;*
(iii) *if x and y are torsion elements of H of orders $o(x)$ and $o(y)$ (respectively) in H, then the order of $[x, y^\varphi]$ in $\nu(H)$ divides the greatest common divisor of $o(x)$ and $o(y)$.*

LEMMA 8. *Let h_1 and h_2 be elements of a group H. Then in $\nu(H)$:*

$$[h_1^n, h_2^\varphi] \cdot [h_2, (h_1^n)^\varphi] = [h_1, (h_2^n)^\varphi] \cdot [h_2^n, (h_1)^\varphi] = ([h_1, h_2^\varphi][h_2, (h_1)^\varphi])^n.$$

2.2. Structural results. In a recent paper [10], Blyth, Fumagalli, and Morigi investigate the structural properties of the nonabelian tensor square and its related groups from Figure 1. The following theorems summarize their results that we will use in our setting.

THEOREM 9. *Let H be any group whose abelianization is finitely generated by the independent set x_iH', $i = 1, \ldots, n$. Let N be the kernel of the epimorphism $\nabla(H) \to \nabla(H/H')$, and let $E(H)$ be the subgroup of $\nu(H)$ defined by*

$$E(H) = \langle [x_i, x_j^\varphi] \mid 1 \le i < j \le n \rangle [H, (H')^\varphi].$$

Then $\nabla(H) \cap E(H) = N$ and $\nabla(H)E(H) = [H, H^\varphi]$.

THEOREM 10. *Let H be any group.*

(i) *The nonabelian tensor square of any group H is satisfies:*

$$(H \otimes H)/N \cong \nabla(H/H') \times (H \wedge H),$$

where N is the kernel of the epimorphism $\nabla(H)\to\nabla(H/H')$.

(ii) *If H/H' has no elements of order 2 then*

$$\nabla(H)\cong\nabla(H/H') \quad \textit{and} \quad \Gamma(H/H')\cong\nabla(H/H').$$

(iii) *If H/H' has no elements of order 2 then $J_2(H)$ is isomorphic to the direct product $\nabla(H)\times H_2(H)$.*

(iv) *If A is any finitely generated abelian group with independent generating set $\{a_1,a_2,\ldots,a_n\}$ then $A\otimes A\cong\nabla(A)\times(A\wedge A)$ where the independent generators of $\nabla(A)$ are the images of $a_i\otimes a_i$ for $i=1,\ldots,n$ and of $(a_i\otimes a_j)(a_j\otimes a_i)$ for all $1\leq i<j\leq n$ and the independent generators of $A\wedge A$ are the images of $a_i\otimes a_j$ for all $1\leq i<j\leq n$.*

The following properties of Γ, the Whitehead universal quadratic functor, can be found in [11]. For A and B abelian groups we have:

$$\Gamma(A\times B)\cong\Gamma(A)\times\Gamma(B)\times(A\otimes B) \tag{11}$$

$$\Gamma(C_n)\cong\begin{cases} C_n & \text{if } n \text{ is odd}\\ C_{2n} & \text{if } n \text{ is even,}\end{cases} \tag{12}$$

where C_n is the cyclic group of order n. We interpret C_0 to be the infinite cyclic group.

Let $G=G_p(\alpha,\beta,\gamma;\rho,\sigma)$ be a 2-generator p-group of class 2 with presentation as in Theorem 1. The abelianization of G depends only on α and β and we have $G/G'\cong\langle aG'\rangle\times\langle bG'\rangle\cong C_{p^\alpha}\times C_{p^\beta}$. By (11) and (12) it follows that

$$\Gamma(G/G')\cong\begin{cases} C_{p^\alpha}\times C_{p^\beta}\times C_{p^\beta} & \text{if } p>2,\\ C_{p^{\alpha+1}}\times C_{p^{\beta+1}}\times C_{p^\beta} & \text{if } p=2.\end{cases} \tag{13}$$

Set $\bar{a}=aG'$ and $\bar{b}=bG'$. Then by Theorem 10(iv) we can compute $\nabla(G/G')$ and the exterior and tensor square of G/G':

$$\nabla(G/G')\cong\langle\bar{a}\otimes\bar{a}\rangle\times\langle\bar{b}\otimes\bar{b}\rangle\times\big\langle(\bar{a}\otimes\bar{b})(\bar{b}\otimes\bar{a})\big\rangle\cong C_{p^\alpha}\times C_{p^\beta}\times C_{p^\beta}, \tag{14}$$

$$(G/G')\wedge(G/G')\cong\langle\bar{a}\otimes\bar{b}\rangle\cong C_{p^\beta}, \tag{15}$$

$$(G/G')\otimes(G/G')\cong C_{p^\alpha}\times C_{p^\beta}\times C_{p^\beta}\times C_{p^\beta}. \tag{16}$$

2.3. Tensor calculations. Let $G=G_p(\alpha,\beta,\gamma;\rho,\sigma)$ be any 2-generator p-group of class 2 with presentation given in Theorem 1. The nonabelian tensor square $G\otimes G$ is abelian; this follows from Theorem 5(ii), as $\nu(G)$ is nilpotent of class at most 3, so $\nu(G)'$ is abelian; as $[G,G^\varphi]\leq\nu(G)'$, we deduce that $G\otimes G$ is indeed abelian. The elements a, b, and $[a,b]$ form a polycyclic generating set for G. By Proposition 20 of [8] we have that $[G,G^\varphi]$ is generated by

$$\begin{gathered}[a,a^\varphi],\ [b,b^\varphi],\ [a,b^\varphi][b,a^\varphi],\\ [a,b^\varphi],\ [a,[a,b]^\varphi],\ [b,[a,b]^\varphi],\ [[a,b],a^\varphi],\ [[a,b],b^\varphi].\end{gathered}$$

By Lemma 6(ii) we have

$$[[a,b],a^\varphi]=[a,[a,b]^\varphi]^{-1} \text{ and } [[a,b],b^\varphi]=[b,[a,b]^\varphi]^{-1}$$

and the following lemma is proven:

LEMMA 17. *The nonabelian tensor square of $G = G_p(\alpha, \beta, \gamma; \rho, \sigma)$ is an abelian group generated by*

$$[a, a^\varphi],\ [b, b^\varphi],\ [a, b^\varphi][b, a^\varphi],\ [a, b^\varphi],\ [a, [a, b]^\varphi],\ \textit{and } [b, [a, b]^\varphi].$$

We now prove a series of lemmas that provide relations between the generators of $[G, G^\varphi]$ listed in Lemma 17.

LEMMA 18. *Let $G = G_p(\alpha, \beta, \gamma; \rho, \sigma)$ be any 2-generator p-group of class 2 with presentation given in Theorem 1. Then*

(i) $[a, ([a, b]^{p^\rho})^\varphi] = [a, (a^{p^\alpha})^\varphi] = [a, a^\varphi]^{p^\alpha} = [a^{p^\alpha}, a^\varphi] = [a, ([a, b]^{p^\rho})^\varphi]^{-1}$,

(ii) $[b, ([a, b]^{p^\sigma})^\varphi] = [b, (b^{p^\beta})^\varphi] = [b, b^\varphi]^{p^\beta} = [b^{p^\beta}, b^\varphi] = [b, ([a, b]^{p^\sigma})^\varphi]^{-1}$,

(iii) $([a, b^\varphi][b, a^\varphi])^{p^\beta} = 1$.

PROOF. The identities (i) and (ii) follow from Lemma 7(i) and the relations of G. Using Lemma 8 and Lemma 6(ii) we have

$$([a, b^\varphi][b, a^\varphi])^{p^\beta} = [a, (b^{p^\beta})^\varphi][b^{p^\beta}, a^\varphi] = [a, ([a, b]^{p^\sigma})^\varphi][[a, b]^{p^\sigma}, a^\varphi] = 1.$$

□

COROLLARY 19. *Let $G = G_p(\alpha, \beta, \gamma; \rho, \sigma)$ be any 2-generator p-group of class 2 with presentation given in Theorem 1. Then $[a, a^\varphi]^{2p^\alpha} = 1$ and $[b, b^\varphi]^{2p^\beta} = 1$.*

The following useful general expansion formula can be found in [2].

LEMMA 20. *Let H be a group of nilpotency class 2. For any x and y in H and any integers n and m,*

$$[x^n, (y^m)^\varphi] = [x, y^\varphi]^{nm}[y, [x, y]^\varphi]^{n\binom{m}{2}}[x, [x, y]^\varphi]^{m\binom{n}{2}}$$

in $\nu(H)$, where $\binom{k}{2} = \frac{k(k-1)}{2}$ for every integer k.

Using Lemma 20 we obtain the following identities in G.

LEMMA 21. *Let $G = G_p(\alpha, \beta, \gamma; \rho, \sigma)$ be any 2-generator p-group of class 2 with presentation given in Theorem 1. Then:*

$$[a, [a, b]^\varphi]^{p^\sigma} = [a, (b^{p^\beta})^\varphi] = \begin{cases} [a, b^\varphi]^{p^\beta} & \textit{if } p > 2, \\ [a, b^\varphi]^{p^\beta}[b, [a, b]^\varphi]^{-p^{\beta-1}} & \textit{if } p = 2. \end{cases} \tag{22}$$

$$[b, [a, b]^\varphi]^{-p^\rho} = [a^{p^\alpha}, b^\varphi] = \begin{cases} [a, b^\varphi]^{p^\alpha} & \textit{if } p > 2, \\ [a, b^\varphi]^{p^\alpha}[a, [a, b]^\varphi]^{-p^{\alpha-1}} & \textit{if } p = 2. \end{cases} \tag{23}$$

PROOF. By Lemma 20 and applying identities found in Lemmas 6, 7, and 8, the following expansions hold in $\nu(G)$:

$$\begin{aligned} [a, (b^{p^\beta})^\varphi] &= [a, b^\varphi]^{p^\beta}[b, [a, b]^\varphi]^{\binom{p^\beta}{2}}[a, [a, b]^\varphi]^{p^\beta\binom{1}{2}} \\ &= [a, b^\varphi]^{p^\beta}[b, [a, b]^\varphi]^{\binom{p^\beta}{2}}. \end{aligned}$$

If $p > 2$, then p^β divides $\binom{p^\beta}{2}$, and since $\gamma \le \beta$ we have

$$[b, [a, b]^\varphi]^{\binom{p^\beta}{2}} = [b, ([a, b]^{p^\beta})^\varphi]^{(p^\beta-1)/2} = 1;$$

if, on the other hand, $p = 2$, then $\binom{p^\beta}{2} = p^{\beta-1}(p^\beta - 1)$. Again, since $\gamma \le \beta$, we have

$$[b, [a, b]^\varphi]^{\binom{p^\beta}{2}} = [b, ([a, b]^{p^\beta-1})^\varphi]^{p^{\beta-1}} = [b, ([a, b]^{-1})^\varphi]^{p^{\beta-1}} = [b, [a, b]^\varphi]^{-p^{\beta-1}};$$

which yields (22) once we note that $[a,(b^{p^\beta})^\varphi] = [a,([a,b]^{p^\sigma})^\varphi] = [a,[a,b]^\varphi]^{p^\sigma}$. The equalities in (23) are proven in the same manner. □

The natural homomorphism $\mu : G \to G/G'$ induces the epimorphism

$$f : [G, G^\varphi] \to (G/G') \otimes (G/G') \tag{24}$$

with $[x, y^\varphi] \mapsto \mu(x) \otimes \mu(y)$ for all x and y in G. The epimorphism f allows us to put lower bounds on the orders of the generators of $G \otimes G$.

LEMMA 25. *Let $G = G_p(\alpha,\beta,\gamma;\rho,\sigma)$ be a 2-generator p-group of class 2 with presentation as in Theorem 1. Then the generators of $[G, G^\varphi]$ have the following order bounds.*

(i) *The generator $[a, a^\varphi]$ has order a multiple of p^α and dividing $2p^\alpha$.*
(ii) *The generator $[b, b^\varphi]$ has order a multiple of p^β and dividing $2p^\beta$.*
(iii) *The generator $[a, b^\varphi]$ has order a multiple of p^β and dividing $2p^\alpha$.*
(iv) *The generator $[a, b^\varphi][b, a^\varphi]$ has order p^β.*
(v) *The generator $[a, [a, b]^\varphi]$ has order dividing $2p^\rho$ and dividing p^γ.*
(vi) *The generator $[b, [a, b]^\varphi]$ has order dividing $2p^{\min(\rho,\sigma)}$ and dividing p^γ.*

PROOF. By Corollary 19 the orders of $[a, a^\varphi]$ and $[b, b^\varphi]$ divide $2p^\alpha$ and $2p^\beta$ respectively, and their images $f([a, a^\varphi])$ and $f([b, b^\varphi])$ have orders p^α and p^β by (16), where f is defined in (24).

The order of $f([a, b^\varphi])$ and $f([a, b^\varphi][b, a^\varphi])$ is p^β by (16) hence the order of $[a, b^\varphi]$ in $[G, G^\varphi]$ is a multiple p^β, and the order of $[a, b^\varphi][b, a^\varphi]$ is exactly p^β since by Lemma 18(iii) $[a, b^\varphi][b, a^\varphi]$ has order dividing p^β.

The orders of $[a, [a, b]^\varphi]$ and $[b, [a, b]^\varphi]$ both divide p^γ, since $[a, [a, b]^\varphi]^{p^\gamma} = [a, ([a, b]^{p^\gamma})^\varphi] = [a, 1] = 1$, and similarly for $[b, [a, b]^\varphi]$. The orders of $[a, [a, b]^\varphi]$ and $[b, [a, b]^\varphi]$ divide $2p^\rho$ and $2p^\sigma$ respectively, since by Lemma 18(i) and (ii)

$$1 = [a, a^\varphi]^{2p^\alpha} = [a, ([a, b]^{p^\rho})^\varphi]^2 = [a, [a, b]^\varphi]^{2p^\rho}$$
$$1 = [b, b^\varphi]^{2p^\sigma} = [b, ([a, b]^{p^\sigma})^\varphi]^2 = [b, [a, b]^\varphi]^{2p^\sigma}.$$

Next we show the order of $[b, [a, b]^\varphi]$ divides $2p^{\min(\rho,\sigma)}$. If $\sigma \le \rho$, then we are done. If $\rho < \sigma$, we want to show that $[b, [a, b]^\varphi]^{2p^\rho} = 1$. From (23) we have that, since $\beta \le \alpha$,

$$\begin{aligned}
[b, [a, b]^\varphi]^{2p^\rho} &= \begin{cases} [a, b^\varphi]^{-2p^\alpha} & \text{if } p > 2, \\ [a, b^\varphi]^{-2p^\alpha}[a, [a, b]^\varphi]^{-2p^{\alpha-1}} & \text{if } p = 2, \end{cases} \\
&= [a, [a, b]^\varphi]^{-2p^{\sigma+\alpha-\beta}} = 1,
\end{aligned}$$

where the last equality follows from (v) since we are assuming that $\rho < \sigma$. Thus, the order of $[b, [a, b]^\varphi]$ divides both $2p^\sigma$ and $2p^\rho$, and so divides $2p^{\min(\rho,\sigma)}$.

Finally, that the order of $[a, b^\varphi]$ divides $2p^\alpha$ now follows from (23), (v), and (vi). □

The kernel N of the epimorphism $\nabla(G) \to \nabla(G/G')$ plays a role in the structure of $G \otimes G$ as seen in Theorem 10.

LEMMA 26. *Let $G = G_p(\alpha,\beta,\gamma;\rho,\sigma)$ be a 2-generator p-group of class 2 with presentation as in Theorem 1. Then the subgroup N of $[G, G^\varphi]$ which is the kernel of the epimorphism $\nabla(G) \to \nabla(G/G')$ is generated by $[a, [a, b]^\varphi]^{p^\rho}$ and $[b, [a, b]^\varphi]^{p^\sigma}$.*

PROOF. The generators of $\nabla(G)$ are $[a,a^\varphi]$, $[b,b^\varphi]$ and $[a,b^\varphi][b,a^\varphi]$ by [8, Lemma 17]. These generators have nontrivial images under the map f of (24). However, $1=f([a,a^\varphi]^{p^\alpha})=f([a,([a,b]^{p^\rho})^\varphi])=f([a,[a,b]^\varphi]^{p^\rho})$ and $1=f([b,a^\varphi]^{p^\beta})=f([b,([a,b]^{p^\sigma})^\varphi])=f([a,[a,b]^\varphi]^{p^\sigma})$. □

2.4. Crossed pairings. The analysis of the previous subsection does not always provide enough information to determine the exact order of (or relations among) some of the generators of $G\otimes G$, where G is the group $G_p(\alpha,\beta,\gamma;\rho,\sigma)$ with presentation as in Theorem 1. Could the order of $[a,[a,b]^\varphi]$ be strictly less than p^ρ, for example?

As discussed in the introduction, we will solve this problem below by using crossed pairings to define maps from $G\otimes G$ into finite cyclic groups, thus yielding lower bounds for the orders when necessary. In general, verifying that a map is a crossed pairing is laborious; fortunately, in our setting we can avoid much of the work by making some prior considerations.

Let $F(2,2)$ be the free nilpotent group of class two 2 and rank 2, with generators x and y. Every element of $F(2,2)$ can be written uniquely in the form $x^n y^m[x,y]^q$ with n, m, and q integers. From Theorem 3.2 in [2], we know $[F(2,2),(F(2,2))^\varphi]$ is free abelian of rank 6, freely generated by $[x,x^\varphi]$, $[y,y^\varphi]$, $[x,y^\varphi]$, $[y,x^\varphi]$, $[x,[x,y]^\varphi]$, and $[y,[x,y]^\varphi]$. We also have the following equality in $\nu(F(2,2))$, which can be verified directly using commutator calculus (or see Eqs 3.2.1–3.2.3 in [2]):

$$(27)\qquad \begin{aligned}&[x^m y^n[x,y]^q,(x^r y^s[x,y]^t)^\varphi]\\ &=[x,x^\varphi]^{mr}[y,y^\varphi]^{ns}[x,y^\varphi]^{ms}[y,x^\varphi]^{nr}\\ &\quad [x,[x,y]^\varphi]^{mt-rq+mrs-mnr+n\binom{r}{2}-s\binom{m}{2}}\\ &\quad [y,[x,y]^\varphi]^{nt-sq+r\binom{n}{2}-m\binom{s}{2}}\end{aligned}$$

We can now define six homomorphism from $F(2,2)\otimes F(2,2)$ to the additive infinite cyclic group $\mathbb{Z}$ (equivalently, crossed pairings from $F(2,2)\times F(2,2)$ to $\mathbb{Z}$) by taking the projection onto each of the generators. Thus, we have the following crossed pairings:

$$\begin{aligned}\phi_{xx}(x^m y^n[x,y]^q,x^r y^s[x,y]^t)&=mr,\\ \phi_{yy}(x^m y^n[x,y]^q,x^r y^s[x,y]^t)&=ns,\\ \phi_{xy}(x^m y^n[x,y]^q,x^r y^s[x,y]^t)&=ms,\\ \phi_{yx}(x^m y^n[x,y]^q,x^r y^s[x,y]^t)&=nr,\\ \phi_{xxy}(x^m y^n[x,y]^q,x^r y^s[x,y]^t)&=mt-rq+mrs-mnr+n\binom{r}{2}-s\binom{m}{2},\\ \phi_{yxy}(x^m y^n[x,y]^q,x^r y^s[x,y]^t)&=nt-sq+r\binom{n}{2}-m\binom{s}{2}.\end{aligned}$$

Trivially, any integer linear combination of these crossed pairings will again be a crossed pairing, as will the composition of any such integer linear combination with the canonical projection $\pi_\mu\colon \mathbb{Z}\to\mathbb{Z}_\mu$ for any integer μ, where $\mathbb{Z}_\mu$ represents the additive cyclic group of integers modulo μ.

Now let $G=G_p(\alpha,\beta,\gamma;\rho,\sigma)$ be any 2-generator p-group of class 2 with presentation as in Theorem 1. Every element of G can be written as $a^m b^n[a,b]^q$, with $0\le m<p^\alpha$ and $0\le n<p^\beta$. Given two elements $g=a^m b^n[a,b]^q$ and $h=a^r b^s[a,b]^t$

written this way, the equality $g = h$ holds if and only if $q \equiv t \pmod{p^\gamma}$, $m = r$, and $n = s$. With this normal form, the multiplication rule in G is given by:

$$\left(a^m b^n [a,b]^q\right)\left(a^r b^s [a,b]^r\right) = a^{m+r-u(m,r)} b^{n+s-v(n,s)} [a,b]^{q+t-rn+w(m,n,r,s)},$$

where

$$\begin{aligned}
u(m,r) &= \begin{cases} 0 & \text{if } m+r < p^\alpha, \\ p^\alpha & \text{if } m+r \geq p^\alpha. \end{cases}\\
v(n,s) &= \begin{cases} 0 & \text{if } n+s < p^\beta, \\ p^\beta & \text{if } n+s \geq p^\beta. \end{cases}\\
w(m,n,r,s) &= \begin{cases} 0 & \text{if } m+r < p^\alpha \text{ and } n+s < p^\beta, \\ p^\rho & \text{if } m+r \geq p^\alpha \text{ and } n+s < p^\beta, \\ p^\sigma & \text{if } m+r < p^\alpha \text{ and } n+s \geq p^\beta, \\ p^\rho + p^\sigma & \text{if } m+r \geq p^\alpha \text{ and } n+s \geq p^\beta. \end{cases}
\end{aligned}$$

We will normally write simply u, v, and w, with the arguments being understood.

We also have a map $\psi\colon G \to F(2,2)$ given by mapping the element $a^m b^n [a,b]^q$ with $0 \leq m < p^\alpha$ and $0 \leq n < p^\beta$, to $x^m y^n [x,y]^q$. This map is of course not well-defined, as the same element has expressions with several different values of q. However, if we have a map $\pi\colon F(2,2) \to L$ and the value of π on $x^m y^n [x,y]^q$ only depends on the congruence class of q modulo p^γ, then the composition $\pi \circ \psi$ will be a well-defined map.

In particular, let ϕ be an integer linear combination of the crossed pairings defined above,

$$\phi = c_1\phi_{xx} + c_2\phi_{yy} + c_3\phi_{xy} + c_4\phi_{yx} + c_5\phi_{xxy} + c_6\phi_{yxy}, \quad c_i \in \mathbb{Z},$$

and let $\pi_\mu \circ \phi\colon F(2,2) \times F(2,2) \to \mathbb{Z}_\mu$ be the crossed pairing obtained by composing ϕ with the canonical projection onto the integers modulo μ. If we let Ψ be the map $G \times G \to F(2,2) \times F(2,2)$ induced by ψ, then $\Phi\colon G \times G \to \mathbb{Z}_\mu$, defined by $\Phi = \pi_\mu \circ \phi \circ \Psi$ is well-defined if and only if the value of $\Phi(a^m b^n [a,b]^q, a^r b^s [a,b]^t)$ depends only on the congruence classes of q and t modulo p^γ, and will be a crossed pairing if and only if for all elements $a^m b^n [a,b]^q$, $a^r b^s [a,b]^t$, and $a^j b^k [a,b]^\ell$ in G, we have:

$$\begin{aligned}
(28) \qquad & \pi_\mu \circ \phi(x^{m+j} y^{n+k} [x,y]^{q+\ell-jn}, x^r y^s [x,y]^t)\\
&= \pi_\mu \circ \phi(x^{m+j-u} y^{n+k-v} [x,y]^{q+\ell-jn+w}, x^r y^s [x,y]^t),
\end{aligned}$$

and

$$\begin{aligned}
(29) \qquad & \pi_\mu \circ \phi(x^m y^n [x,y]^q, x^{r+j} y^{s+k} [x,y]^{t+\ell-js})\\
&= \pi_\mu \circ \phi(x^m y^n [x,y]^q, x^{r+j-u} y^{s+k-v} [x,y]^{t+\ell-js+w}).
\end{aligned}$$

These equalities reduce, in turn, to showing that certain expressions involving u, v, and w are congruent to 0 modulo μ.

Choosing the right value of μ and the appropriate linear combination ϕ is usually fairly straightforward: in most cases, the "obvious" choices will work, as we will see below. However, there are a few situations where we encounter some possibly surprising complications. We anticipated these by doing trial runs with GAP to find the abelian invariants of the tensor squares of all 2-generator groups of a given order p^n and class two; Theorem 1 gives an easy way to run through them all and compute these invariants. Once we knew what the answer should be, verifying it was straightforward; this is the source of some of the possibly mysterious-looking choices we will make in the next two sections.

3. Computing the nonabelian tensor squares

In this section, given the group $G = G_p(\alpha, \beta, \gamma; \rho, \sigma)$ we list the abelian invariants of its nonabelian tensor square in terms of the canonical parameters α, β, γ, ρ, and σ. We start with the odd prime case.

3.1. The p odd case. Throughout this section we let G be an arbitrary 2-generator p-group of class 2, given by $G = G_p(\alpha, \beta, \gamma; \rho, \sigma)$ as in Theorem 1 and $p > 2$. By Theorem 10(ii) and (iii) we have

$$G \otimes G \cong \nabla(G) \times G \wedge G \qquad \text{and} \qquad \nabla(G) \cong \Gamma(G/G'),$$

as G/G' has no elements of order 2. Since G/G' is isomorphic to $C_{p^\alpha} \times C_{p^\beta}$ using (11) and (12) we see that

$$\nabla(G) \cong \Gamma(G/G') \cong C_{p^\alpha} \times C_{p^\beta} \times C_{p^\beta}. \tag{30}$$

Therefore it remains to be shown what the abelian invariants of

$$G \wedge G \cong \langle [a, b^\varphi], [a, [a, b]^\varphi], [b, [a, b]^\varphi] \rangle$$

are. To find the orders of these generators, and any possible relations among them, we will use crossed pairings into suitable cyclic groups.

By performing the relevant computations using GAP for several of these groups, it quickly becomes apparent that $G \wedge G$ will have two invariants of order $p^{\min(\rho,\sigma)}$ (which in light of Lemma 25(iv) and (v) are likely to be associated with the generators $[a, [a, b]^\varphi]$ and $[b, [a, b]^\varphi]$ in some manner), and one of order p^β or larger, which ought to be related to $[a, b^\varphi]$. An order larger than p^β occurs when $\rho > \sigma$, which makes sense in view of (22) which shows that $[a, [a, b]^\varphi]^{p^\sigma} = [a, b^\varphi]^{p^\beta}$. We now establish these observation more precisely.

The generator $[b, [a, b]^\varphi]$. As p is an odd prime, Lemma 25(v) shows that $[b, [a, b]^\varphi]$ has order dividing $p^{\min(\rho,\sigma)}$. As noted above, computations with GAP suggest that this will be its order in $G \otimes G$, and so we expect this generator to be independent of the rest as well. If this is the case, then we ought to be able to set up a crossed pairing into a cyclic group of order $p^{\min(\rho,\sigma)}$ that maps $[b, [a, b]^\varphi]$ to a generator of the group, and maps all other generators to the identity; the clear choice is to use the crossed pairing ϕ_{yxy} and the setup described in Section 2.4. Indeed, we have:

LEMMA 31. *Let $G = G_p(\alpha, \beta, \gamma; \rho, \sigma)$ be a 2-generator p-group with presentation as in Theorem 1 with $p > 2$. Then $[b, [a, b]^\varphi]$ has order $p^{\min(\rho,\sigma)}$, and is independent of the other five generators of $G \otimes G$.*

PROOF. Let $\Phi \colon G \times G \to \mathbb{Z}_{p^{\min(\rho,\sigma)}}$ be given by $\Phi = \pi_{p^{\min(\rho,\sigma)}} \circ \phi_{yxy} \circ \Psi$.

Because $\rho, \sigma \leq \gamma$, the map is well-defined. To verify that it is in fact a crossed pairing, we need to verify (28) and (29). That is if $0 \leq m, j, r < p^\alpha$, $0 \leq n, k, s < p^\beta$, and q, ℓ, t are integers, then we need to show that

$$(n + k)t - s(q + \ell - nj) + r\binom{n+k}{2} - (m + j)\binom{s}{2}$$

and

$$(n+k-v)t - s(q+\ell-nj-w) + r\binom{n+k-v}{2} - (m+j-u)\binom{s}{2}$$

are congruent modulo $p^{\min(\rho,\sigma)}$. Expanding, and using the fact that for all a and b we have $\binom{a-b}{2} = \binom{a}{2} - \binom{b}{2} - b(a-b)$, the equality of these two expressions modulo $p^{\min(\rho,\sigma)}$ is equivalent to showing that

$$-vt + ws - r\binom{v}{2} - rv(n+k-v) + u\binom{s}{2} \equiv 0 \pmod{p^{\min(\rho,\sigma)}}$$

holds for all possible values of u, v, and w. There are four possible combinations of values: $u=v=w=0$; $u=p^\alpha$, $v=0$, $w=p^\rho$; $u=0$, $v=p^\beta$, $w=p^\sigma$; and $u=p^\alpha$, $v=p^\beta$, $w=p^\rho+p^\sigma$.

Note, however, that each of u, v, and w are congruent to 0 modulo $p^{\min(\rho,\sigma)}$ in all cases, from which it follows that this congruence holds. Likewise, (29) follows because $u\equiv v\equiv w\equiv 0 \pmod{p^{\min(\rho,\sigma)}}$, and so we conclude that we do indeed have a crossed pairing $G\times G\to \mathbb{Z}_{p^{\min(\rho,\sigma)}}$. This crossed pairing defines a unique homomorphism Φ^* from $[G,G^\varphi]$ to $\mathbb{Z}_{p^{\min(\rho,\sigma)}}$ by $\Phi^*([g,h^\varphi]) = \Phi(g,h)$.

Now we simply note that $\Phi^*([b,[a,b]^\varphi]) = 1 \mod p^{\min(\rho,\sigma)}$; therefore, we conclude that the order of $[b,[a,b]^\varphi]$ is a multiple of $p^{\min(\rho,\sigma)}$. Since we already know that $[b,[a,b]^\varphi]^{p^{\min(\rho,\sigma)}} = 1$ in $\nu(G)$, we therefore obtain that the order is exactly $p^{\min(\rho,\sigma)}$.

In addition, note that each of the other generators $[a,a^\varphi]$, $[b,b^\varphi]$, $[a,b^\varphi][b,a^\varphi]$, $[a,b^\varphi]$, and $[a,[a,b]^\varphi]$ lie in the kernel of Φ^*; since Φ^* restricts to the identity on the subgroup generated by $[b,[a,b]^\varphi]$, we also conclude that $[b,[a,b]^\varphi]$ is independent of the other generators of $[G,G^\varphi]$. □

The generators $[a,b^\varphi]$ **and** $[a,[a,b]^\varphi]$**.** We have already shown that the order of $[a,[a,b]^\varphi]$ divides p^ρ, and that the order of $[a,b^\varphi]$ must be a multiple of p^β. Moreover, since $[a,b^\varphi]^{p^\beta} = [a,[a,b]^\varphi]^{p^\sigma}$ holds in $G\otimes G$, it follows that if $\rho\leq\sigma$ then $[a,b^\varphi]$ has order dividing p^β, and hence the order is exactly p^β. Verifying that the order of $[a,[a,b]^\varphi]$ is exactly p^ρ in this case can be done as above, by using the crossed pairing ϕ_{xxy} and mapping to $\mathbb{Z}_{p^\rho}$.

On the other hand, if $\rho>\sigma$, then we deduce that $[a,b^\varphi]$ must have order dividing $p^{\beta+\rho-\sigma}$. However, the order may be less than this quantity: we also have that $[a,b^\varphi]^{p^\alpha} = [b,[a,b]^\varphi]^{-p^\rho}$ by (23), and so the order of $[a,b^\varphi]$ must be a divisor of p^α. That is, when $\rho>\sigma$ the order of $[a,b^\varphi]$ must be a divisor of $p^{\beta+\tau}$ with $\tau=\min(\rho-\sigma,\alpha-\beta)$. Note that we can only have $\rho>\sigma$ when $\alpha>\beta$, by Theorem 1, and that this theorem also indicates a qualitative difference between the case $\sigma<\rho<\sigma+\alpha-\beta$ and the case $\sigma+\alpha-\beta\leq\rho$. We expect (and indeed, computations with GAP support) that the order of $[a,b^\varphi]$ when $\rho>\sigma$ will be $p^{\beta+\tau}$, and that the relation given in (22) will be the smallest nontrivial relation between $[a,b^\varphi]$ and $[a,[a,b]^\varphi]$ in $G\otimes G$.

Assuming that this is the case, then in order to get the second abelian invariant of order $p^{\min(\rho,\sigma)} = p^\sigma$, we will want to replace the generator $[a,[a,b]^\varphi]$ with the generator $[a,b^\varphi]^{-p^{\beta-\sigma}}[a,[a,b]^\varphi]$. As we are also using $[a,b^\varphi][b,a^\varphi]$ instead of simply $[a,b^\varphi]$, this means that we would have to rework (27) so the expression is in terms of this new basis. In such an expression, the exponent of $[a,b^\varphi]$ will not be given by ϕ_{xy}, but rather by $\phi_{xy}-\phi_{yx}+p^{\beta-\sigma}\phi_{xxy}$. So what we should try is to use this latter crossed pairing, and then map to $\mathbb{Z}_{p^{\beta+\tau}}$ (the expected order of $[a, b^{\varphi}]$) in order establish the orders and relations when $\rho > \sigma$. And indeed, this does take care of the situation.

LEMMA 32. *Let $G = G_p(\alpha, \beta, \gamma; \rho, \sigma)$ be a 2-generator p-group with presentation as in Theorem 1 with $p > 2$.*

(i) *If $\rho \leq \sigma$, then the order of $[a, [a, b]^{\varphi}]$ is p^{ρ}, the order of $[a, b^{\varphi}]$ is p^{β}, and each is independent from the other generators of $G \otimes G$.*
(ii) *If $\rho > \sigma$, then the order of $[a, [a, b]^{\varphi}]$ is $p^{\sigma+\tau}$ and the order of $[a, b^{\varphi}]$ is $p^{\beta+\tau}$, where $\tau = \min(\alpha - \beta, \rho - \sigma)$. The two satisfy $[a, b^{\varphi}]^{p^{\beta}} = [a, [a, b]^{\varphi}]^{p^{\sigma}}$, and this is the only nontrivial relation between them and the other generators of $G \otimes G$.*

PROOF. For (i), we assume $\rho \leq \sigma$ and we consider $\Phi\colon G \times G \to \mathbb{Z}_{p^{\rho}}$ given by $\Phi = \pi_{p^{\rho}} \circ \phi_{xxy} \circ \Psi$. As in Lemma 31, because $\rho \leq \sigma \leq \gamma$ the map is well defined, and since $u \equiv v \equiv w \equiv 0 \pmod{p^{\rho}}$ for all possible values of u, v, and w, the map is a crossed pairing. The induced homomorphism $\Phi^*\colon G \otimes G \to \mathbb{Z}_{p^{\rho}}$ maps $[a, [a, b]^{\varphi}]$ to $1 \mod p^{\rho}$, and every other generator to 0, establishing the order and the independence for $[a, [a, b]^{\varphi}]$. We also saw above that when $\rho \leq \sigma$ the order of $[a, b^{\varphi}]$ will be p^{β}, and the independence follows from its independence when mapping to $G^{\rm ab} \otimes G^{\rm ab}$, and the independence of both $[a, [a, b]^{\varphi}]$ and $[b, [a, b]^{\varphi}]$.

For (ii), as discussed above we let $\phi = \phi_{xy} - \phi_{yx} + p^{\beta-\sigma}\phi_{xxy}$ and then take $\Phi = \pi_{p^{\beta+\tau}} \circ \phi \circ \Psi$, with $\tau = \min(\rho - \sigma, \alpha - \beta)$.

First, we note that Φ is well-defined: given $g = a^m b^n [a, b]^q$ and $h = a^r b^s [a, b]^t$, neither ϕ_{xy} nor ϕ_{yx} depend on the congruence classes of q or t modulo p^{γ}; and while ϕ_{xxy} does involve q and t, we are multiplying them by $p^{\beta-\sigma}$. Since we have that $\beta + \tau \leq \beta + \rho - \sigma \leq \beta + \gamma - \sigma$, it follows that $p^{\beta-\sigma}q$ and $p^{\beta-\sigma}t$ are well-defined modulo $p^{\beta+\tau}$, so Φ is well-defined.

To verify that Φ defines a crossed pairing, we note that $\tau + \sigma \leq \rho \leq \beta$, so $2\beta - \sigma \geq \beta + \tau$. Now verifying (28) amounts to proving that

$$
\begin{aligned}
&-us + vr + p^{\beta-\sigma}\big(-ut - rw - urs + ur(n + k - v) + (m + j)vr\big) \\
&\qquad + p^{\beta-\sigma}\left(-v\binom{r}{s} + s\binom{u}{2} + su(m + j - u)\right) \equiv 0 \pmod{p^{\beta+\tau}}
\end{aligned}
\tag{33}
$$

for the different possible values of u, v, and w. Note $u \equiv p^{\beta-\sigma}v \equiv 0 \pmod{p^{\beta+\tau}}$ in all cases: for $\beta + \tau \leq \beta + (\alpha - \beta) = \alpha$, and $2\beta - \sigma \geq \beta + \tau$; so none of these summands will affect the value of the left side of expression (33) modulo $p^{\beta+\tau}$. This leaves only $vr - p^{\beta-\sigma}rw$; when $v = 0$, the value of w is either 0 or p^{ρ}, and in both cases we have $-p^{\beta-\sigma}rw \equiv 0 \pmod{p^{\beta+\tau}}$ (since $\beta + \rho - \sigma \geq \beta + \tau$). So the only case to worry about is when $v = p^{\beta}$ and $w = p^{\sigma}$ or $p^{\sigma} + p^{\rho}$, and in both cases we obtain $vr - p^{\beta-\sigma}rw \equiv p^{\beta}r - p^{\beta}r \equiv 0 \pmod{p^{\beta+\tau}}$.

To verify (29), we likewise need to verify that for all values of u, v, and w,

$$
\begin{aligned}
&-mv + nu + p^{\beta-\sigma}\big(mw + uq - mu(s + k - v) + mnu\big) \\
&\qquad + p^{\beta-\sigma}\left(-n\binom{u}{2} - nu(r + j - u) + v\binom{m}{2}\right) \equiv 0 \pmod{p^{\beta+\sigma}}
\end{aligned}
$$

holds. Once again, this reduces to verifying that $-mv + p^{\beta-\sigma}mw \equiv 0 \pmod{p^{\beta+\tau}}$ always holds, and this follows as it did above.

Thus, Φ is indeed a crossed pairing, and we obtain our desired induced map $\Phi^*\colon G\otimes G \to \mathbb{Z}_{p^{\beta+\tau}}$, mapping $[a,b^{\varphi}]$ to the generator 1 of $\mathbb{Z}_{p^{\beta+\tau}}$, $[a,[a,b]^{\varphi}]$ to $p^{\beta-\sigma}$, and all other generators to 0. We deduce that the order of $[a,b^{\varphi}]$ is exactly $p^{\beta+\tau}$, since we already knew it divides $p^{\beta+\tau}$; that the order of $[a,[a,b]^{\varphi}]$ is $p^{\sigma+\tau}$, and that there is no relation among the generators other than $[a,b^{\varphi}]^{p^{\beta}} = [a,[a,b]^{\varphi}]^{p^{\sigma}}$. This establishes the lemma. $\square$

The tensor square and exterior square for odd primes. From the analysis in the previous two parts, we obtain the desired description of the exterior square of $G = G_p(\alpha,\beta,\gamma;\rho,\sigma)$ when p is odd:

PROPOSITION 34. *Let $G = G_p(\alpha,\beta,\gamma;\rho,\sigma)$ be any 2-generator p-group with presentation as in Theorem 1 and $p>2$. Then*

$$G\wedge G \cong \begin{cases} C_{p^{\beta}}\times C_{p^{\rho}}\times C_{p^{\rho}} & \text{if } \rho\leq\sigma\\ C_{p^{\beta+\tau}}\times C_{p^{\sigma}}\times C_{p^{\sigma}} & \text{if } \rho>\sigma\end{cases} \tag{35}$$

where $\tau = \min(\alpha-\beta,\rho-\sigma)$. The cyclic factors correspond to the subgroup generated by $[a,b^{\varphi}]$, $[a,[a,b]^{\varphi}]$, and $[b,[a,b]^{\varphi}]$, respectively, when $\rho\leq\sigma$; and to the subgroups generated by $[a,b^{\varphi}]$, $[a,b^{\varphi}]^{-p^{\beta-\sigma}}[a,[a,b]^{\varphi}]$, and $[b,[a,b]^{\varphi}]$, respectively, when $\rho>\sigma$.

Hence we have the our description for $G\otimes G$.

THEOREM 36. *Let $G = G_p(\alpha,\beta,\gamma;\rho,\sigma)$ be any 2-generator p-group with presentation as in Theorem 1 and $p>2$. Then*

$$G\otimes G \cong \begin{cases} C_{p^{\alpha}}\times C_{p^{\beta}}\times C_{p^{\beta}}\times C_{p^{\beta}}\times C_{p^{\rho}}\times C_{p^{\rho}} & \text{if } \rho\leq\sigma\\ C_{p^{\alpha}}\times C_{p^{\beta}}\times C_{p^{\beta}}\times C_{p^{\beta+\tau}}\times C_{p^{\sigma}}\times C_{p^{\sigma}} & \text{if } \rho>\sigma\end{cases} \tag{37}$$

where $\tau = \min(\alpha-\beta,\rho-\sigma)$. The invariant cyclic factors can be taken to be the subgroups generated by:

(i) *$[a,a^{\varphi}]$, $[b,b^{\varphi}]$, $[a,b^{\varphi}][b,a^{\varphi}]$, $[a,b^{\varphi}]$, $[a,[a,b]^{\varphi}]$, and $[b,[a,b]^{\varphi}]$, respectively, if $\rho\leq\sigma$;*

(ii) *$[a,a^{\varphi}]$, $[b,b^{\varphi}]$, $[a,b^{\varphi}][b,a^{\varphi}]$, $[a,b^{\varphi}]$, $[a,b^{\varphi}]^{p^{\beta-\sigma}}[a,[a,b]^{\varphi}]$, and $[b,[a,b]^{\varphi}]$, respectively, if $\rho>\sigma$.*

3.2. The $p=2$ case. As can be expected, the $p=2$ case presents more difficulties; here Theorem 9 does not give us a splitting of $G\otimes G$ into $\nabla(G)\times G\wedge G$, the orders of $[a,a^{\varphi}]$ and $[b,b^{\varphi}]$ are not completely determined by the analysis in Section 2.3, and the identities found there are generally more complex (e.g., compare the $p>2$ and the $p=2$ cases of (22) and (23)).

Let $G = G_2(\alpha,\beta,\gamma;\rho,\sigma)$ be a 2-generator 2-group of class 2, with presentation as in Theorem 1. We know the order of $[a,b^{\varphi}][b,a^{\varphi}]$ is 2^{β}; and that we have a central extension

$$1 \longrightarrow N \longrightarrow G\otimes G \stackrel{\psi}{\longrightarrow} \nabla(G/G')\times G\wedge G \longrightarrow 1,$$

where N is the kernel of the mapping $\nabla(G)\to\nabla(G/G')$. Since we have:

$$[a,a^{\varphi}]^{2^{\alpha}} = [a,[a,b]^{\varphi}]^{2^{\rho}},\quad [b,b^{\varphi}]^{2^{\beta}} = [b,[a,b]^{\varphi}]^{2^{\sigma}},$$
$$[a,[a,b]^{\varphi}]^{2^{\rho+1}} = 1,\quad \text{and}\quad [b,[a,b]^{\varphi}]^{2^{\sigma+1}} = 1,$$

we know that N is generated by $[a,[a,b]^{\varphi}]^{2^{\rho}}$ and $[b,[a,b]^{\varphi}]^{2^{\sigma}}$, and is an elementary abelian 2-group.

Thus, the order of $[a, a^\varphi]$ will be 2^α when $[a,[a,b]^\varphi]$ is of order dividing 2^ρ, and will be $2^{\alpha+1}$ if and only if the order of $[a,[a,b]^\varphi]$ is $2^{\rho+1}$ (we know from Lemma 25(v) that the order of the latter divides $2^{\rho+1}$). And likewise the order of $[b, b^\varphi]$ is 2^β if and only if the order of $[b,[a,b]^\varphi]$ divides 2^σ, and will be $2^{\beta+1}$ if and only if the order of $[b,[a,b]^\varphi]$ is exactly $2^{\sigma+1}$.

We also know that $[a,b^\varphi][b,a^\varphi]$ is independent of the other generators; that $[a,a^\varphi]$ and $[b,b^\varphi]$ are independent of each other except perhaps when their orders are $2^{\alpha+1}$ and $2^{\beta+1}$, respectively; and that $[a, b^\varphi]$ is independent of the other generators when its order is 2^β.

These are almost all the new considerations we need to make; other than the adjustments to the orders of $[a,a^\varphi]$ and $[b,b^\varphi]$, we expect the invariants to follow the same pattern as for odd p. And for the most part they do; however, there are a couple of subtleties, as we will see below, which we discovered through some sample computations in GAP. The perhaps "mysterious" maps we will use to establish the orders of the generators come from doing a suitable change of basis to obtain the invariants.

The order of $[a,a^\varphi]$**.** Let $G = G_2(\alpha,\beta,\gamma;\rho,\sigma)$ be a 2-generator 2-group of class 2, with presentation as in Theorem 1. We know from general considerations that the order of $[a,a^\varphi]$ is either 2^α or $2^{\alpha+1}$, and since $[a,a^\varphi]^{2^\alpha} = [a,[a,b]^\varphi]^{2^\rho}$, the order is $2^{\alpha+1}$ if and only if the order of $[a,[a,b]^\varphi]^{2^\rho}$ is $2^{\rho+1}$. In particular, if $\rho=\gamma$, then the order of $[a,a^\varphi]$ is 2^α. In fact, this is the only case in which the order is 2^α. In the proposition below, $\delta_{\rho\gamma}$ is Kronecker's delta, with value 1 if $\rho=\gamma$, and 0 otherwise.

PROPOSITION 38. *Let $G = G_2(\alpha,\beta,\gamma;\rho,\sigma)$ be a 2-generator 2-group of class 2 with presentation as in Theorem 1. Then the order of $[a,a^\varphi]$ is $2^{\alpha+1-\delta_{\rho\gamma}}$, and $[a,a^\varphi]$ and $[b,b^\varphi]$ are independent.*

PROOF. As noted above, the order of $[a,a^\varphi]$ is 2^α when $\rho=\gamma$. So suppose that $\rho<\gamma$, and we will show the order to be exactly $2^{\alpha+1}$.

Consider $\Phi = \pi_{2^{\alpha+1}}\circ\phi\circ\Psi$, where ϕ is the map

$$\phi = \phi_{xx} + 2^{\alpha-\rho}\phi_{xxy} + 2^{\alpha-\beta+\sigma-\rho}\big(\phi_{xy}-\phi_{yx}\big).$$

Note first that the exponent $\alpha-\beta+\sigma-\rho$ is nonnegative. Indeed, if $\sigma\geq\rho$, then this is immediate; if $\sigma<\rho$, then because $\rho<\gamma$ we must be in Family 1(c) from Theorem 1, so $0\leq\sigma<\rho<\sigma+\alpha-\beta$, from which $\alpha-\beta+\sigma-\rho>0$ follows. Note, moreover, that the only case in which the exponent is 0 is when $\alpha=\beta$ and $\sigma=\rho$, which given the condition $\rho<\gamma$ can only occur in Family 3(b) of Theorem 1, that is, when $\alpha=\beta=\gamma=\rho+1=\sigma+1$.

Next, note that the map Φ is well defined, since $\alpha+\gamma-\rho\geq\alpha+1$; as the only occurrences of q or t when evaluating $\Phi(a^mb^n[a,b]^q, a^rb^s[a,b]^t)$ are in the expression for ϕ_{xxy}, this means that the value modulo $2^{\alpha+1}$ will remain unchanged if we modify q or t by adding multiples of 2^γ.

We claim that Φ defines a crossed pairing $G\times G\to\mathbb{Z}_{2^{\alpha+1}}$. Since $\alpha-\rho$ and $\beta-\rho$ are both positive, we have $2^{\alpha-\rho}u\equiv 2^{\alpha-\rho}v\equiv 0 \pmod{2^{\alpha+1}}$, so verifying (28) and (29) is equivalent to verifying that:

$$-ur + 2^{\alpha-\rho}\left(-rw + s\binom{u}{2}\right) + 2^{\alpha-\beta+\sigma-\rho}\big(-us+vr\big)\equiv 0 \pmod{2^{\alpha+1}},$$

and

$$-mu + 2^{\alpha-\rho}\left(mw - n\binom{u}{2}\right) + 2^{\alpha-\beta+\sigma-\rho}(-mv + un) \equiv 0 \pmod{2^{\alpha+1}}$$

hold for all possible values of u, v, and w. If $u = 2^{\alpha}$, $v = 0$, and $w = 2^{\rho}$, then $-ur - 2^{\alpha-\rho}rw$ and $-mu + 2^{\alpha-\rho}mw$ are both congruent to zero modulo $2^{\alpha+1}$; if $\alpha - \beta + \sigma - \rho = 0$, then we have $\alpha = \beta = \gamma$, $\rho = \sigma = \gamma - 1$, as noted above, so:

$$2^{\alpha-\rho}s\binom{2^{\alpha}}{2} - 2^{\alpha}s \equiv 2^{\alpha}(2^{\alpha}-1)s - 2^{\alpha}s \equiv 0 \pmod{2^{\alpha+1}},$$

so the first congruence holds, and

$$-2^{\alpha-\rho}n\binom{2^{\alpha}}{2} + 2^{\alpha}n = 2^{\alpha}(2^{\alpha}-1)n + 2^{\alpha}n \equiv 0 \pmod{2^{\alpha+1}},$$

so the second congruence also holds. If $\alpha - \beta + \sigma - \rho > 0$, on the other hand, then $2^{\alpha-\beta+\sigma-\rho}u \equiv 0 \pmod{2^{\alpha+1}}$. We must have $\alpha > \beta$, or else $\alpha = \beta$ and either $\beta > \gamma = \sigma > \rho$ or $\beta = \gamma = \sigma$ and $\rho < \gamma - 1$. In all cases we have $\alpha - \rho \geq 2$, so $2^{\alpha-\rho}\binom{u}{2}s \equiv 0 \pmod{2^{\alpha+1}}$, and so both congruences will also hold.

If $u = 0$, $v = 2^{\beta}$, and $w = 2^{\sigma}$, the congruences become

$$-2^{\alpha-\rho+\sigma}r + 2^{\alpha+\sigma-\rho}r \equiv 0 \pmod{2^{\alpha+1}},$$
$$2^{\alpha-\rho+\sigma}m - 2^{\alpha+\sigma-\rho}m \equiv 0 \pmod{2^{\alpha+1}}.$$

Finally, the congruences hold when $u = 2^{\alpha}$, $v = 2^{\beta}$, and $w = 2^{\rho} + 2^{\sigma}$, as we simply get the sum of the congruences in the previous two cases.

Thus, we deduce that Φ is a crossed pairing, and induces a homomorphism $\Phi^*\colon [G, G^{\varphi}] \to \mathbb{Z}_{2^{\alpha+1}}$. This homomorphism maps $[a, a^{\varphi}]$ to the class of 1 modulo $2^{\alpha+1}$, which shows that the order of $[a, a^{\varphi}]$ is a multiple of $2^{\alpha+1}$, and hence exactly equal to $2^{\alpha+1}$.

Noting that the map Φ^* maps $[b, b^{\varphi}]$ to 0 shows the independence in this case; when the order of $[a, a^{\varphi}]$ is 2^{α}, the independence from $[b, b^{\varphi}]$ follows by mapping to $G^{\mathrm{ab}} \otimes G^{\mathrm{ab}}$. □

The generator $[b, b^{\varphi}]$**.** The case of $[b, b^{\varphi}]$ is similar to that of $[a, a^{\varphi}]$, though because a and b play not-quite-symmetric roles, the function we consider is slightly more complicated.

If $\sigma = \gamma$ then the order of $[b, b^{\varphi}]$ is 2^{β}, since then $[b, [a, b]^{\varphi}]^{2^{\sigma}} = 1$. We claim that the order of $[b, b^{\varphi}]$ will be $2^{\beta+1}$ whenever $\sigma < \gamma$. Again letting $\delta_{\sigma\gamma}$ denote Kronecker's delta, we have:

PROPOSITION 39. *Let $G = G_2(\alpha, \beta, \gamma; \rho, \sigma)$ be a 2-generator 2-group of class 2 with presentation as in Theorem 1. Then the order of $[b, b^{\varphi}]$ is $2^{\beta+1-\delta_{\sigma\gamma}}$.*

PROOF. It remains to be shown that if $\sigma < \gamma$, then the order of $[b, b^{\varphi}]$ is a multiple of $2^{\beta+1}$. To that end, we consider the map $\Phi = \pi_{2^{\beta+1}} \circ \phi \circ \Psi$, where $\phi = \phi_{yy} + 2^{\beta-\sigma}\phi_{yxy} + 2^{1-\delta_{(\beta-1)\sigma}}(\phi_{xy} - \phi_{yx})$, with $\delta_{(\beta-1)\sigma} = 1$ if $\sigma = \beta - 1$, and 0 otherwise. Note that for any specific choice of β and σ, ϕ is an integer linear combination of the crossed pairings for $F(2,2)$, and therefore is itself a crossed pairing. Note as well that since $\gamma > \sigma$, the map Φ is well defined since the only occurrence of q or t in evaluating $\Phi(a^m b^n [a,b]^q, a^r b^s [a,b]^t)$ occurs in the expression of ϕ_{yxy}, and is multiplied by $2^{\beta-\sigma}$.

For simplicity, write $i = 1 - \delta_{(\beta-1)\sigma}$. Since $2u \equiv 2v \equiv 0 \pmod{2^{\beta+1}}$ for all values of u and v, verifying (28) and (29) reduces to verifying that

$$\begin{aligned} -vs + 2^{\beta-\sigma}\left(-sw - r\binom{v}{2}\right) + 2^i(-us+vr) &\equiv 0 \pmod{2^{\beta+1}} \\ -nv + 2^{\beta-\sigma}\left(nw + m\binom{v}{2}\right) + 2^i(-mv+nu) &\equiv 0 \pmod{2^{\beta+1}} \end{aligned}$$

both hold, and we can do this by looking separately at the cases $u = 2^\alpha$, $v = 0$, $w = 2^\rho$; and $u = 0$, $v = 2^\beta$, $w = 2^\sigma$; as the third case, $u = 2^\alpha$, $v = 2^\beta$, $w = 2^\rho + 2^\sigma$ will follow from these two.

If $u = 2^\alpha$, $v = 0$, and $w = 2^\rho$, then the left hand side of the first congruence is $-2^{\beta+\rho-\sigma}s - 2^{\alpha+i}s$, and the second evaluates to $2^{\beta+\rho-\sigma}n + 2^{\alpha+i}n$. If $\rho > \sigma$, then $\alpha > \beta$, so all summands are congruent to 0 modulo $2^{\beta+1}$, regardless of the value of i, and both congruences hold. If $\rho = \sigma < \gamma$, then we must have $\rho = \sigma = \gamma - 1$, and $\alpha = \beta = \gamma$, so $i = 1 - \delta_{(\beta-1)\sigma} = 0$, and the left hand side of the first congruence is $-2^\beta s - 2^\beta s = -2^{\beta+1}s$, while the left hand side of the second congruence is likewise equal to $2^{\beta+1}n$; so again the congruences hold.

If $u = 0$, $v = 2^\beta$, and $w = 2^\sigma$, the first congruence evaluates to

$$-2^\beta s - 2^\beta s + 2^{\beta-\sigma+\beta-1}r + 2^{\beta+i}r \equiv 2^{2\beta-\sigma-1}r + 2^{\beta+i}r \pmod{2^{\beta+1}},$$

while the second yields

$$-2^\beta n + 2^\beta n - 2^{\beta-\sigma+\beta-1}m - 2^{\beta+i}m \equiv -2^{2\beta-\sigma-1}m - 2^{\beta+i}m \pmod{2^{\beta+1}}.$$

If $\sigma = \beta - 1$, then $i = 0$, so the first congruence yields $2^\beta r + 2^\beta r \equiv 0 \pmod{2^{\beta+1}}$, while the second gives $-2^\beta m - 2^\beta m \equiv 0 \pmod{2^{\beta+1}}$, hence the congruences hold. If $\sigma < \beta - 1$, then $i = 1$ and $2\beta - \sigma - 1 \geq \beta + 1$, so both congruences hold again. And we cannot have $\sigma > \beta - 1$, as that would force $\sigma = \gamma = \beta$, contrary to our hypothesis that $\sigma < \gamma$.

Thus the crossed pairing induces a map $\Phi^*\colon [G, G^\varphi] \to \mathbb{Z}_{2^{\beta+1}}$ that sends $[b, b^\varphi]$ to the class of 1 modulo $2^{\beta+1}$, proving that $[b, b^\varphi]$ has order a multiple of $2^{\beta+1}$ in this case, which establishes the proposition. □

The generator $[b, [a,b]^\varphi]$. If $\sigma < \gamma$ then the order of $[b, b^\varphi]$ is $2^{\beta+1}$; since $[b, b^\varphi]^{2^\beta} = [b, [a,b]^\varphi]^{2^\sigma}$, this means that the order of $[b, [a,b]^\varphi]$ must be $2^{\sigma+1}$ in this case. So we only need to figure out the order when $\sigma = \gamma$. Since we will have $\rho \leq \sigma$, from our experience from the $p > 2$ case we expect the order to be 2^ρ. And indeed, this is what we have.

PROPOSITION 40. *Let $G = G_2(\alpha, \beta, \gamma; \rho, \sigma)$ be a 2-generator 2-group of class 2, with presentation as in Theorem 1. Then the order of $[b, [a,b]^\varphi]$ is $2^{\sigma+1}$ if $\sigma < \gamma$, and 2^ρ if $\sigma = \gamma$.*

PROOF. The result follows from previous sections if $\sigma < \gamma$, so assume that $\sigma = \gamma$. That the order divides 2^ρ is a consequence of the identities we already have; this is clear if $\rho = \sigma = \gamma$. And if $\rho < \sigma$, then since $[a, [a,b]^\varphi]^{2^{\rho+1}} = 1$ we have from

(22) and (23):

$$\begin{aligned}
[b,[a,b]^{\varphi}]^{-2^{\rho}} &= [a,b^{\varphi}]^{2^{\alpha}}[a,[a,b]^{\varphi}]^{-2^{\alpha-1}}\\
&= \left([a,b^{\varphi}]^{2^{\beta}}\right)^{2^{\alpha-\beta}}[a,[a,b]^{\varphi}]^{-2^{\alpha-1}}\\
&= [a,[a,b]^{\varphi}]^{2^{\sigma+\alpha-\beta}}[b,[a,b]^{\varphi}]^{2^{\alpha-1}}[a,[a,b]^{\varphi}]^{-2^{\alpha-1}}\\
&= [b,[a,b]^{\varphi}]^{2^{\alpha-1}}[a,[a,b]^{\varphi}]^{-2^{\alpha-1}},
\end{aligned}$$

where the last equality holds since $\sigma+\alpha-\beta>\rho$. If $\alpha-1\geq\gamma$, then both terms in the last equality are trivial, so $[b,[a,b]^{\varphi}]^{2^{\rho}}=1$. If $\alpha-1<\gamma$, then we must have $\alpha=\gamma$, and since $\rho<\sigma$, we must be in family 3(a) from Theorem 1, that is, the group $G_2(\gamma,\gamma,\gamma;\rho,\gamma)$ with $\rho<\gamma-1$. In that case, $\alpha-1>\rho$, so $[a,[a,b]^{\varphi}]^{-2^{\alpha-1}}=1$, and we obtain that $[b,[a,b]^{\varphi}]^{-2^{\rho}}=[b,[a,b]^{\varphi}]^{2^{\alpha-1}}$. Therefore, $[b,[a,b]^{\varphi}]^{2^{\rho}(2^{\alpha-\rho-1}+1)}=1$, and since we are assuming that $\rho<\alpha-1$, the order of $[b,[a,b]^{\varphi}]$ divides 2^{ρ}.

So it only remains to be shown that the order of $[b,[a,b]^{\varphi}]$ is a multiple of 2^{ρ} as well. Consider the map $\Phi=\pi_{2^{\rho+1}}\circ\phi\circ\Psi$, where $\phi=\phi_{xy}-\phi_{yx}+2(\phi_{xxy}+\phi_{yxy})$. This map is well-defined, since $\rho\leq\gamma$. Note also that $2u\equiv 2v\equiv 2w\equiv 0 \pmod{2^{\rho+1}}$. Therefore, establishing (28) and (29) reduces to the easy congruences:

$$\begin{aligned}
-us+vr+(u^2-u)s-(v^2-v)r &\equiv u^2s-v^2r\equiv 0 \pmod{2^{\rho+1}},\\
-mv+nu+(u^2-u)n+(v^2-v)m &\equiv u^2n+v^2m\equiv 0 \pmod{2^{\rho+1}},
\end{aligned}$$

which trivially hold since $u^2\equiv v^2\equiv 0 \pmod{2^{\rho+1}}$. Thus we have a crossed pairing, and an induced map Φ^* that maps $[b,[a,b]^{\varphi}]$ to the class of 2 modulo $2^{\rho+1}$, proving that the order of $[b,[a,b]^{\varphi}]$ must be a multiple of 2^{ρ}. This proves the proposition. □

The generators $[a,b^{\varphi}]$ and $[a,[a,b]^{\varphi}]$. For the final two generators we expect a modification of the result for $p>2$ along the lines we have seen above: when $\rho<\gamma$, we know the order of $[a,[a,b]^{\varphi}]$ is $2^{\rho+1}$; and if $\rho=\gamma$, the order should be the smaller of $2^{\sigma+\alpha-\beta}$ and 2^{γ}. From this, we can deduce the order of $[a,b^{\varphi}]$; this order is given by essentially the same formula as it did before, with one notable change: when $\sigma=\rho=\beta$, the order is $2^{\beta+1}$, as we will see below.

PROPOSITION 41. *Let $G=G_2(\alpha,\beta,\gamma;\rho,\sigma)$ be a 2-generator 2-group of class 2 with presentation as in Theorem 1. If $\rho<\gamma$, then the order of $[a,[a,b]^{\varphi}]$ is $2^{\rho+1}$. If $\rho=\gamma$, then the order of $[a,[a,b]^{\varphi}]$ is $2^{\sigma+\tau}$, where $\tau=\min(\rho-\sigma,\alpha-\beta)$.*

PROOF. Since $[a,a^{\varphi}]^{2^{\alpha}}=[a,[a,b]^{\varphi}]^{2^{\rho}}$, from Proposition 38 we deduce that if $\rho<\gamma$, then the order of $[a,[a,b]^{\varphi}]$ is exactly $2^{\rho+1}$. It only remains then to establish the order when $\rho=\gamma$.

Assume then that $\rho=\gamma$. We first prove that the order of $[a,[a,b]^{\varphi}]$ will divide $2^{\sigma+\tau}$; to that end, we consider the two possible values of τ. If $\tau=\rho-\sigma\leq\alpha-\beta$, then $\sigma+\tau=\gamma$, and so $[a,[a,b]^{\varphi}]^{2^{\sigma+\tau}}=[a,[a,b]^{\varphi}]^{2^{\gamma}}=1$, by Lemma 25(v). On the other hand, if $\tau=\alpha-\beta<\rho-\sigma$, then we note that $\rho-\sigma>0$ implies that $\alpha>\beta$ (by the presentation in Theorem 1; we can only have $\rho>\sigma$ if $\alpha>\beta$), so $\rho-\sigma=\gamma-\sigma>1$. Thus, $\sigma<\gamma-1<\beta$, so $\sigma<\beta-1$. From Proposition 40 we therefore have that $[b,[a,b]^{\varphi}]^{2^{\beta-1}}=1$, so from (22) we obtain obtain

$$[a,b^{\varphi}]^{2^{\beta}}=[a,[a,b]^{\varphi}]^{2^{\sigma}}[b,[a,b]^{\varphi}]^{2^{\beta-1}}=[a,[a,b]^{\varphi}]^{2^{\sigma}}.$$

As $\sigma+\tau=\sigma+\alpha-\beta$ under our current assumption, we have:

$$[a,[a,b]^{\varphi}]^{2^{\sigma+\tau}}=[a,[a,b]^{\varphi}]^{2^{\sigma+\alpha-\beta}}=[a,b^{\varphi}]^{2^{\beta+\alpha-\beta}}=[a,b^{\varphi}]^{2^{\alpha}}.$$

Finally, using (23), that $\alpha - 1 \geq \gamma = \rho$, and Lemma 25(v) and (vi) we conclude that:

$$[a,[a,b]^{\varphi}]^{2^{\sigma+\tau}} = [a,b^{\varphi}]^{2^{\alpha}} = [b,[a,b]^{\varphi}]^{-2^{\rho}}[a,[a,b]^{\varphi}]^{2^{\alpha-1}} = 1.$$

Therefore, both in the case when $\tau = \rho - \sigma$ and when $\tau = \alpha - \beta$, we have that the order of $[a,[a,b]^{\varphi}]$ divides $2^{\sigma+\tau}$.

To show that the order of $[a,[a,b]^{\varphi}]$ is also a multiple of $2^{\sigma+\tau}$, we note first that if $\sigma = \rho = \gamma$, then $\tau = 0$; in this case, the map Φ used in the proof of Proposition 40 is a crossed pairing (the only assumption there was that $\rho \leq \sigma = \gamma$), and maps $[a,[a,b]^{\varphi}]$ to the class of 2 modulo $2^{\rho+1}$, which finishes the proof that the order of $[a,[a,b]^{\varphi}]$ is exactly $2^{\rho} = 2^{\sigma+\tau}$. Thus, we may assume that $\sigma < \rho = \gamma$, which means $\alpha > \beta$. We consider then $\Phi = \pi_{2^{\beta+\tau}} \circ \phi \circ \Psi$, where $\phi = \phi_{xy} - \phi_{yx} + 2^{\beta-\sigma}\phi_{xxy}$. The map is well-defined, since $\beta + \gamma - \sigma \geq \beta + \tau$. Since $\alpha \geq \beta + \tau$, $\beta + \rho - \sigma \geq \beta + \tau$, and $\beta - \sigma > 0$, it follows that $u \equiv 2^{\beta-\sigma}\binom{u}{2} \equiv 2^{\beta-\sigma}v \equiv 0 \pmod{2^{\beta+\tau}}$, so verifying (28) and (29) reduces to checking that:

$$vr - 2^{\beta-\sigma}rw \equiv 0 \pmod{2^{\beta+\tau}} \quad \text{and} \quad -mv + 2^{\beta-\sigma}mw \equiv 0 \pmod{2^{\beta+\tau}}.$$

It is now trivial that both congruences hold for all possible values of u, v, and w, and therefore Φ defines a crossed pairing. The induced map Φ^* maps $[a,[a,b]^{\varphi}]$ to the class of $2^{\beta-\sigma}$ modulo $2^{\beta+\tau}$, proving that the order of $[a,[a,b]^{\varphi}]$ is also equal to $2^{\sigma+\tau}$ in this case. This finishes the proof. □

Proposition 42. *Let $G = G_2(\alpha,\beta,\gamma;\rho,\sigma)$ be a 2-generator 2-group of class 2, with presentation as in Theorem 1. Then the order of $[a,b^{\varphi}]$ is:*

$$\bigl|[a,b^{\varphi}]\bigr| = \begin{cases} 2^{\beta} & \text{if } \rho < \sigma;\\ 2^{\beta+\rho-\sigma+1} & \text{if } \sigma \leq \rho < \gamma;\\ 2^{\beta+\tau} & \text{if } \sigma < \rho = \gamma;\\ 2^{\beta+\delta_{\sigma\beta}} & \text{if } \sigma = \rho = \gamma; \end{cases}$$

where $\tau = \min(\alpha-\beta, \rho-\sigma)$ and δ_{ij} is Kronecker's delta.

Proof. By Lemma 25(iii), we know that the order of $[a,b^{\varphi}]$ is a multiple of 2^{β} and divides $2^{\alpha+1}$; and by (22) we have that

$$[a,b^{\varphi}]^{2^{\beta}} = [a,[a,b]^{\varphi}]^{2^{\sigma}}[b,[a,b]^{\varphi}]^{2^{\beta-1}}.$$

Suppose first that $\rho < \sigma$. Note that we cannot have $\sigma < \gamma$, since there is no canonical choice of values in Theorem 1 with $\rho < \sigma < \gamma$; thus, $\sigma = \gamma$ and the order of $[b,[a,b]^{\varphi}]$ is 2^{ρ} by Proposition 40, and $\rho \leq \beta - 1$, so from the equation above we obtain that $[a,b^{\varphi}]^{2^{\beta}} = [a,[a,b]^{\varphi}]^{2^{\sigma}}$. Since $\rho < \gamma$, Proposition 41 yields that $[a,[a,b]^{\varphi}]^{2^{\sigma}} = 1$, so the order of $[a,b^{\varphi}]$ divides 2^{β}. We deduce that in this situation the order of $[a,b^{\varphi}]$ is 2^{β}, as desired.

Consider next the case when $\alpha = \beta = \gamma = \rho + 1 = \sigma + 1$; that is, we are in the second case of the theorem with some added hypotheses, and this is the only situation in which we have $\sigma = \rho < \gamma$. We need to show that the order of $[a,b^{\varphi}]$ in this case is $2^{\beta+\rho-\sigma+1} = 2^{\beta+1}$. We have that:

$$[a,b^{\varphi}]^{2^{\beta}} = [a,[a,b]^{\varphi}]^{2^{\gamma-1}}[b,[a,b]^{\varphi}]^{2^{\gamma-1}} = [a,a^{\varphi}]^{2^{\gamma}}[b,b^{\varphi}]^{2^{\gamma}},$$

where the last equality follows since $\alpha = \beta = \gamma$ and $[b,b^{\varphi}]^{2^{\beta}} = [b,[a,b]^{\varphi}]^{2^{\sigma}}$ and $[a,a^{\varphi}]^{2^{\alpha}} = [a,[a,b]^{\varphi}]^{2^{\rho}}$. Since $\rho = \sigma < \gamma$, by Propositions 38 and 39 the commutators $[a,a^{\varphi}]$ and $[b,b^{\varphi}]$ are both of order $2^{\gamma+1}$ and independent. Therefore,

$[a, b^\varphi]^{2^\beta} \neq 1$. From Lemma 25(v) we also know that the order of $[a, b^\varphi]$ divides $2^{\alpha+1} = 2^{\beta+1}$, and so we conclude that if $\alpha = \beta = \gamma = \rho+1 = \sigma+1$, then the order of $[a, b^\varphi]$ is $2^{\beta+1}$, as claimed.

Next, consider the case with $\sigma < \rho < \gamma$. As in the case with $\rho < \sigma = \gamma$, we have that $[b, [a, b]^\varphi]^{2^{\beta-1}} = 1$, hence $[a, b^\varphi]^{2^\beta} = [a, [a, b]^\varphi]^{2^\sigma}$. Since the order of $[a, [a, b]^\varphi]$ in this situation is $2^{\rho+1}$ by Proposition 41, we deduce that the order of $[a, b^\varphi]$ is $2^{\beta+\rho-\sigma+1}$, which is what we needed to establish.

Next we have the case $\sigma < \rho = \gamma$; then $[a, [a, b]^\varphi]$ has order $2^{\sigma+\tau}$ and $[b, [a, b]^\varphi]$ has order $2^{\sigma+1}$ by Propositions 40 and 41. By (22) we have

$$[a, b^\varphi]^{2^{\beta+\tau}} = [a, [a, b]^\varphi]^{2^{\sigma+\tau}} [b, [a, b]^\varphi]^{2^{\beta+\tau-1}} = 1$$

(since $\beta+\tau-1 \geq \gamma$), so the order of $[a, b^\varphi]$ divides $2^{\beta+\tau}$. We need to show the order is exactly $2^{\beta+\tau}$. If $\beta-1 \geq \sigma+1$ then $[b, [a, b]^\varphi]^{2^{\beta-1}} = 1$, hence $[a, b^\varphi]^{2^\beta} = [a, [a, b]^\varphi]^{2^\sigma}$, and we obtain from Proposition 41 that the order of $[a, b^\varphi]$ is exactly $2^{\beta+\tau}$. On the other hand, if $\beta = \sigma + 1$, then we must have $\sigma + 1 = \rho = \gamma = \beta$ and $\tau = 1$, and we want to show that the order of $[a, b^\varphi]$ is exactly $2^{\beta+1}$. To that end, consider the crossed pairing $\Phi = \pi_{2^{\beta+1}} \circ \phi \circ \Psi$ from Proposition 39, with

$$\phi = \phi_{yy} + 2^{\beta-\sigma}\phi_{yxy} + 2^{1-\delta_{(\beta-1)\sigma}}(\phi_{xy} - \phi_{yx}) = \phi_{yy} + 2\phi_{yxy} + \phi_{xy} - \phi_{yx}.$$

Then we have that the induced homomorphism Φ^* maps $[a, b^\varphi]$ to the class of 1 modulo $2^{\beta+1}$, and therefore the order of $[a, b^\varphi]$ is a multiple of $2^{\beta+1} = 2^{\beta+\tau}$. Since we already know that the order of $[a, b^\varphi]$ divides $2^{\beta+\tau}$, we again conclude that the order is exactly $2^{\beta+\tau}$, as desired.

Thus, we are left with the case $\sigma = \rho = \gamma$, where we need to show that the order of $[a, b^\varphi]$ is exactly $2^{\beta+\delta_{\sigma\beta}}$. If $\sigma < \beta$, then $\beta - 1 \geq \gamma$ so $[a, b^\varphi]^{2^\beta} = [a, [a, b]^\varphi]^{2^\sigma} = 1$, by Proposition 41. Thus, the order of $[a, b^\varphi]$ is indeed $2^{\beta+\delta_{\sigma\beta}}$. If $\sigma = \beta$ then $\tau = 0$, and so by Proposition 41 the order of $[a, [a, b]^\varphi]$ is 2^σ. Thus we have that $[a, b^\varphi]^{2^\beta} = [b, [a, b]^\varphi]^{2^{\beta-1}} = [b, [a, b]^\varphi]^{2^{\rho-1}}$. By Proposition 40, the order of $[b, [a, b]^\varphi]$ in this case is 2^ρ, so the order of $[a, b^\varphi]$ is indeed $2^{\beta+\delta_{\sigma\beta}}$, as claimed. ☐

Orders of the generators of $G \wedge G$. In the $p > 2$ case $G \otimes G$ splits and $G \wedge G$ is isomorphic to the subgroup of $G \otimes G$ generated by $[a, b^\varphi]$, $[a, [a, b]^\varphi]$, and $[b, [a, b]^\varphi]$. Hence we found the orders of these generators directly by our analysis of $G \otimes G$. This is not the case in general when $p = 2$ and we must compute the images of these generators under the epimorphism

$$\theta : G \otimes G \to G \wedge G \tag{43}$$

found in Figure 1 factoring out $\nabla(G)$. The next result uses the preceding analysis to determine orders of these images.

PROPOSITION 44. *Let $G = G_2(\alpha, \beta, \gamma; \rho, \sigma)$ be any 2-generator 2-group of class 2 with presentation as in Theorem 1. Then the orders of $\theta([a, b^\varphi])$, $\theta([a, [a, b]^\varphi])$ and $\theta([b, [a, b]^\varphi])$ are as follows:*

$$|\theta([a, b^\varphi])| = \begin{cases} 2^\beta & \text{if } \rho < \sigma \\ 2^{\beta+\rho-\sigma} & \text{if } \sigma \leq \rho < \gamma \\ 2^{\beta+\tau} & \text{if } \sigma < \rho = \gamma \\ 2^{\beta+\delta_{\sigma\beta}} & \text{if } \sigma = \rho = \gamma \end{cases}$$

$$\left|\theta([a,[a,b]^\varphi])\right| = \begin{cases} 2^\rho & \textit{if } \rho < \sigma \\ 2^\rho & \textit{if } \sigma \le \rho < \gamma \\ 2^{\sigma+\tau} & \textit{if } \sigma < \rho = \gamma \\ 2^\sigma & \textit{if } \sigma = \rho = \gamma \end{cases}$$

$$\left|\theta([b,[a,b]^\varphi])\right| = 2^{\min(\rho,\sigma)}$$

where θ is as defined in (43), $\tau = \min(\alpha-\beta, \rho-\sigma)$, *and δ_{ij} is Kronecker's delta.*

PROOF. The general observation guiding us is that if $[a, a^\varphi]$ has order $2^{\alpha+1}$ then $[a,a^\varphi]^{2^\alpha} = [a,[a,b]^\varphi]^{2^\rho}$. Hence $[a,[a,b]^\varphi]$ has order $2^{\rho+1}$ in $G \otimes G$ and its image in $G \wedge G$ has order 2^ρ. On the other hand if $[a,a^\varphi]$ has order 2^α then no power of $[a,[a,b]^\varphi]$ can be in $\nabla(G)$ and the order of $\theta([a,[a,b]^\varphi])$ is the order of $[a,[a,b]^\varphi]$. A similar analysis holds for $[b,b^\varphi]$ and $[b,[a,b]^\varphi]$. We use Propositions 38 through 42 to determine the orders of $[a,a^\varphi]$, $[b,b^\varphi]$, $[a,[a,b]^\varphi]$, $[b,[a,b]^\varphi]$, and $[a,b^\varphi]$ respectively. We will use Lemma 21 to determine if any powers of $[a,b^\varphi]$ are in $\nabla(G)$.

Suppose $\rho < \sigma$. Then $[a,b^\varphi]$ has order 2^β. But this is the lower bound for the order of $\theta([a,b^\varphi])$ as its image in $(G/G') \wedge (G/G')$ has order 2^β. Hence $\theta([a,b^\varphi])$ has order 2^β. The orders of $[a,a^\varphi]$, $[b,b^\varphi]$, and $[b,[a,b]^\varphi]$ are $2^{\alpha+1}$, 2^β and 2^ρ respectively. Hence $\theta([a,[a,b]^\varphi])$ and $\theta([b,[a,b]^\varphi])$ each have order 2^ρ.

Suppose $\sigma \le \rho < \gamma$. The orders of $[a,a^\varphi]$, $[b,b^\varphi]$, $[a,[a,b]^\varphi]$ and $[b,[a,b]^\varphi]$ are $2^{\alpha+1}$, $2^{\beta+1}$, $2^{\rho+1}$, and $2^{\sigma+1}$ respectively. Hence $\theta([a,[a,b]^\varphi])$ and $\theta([b,[a,b]^\varphi])$ have orders 2^ρ and 2^σ respectively. The order of $[a,b^\varphi]$ is $2^{\beta+\rho-\sigma+1}$. But by (22)

$$\begin{aligned}
[a,b^\varphi]^{2^{\beta+\rho-\sigma}} &= [a,[a,b]^\varphi]^{2^{\sigma+\rho-\sigma}}[b,[a,b]^\varphi]^{2^{\beta-1+\rho-\sigma}} \\
&= [a,[a,b]^\varphi]^{2^\rho}[b,[a,b]^\varphi]^{2^{\beta-1+\rho-\sigma}} \\
&= [a,a^\varphi]^{2^\alpha}[b,[a,b]^\varphi]^{2^{\beta-1+\rho-\sigma}}.
\end{aligned}$$

If $(\rho-\sigma) > 0$ then $\beta-1+\rho-\sigma \ge \gamma$ and $[b,[a,b]^\varphi]^{2^{\beta-1+\rho-\sigma}}$ is trivial. If $(\rho-\sigma)=0$ then $\beta-1 \ge \sigma$ as $\beta \ge \gamma$. Hence $[b,[a,b]^\varphi]^{2^{\beta-1+\rho-\sigma}}$ is either trivial or an element of $\nabla(G)$. Therefore $\theta([a,b^\varphi])$ has order $2^{\beta+\rho-\sigma}$.

Suppose $\sigma < \rho = \gamma$. The orders of $[a,a^\varphi]$, $[b,b^\varphi]$, $[a,[a,b]^\varphi]$ and $[b,[a,b]^\varphi]$ are 2^α, $2^{\beta+1}$, $2^{\sigma+\tau}$, and $2^{\sigma+1}$ respectively. Hence $\theta([a,[a,b]^\varphi])$ and $\theta([b,[a,b]^\varphi])$ have orders $2^{\sigma+\tau}$ and 2^σ respectively. Since no power of $[a,b^\varphi]$ is in $\nabla(G)$, the order of $\theta([a,b^\varphi])$ is $2^{\beta+\tau}$.

Finally suppose that $\sigma = \rho = \gamma$. Then $[a,a^\varphi]$ and $[b,b^\varphi]$ have order 2^α and 2^β respectively and $\nabla(G) \cong \nabla(G/G')$. Therefore $G \otimes G$ splits and the orders of $\theta([a,b^\varphi])$, $\theta([a,[a,b]^\varphi])$ and $\theta([b,[a,b]^\varphi])$ are $2^{\beta+\delta_{\sigma\beta}}$, $2^{\sigma+\tau}$ and 2^σ respectively with $\tau = 0$ since $\rho = \sigma$. □

The tensor and exterior squares for $p = 2$. From the analysis above, we obtain the structure of $G \wedge G$ and $G \otimes G$ when $p = 2$. Writing down a basis for $G \otimes G$ is not nearly as nice as in the $p > 2$ case, as might be guessed from both (22) and (23), so rather than write them out in the statements, we describe how to obtain such a basis. The first three generators, $[a,a^\varphi]$, $[b,b^\varphi]$, and $[a,b^\varphi][b,a^\varphi]$ can always be taken to be the first three basis elements.

When the order of $[a,b^{\varphi}]$ is 2^{β}, we may simply take the last three basis elements to be $[a,b^{\varphi}]$, $[a,a^{\varphi}]^{-2^{\alpha-\rho}}[a,[a,b]^{\varphi}]$, and $[b,[a,b]^{\varphi}]$, as the only possible "entanglement" arises if $\rho<\gamma$ between $[a,a^{\varphi}]$ and $[a,[a,b]^{\varphi}]$. If $\rho=\gamma$ we may in fact omit the $[a,a^{\varphi}]^{-2^{\alpha-\rho}}$ factor from the fifth basis element.

When $\sigma=\rho=\gamma=\beta<\alpha$, the required modification arises from the identity $[a,b^{\varphi}]^{2^{\gamma}}=[b,[a,b]^{\varphi}]^{2^{\gamma-1}}$; the basis is then constructed by taking our standard six generators, and replacing $[b,[a,b]^{\varphi}]$ with $[a,b^{\varphi}]^{-2}[b,[a,b]^{\varphi}]$ to obtain an element of order $2^{\gamma-1}$. Similarly, if $\sigma=\rho=\gamma=\alpha$, the problem arises from the fact that $[a,b^{\varphi}]^{2^{\gamma}}$ is equal to both $[a,[a,b]^{\varphi}]^{2^{\gamma-1}}$ and to $[b,[a,b]^{\varphi}]^{2^{\gamma-1}}$. This is solved by replacing each of the last two generators by their product with $[a,b^{\varphi}]^{-2}$.

If $\sigma+1=\rho+1=\gamma=\alpha=\beta$, the difficulty stems from the equality

$$[a,b^{\varphi}]^{2^{\gamma}}=[a,[a,b]^{\varphi}]^{2^{\gamma-1}}[b,[a,b]^{\varphi}]^{2^{\gamma-1}}=[a,a^{\varphi}]^{2^{\gamma}}[b,b^{\varphi}]^{2^{\gamma}}.$$

Here, the three generators $[a,b^{\varphi}]$, $[a,[a,b]^{\varphi}]$, and $[b,[a,b]^{\varphi}]$ are replaced with the elements $[a,a^{\varphi}][b,b^{\varphi}][a,b^{\varphi}]$, $[a,a^{\varphi}]^{2}[a,[a,b]^{\varphi}]$, and $[b,b^{\varphi}]^{2}[b,[a,b]^{\varphi}]$, respectively.

This leaves only Families 1(b) and 1(c) from Theorem 1, where $\sigma<\rho\leq\gamma$. To account for the fact that $\sigma<\gamma$, we must replace $[b,[a,b]^{\varphi}]$ with $[b,b^{\varphi}]^{-2^{\beta-\sigma}}[b,[a,b]^{\varphi}]$. The replacement of $[a,[a,b^{\varphi}]]$ and of $[a,b^{\varphi}]$ is more difficult; essentially, one uses (22) and (23) to obtain expressions for $[a,[a,b]^{\varphi}]^{2^{\sigma}}$ in terms of $[a,b^{\varphi}]$ and $[b,[a,b]^{\varphi}]$, and so multiply $[a,[a,b]^{\varphi}]$ by suitable powers of them to make sure we get a generator of order 2^{σ}; when $\sigma<\rho<\gamma$, one must also multiply $[a,b^{\varphi}]$ by a suitable power of $[a,a^{\varphi}]$ to get an element of order $2^{\beta+\sigma-\rho}$, rather than its "natural" order of $2^{\beta+\sigma-\rho+1}$. Once one does that, we obtain:

PROPOSITION 45. *Let $G=G_2(\alpha,\beta,\gamma;\rho,\sigma)$ be any 2-generator 2-group of class 2 with presentation as in Theorem 1. Then*

$$G\wedge G\cong\begin{cases} C_{2^{\beta}}\times C_{2^{\rho}}\times C_{2^{\rho}} & \text{if } \rho<\sigma;\\ C_{2^{\beta+\rho-\sigma}}\times C_{2^{\sigma}}\times C_{2^{\sigma}} & \text{if } \sigma\leq\rho<\gamma;\\ C_{2^{\beta+\tau}}\times C_{2^{\sigma}}\times C_{2^{\sigma}} & \text{if } \sigma<\rho=\gamma;\\ C_{2^{\beta+\delta_{\sigma\beta}}}\times C_{2^{\gamma-\delta_{\alpha\gamma}}}\times C_{2^{\gamma-\delta_{\beta\gamma}}} & \text{if } \sigma=\rho=\gamma; \end{cases}\tag{46}$$

where $\tau=\min(\rho-\sigma,\alpha-\beta)$, and δ_{ij} is Kronecker's delta.

THEOREM 47. *Let $G=G_2(\alpha,\beta,\gamma;\rho,\sigma)$ be any 2-generator 2-group of class 2 with presentation as in Theorem 1. Then*

$$G\otimes G\cong\begin{cases} C_{2^{\alpha+1}}\times C_{2^{\beta}}\times C_{2^{\beta}}\times C_{2^{\beta}}\times C_{2^{\rho}}\times C_{2^{\rho}} & \text{if } \rho<\sigma=\gamma;\\ C_{2^{\alpha+1}}\times C_{2^{\beta+1}}\times C_{2^{\beta}}\times C_{2^{\beta+\rho-\sigma}}\times C_{2^{\sigma}}\times C_{2^{\sigma}} & \text{if } \sigma\leq\rho<\gamma;\\ C_{2^{\alpha}}\times C_{2^{\beta+1}}\times C_{2^{\beta}}\times C_{2^{\beta+\tau}}\times C_{2^{\sigma}}\times C_{2^{\sigma}} & \text{if } \sigma<\rho=\gamma;\\ C_{2^{\alpha}}\times C_{2^{\beta}}\times C_{2^{\beta}}\times C_{2^{\beta+\delta_{\sigma\beta}}}\times C_{2^{\gamma-\delta_{\alpha\gamma}}}\times C_{2^{\gamma-\delta_{\beta\gamma}}} & \text{if } \sigma=\rho=\gamma; \end{cases}$$

where $\tau=\min(\rho-\sigma,\alpha-\beta)$, and δ_{ij} is Kronecker's delta.

In the $p>2$ case, the map $G\otimes G\to G\wedge G$ always has a natural splitting, thus identifying the exterior square with the subgroup of $G\otimes G$ generated by $[a,b^{\varphi}]$, $[a,[a,b]^{\varphi}]$, and $[b,[a,b]^{\varphi}]$. In the $p=2$ case, this will occur if and only if $\rho=\sigma=\gamma$; otherwise, the subgroup generated by these three generators will intersect the subgroup generated by the other three. Nonetheless, as we will see below, at least in our class we always have an isomorphism $G\otimes G\cong\nabla(G)\times(G\wedge G)$, even though the isomorphism need not arise from a splitting.

4. Other homological functors

In this section we compute the groups $\Gamma(G/G')$, $\nabla(G)$, $J_2(G)$, and $H_2(G)$, where G is a 2-generator p-group of class 2. The computation of Whitehead's universal quadratic functor, $\Gamma(G/G')$, is dependent on p but otherwise it is invariant to α and β. It is a direct calculation using (11) and (12).

THEOREM 48. *Let $G = G_p(\alpha, \beta, \gamma; \rho, \sigma)$ be any 2-generator p-group of class 2 with presentation as in Theorem 1. Then*

$$\Gamma(G/G') \cong \begin{cases} C_{2^{\alpha+1}} \times C_{2^{\beta+1}} \times C_{2^\beta} & \text{if } p = 2 \\ C_{p^\alpha} \times C_{p^\beta} \times C_{p^\beta} & \text{if } p > 2. \end{cases}$$

For the other groups, it is easier to describe them separately for $p > 2$ and $p = 2$.

4.1. The p odd case.

For $p > 2$, we obtain $\nabla(G)$ and $J_2(G)$ using the structure theorems from Subsection 2.2.

THEOREM 49. *Let $G = G_p(\alpha, \beta, \gamma; \rho, \sigma)$ be any 2-generator p-group of class 2 with presentation as in Theorem 1 and $p > 2$. Then*

$$\nabla(G) \cong \Gamma(G/G') \cong C_{p^\alpha} \times C_{p^\beta} \times C_{p^\beta};$$

$$H_2(G) \cong \begin{cases} C_{p^{\beta-\gamma}} \times C_{p^\rho} \times C_{p^\rho} & \text{if } \rho \le \sigma, \\ C_{p^{\beta+\tau-\gamma}} \times C_{p^\sigma} \times C_{p^\sigma} & \text{if } \rho > \sigma, \end{cases}$$

$$J_2(G) \cong \nabla(G) \times H_2(G)$$

where $\tau = \min(\alpha - \beta, \rho - \sigma)$.

PROOF. The structure of $\nabla(G)$ and its isomorphism to $\Gamma(G/G')$ was shown in (30). The split of $J_2(G)$ follows from Theorem 10(iii) since G/G' has no elements of order 2. To find the structure of $H_2(G)$ we compute the kernel of the epimorphism

$$\kappa' : G \wedge G \to G',$$

defined by $x \wedge y \mapsto [x, y]$. An isomorphic copy of the exterior square of G is generated by $[a, b^\varphi]$, $[a, [a, b]^\varphi]$, $[b, [a, b]^\varphi]$. The generators $[a, [a, b]^\varphi]$ and $[b, [a, b]^\varphi]$ of this copy of $G \wedge G$ are mapped to the identity by κ' since G has no nontrivial commutators of weight 3. Since the order of $[a, b]$ in G is p^γ, it follows that $[a, b^\varphi]^{p^\gamma}$ is an element of the kernel of κ'. If $\rho \le \sigma$ then by Lemma 32 the order of $[a, b^\varphi]$ is p^β and by Proposition 34

$$G \wedge G \cong \langle [a, b^\varphi] \rangle \times \langle [a, [a, b]^\varphi] \rangle \times \langle [b, [a, b]^\varphi] \rangle \cong C_{p^\beta} \times C_{p^\rho} \times C_{p^\rho}.$$

Therefore the kernel of κ' is

$$\langle [a, b^\varphi]^{p^\gamma} \rangle \times \langle [a, [a, b]^\varphi] \rangle \times \langle [b, [a, b]^\varphi] \rangle \cong C_{p^{\beta-\gamma}} \times C_{p^\rho} \times C_{p^\rho}.$$

If $\rho > \sigma$ then the order of $[a, b^\varphi]$ is $p^{\beta+\tau}$ where $\tau = \min(\alpha - \beta, \rho - \sigma)$. The nonabelian exterior square of G is isomorphic to the abelian group

$$\left\langle [a, b^\varphi], [a, [a, b]^\varphi], [b, [a, b]^\varphi] \mid [a, b^\varphi]^{p^\beta} = [a, [a, b]^\varphi]^{p^\sigma} \right\rangle \cong C_{p^{\beta+\tau}} \times C_{p^\sigma} \times C_{p^\sigma}.$$

Hence the kernel of κ' is the abelian group

$$\left\langle [a,b^{\varphi}]^{p^{\gamma}}, [a,[a,b]^{\varphi}], [b,[a,b]^{\varphi}] \mid [a,b^{\varphi}]^{p^{\beta}} = [a,[a,b]^{\varphi}]^{p^{\sigma}} \right\rangle \cong C_{p^{\beta+\tau-\gamma}} \times C_{p^{\sigma}} \times C_{p^{\sigma}}.$$

□

4.2. The p even case. The computation of $\nabla(G)$, $H_2(G)$ and $J_2(G)$ for $p=2$ is not aided by the structure results in all cases as in the p odd case. However, similar to the $p>2$ case, our computations are aided by Proposition 44, which gives the orders of the generators of $G\wedge G$.

THEOREM 50. *Let $G = G_2(\alpha,\beta,\gamma;\rho,\sigma)$ be any 2-generator 2-group of class 2 with presentation as in Theorem 1. Then:*

$$\nabla(G) \cong C_{2^{\alpha+1-\delta_{\rho\gamma}}} \times C_{2^{\beta+1-\delta_{\sigma\gamma}}} \times C_{2^{\beta}}$$

$$H_2(G) \cong \begin{cases} C_{2^{\beta-\gamma}} \times C_{2^{\rho}} \times C_{2^{\rho}} & \text{if } \rho<\sigma,\\ C_{2^{\beta+\rho-\sigma-\gamma}} \times C_{2^{\sigma}} \times C_{2^{\sigma}} & \text{if } \sigma\leq\rho<\gamma,\\ C_{2^{\beta+\tau-\gamma}} \times C_{2^{\sigma}} \times C_{2^{\sigma}} & \text{if } \sigma<\rho=\gamma,\\ C_{2^{\beta-\gamma}} \times C_{2^{\sigma}} \times C_{2^{\sigma-\delta_{\alpha\gamma}}} & \text{if } \sigma=\rho=\gamma.\end{cases}$$

and

$$J_2(G) \cong \nabla(G)\times H_2(G),$$

where $\tau=\min(\alpha-\beta,\rho-\sigma)$, δ_{ij} is Kronecker's delta.

PROOF. The subgroup $\nabla(G)$ is independently generated by $[a,a^{\varphi}]$, $[b,b^{\varphi}]$, and $[a,b^{\varphi}][b,a^{\varphi}]$ by Propositions 38 and 39, with orders $2^{\alpha+1-\delta_{\rho\gamma}}$, $2^{\beta+1-\delta_{\sigma\gamma}}$, and 2^{β} respectively.

The structure of $H_2(G)$ is determined by computing the kernel of the mapping $\kappa' : G\wedge G \to G'$ as in the p odd case. The generators of $G\wedge G$ are the images of $[a,[a,b]^{\varphi}]$, $[b,[a,b]^{\varphi}]$, and $[a,b^{\varphi}]$ of $G\otimes G$ under the epimorphism θ (43). We will subscript these generators by θ to indicate their images in the nonabelian exterior square. The generators $[a,[a,b]^{\varphi}]_{\theta}$ and $[b,[a,b]^{\varphi}]_{\theta}$ are mapped to the identity by κ' since G has no nontrivial commutators of weight 3. The order of $[a,b]$ in G is 2^{γ}, hence $[a,b^{\varphi}]^{2^{\gamma}}$ is in the kernel of κ'. It follows that $H_2(G)$ is a homomorphic image of the abelian group generated by

$$([a,b^{\varphi}]_{\theta})^{2^{\gamma}},\quad [a,[a,b]^{\varphi}]_{\theta},\quad [b,[a,b]^{\varphi}]_{\theta} \tag{51}$$

subject to the relations of Lemma 21 under the image of θ:

$$([a,b^{\varphi}]_{\theta})^{2^{\beta}} = ([a,[a,b]^{\varphi}]_{\theta})^{2^{\sigma}}([b,[a,b]^{\varphi}]_{\theta})^{2^{\beta-1}} \tag{52}$$

$$([a,b^{\varphi}]_{\theta})^{2^{\alpha}} = ([a,[a,b]^{\varphi}]_{\theta})^{2^{\alpha-1}}([b,[a,b]^{\varphi}]_{\theta})^{-2^{\rho}}. \tag{53}$$

If $\rho<\sigma$ then the generators of $H_2(G)$ in the order given in (51) have orders $2^{\beta-\gamma}$, 2^{ρ}, and 2^{ρ} respectively. Relations (52) and (53) are trivial; that is, both the left and right hand sides are the identity. Hence $H_2(G)\cong C_{2^{\beta-\gamma}}\times C_{2^{\rho}}\times C_{2^{\rho}}$.

If $\sigma\leq\rho<\gamma$ then the generators of $H_2(G)$ in the order given in (51) have orders $2^{\beta+\rho-\sigma-\gamma}$, 2^{ρ}, and 2^{σ} respectively, and relation (53) is trivial. Since $\sigma<\beta-1$, relation (52) becomes $([a,b^{\varphi}]_{\theta})^{2^{\beta}} = ([a,[a,b]^{\varphi}]_{\theta})^{2^{\sigma}}$. Hence

$$H_2(G)\cong C_{2^{\beta+\rho-\sigma-\gamma}}\times C_{2^{\sigma}}\times C_{2^{\sigma}}.$$

If $\sigma < \rho = \gamma$ then the generators of $H_2(G)$ in the order given in (51) have orders $2^{\beta+\tau-\gamma}$, $2^{\sigma+\tau}$, and 2^σ respectively. Relation (53) is trivial, and as $\sigma < \beta - 1$, relation (52) again becomes $([a, b^\varphi]_\theta)^{2^\beta} = ([a, [a, b]^\varphi]_\theta)^{2^\sigma}$. Hence

$$H_2(G) \cong C_{2^{\beta+\tau-\gamma}} \times C_{2^\sigma} \times C_{2^\sigma}.$$

Next, suppose $\sigma = \rho = \gamma$. Then the generators of $H_2(G)$ in the order given in (51) have orders $2^{\beta-\gamma+\delta_{\sigma\beta}}$, 2^ρ, and 2^σ respectively. If $\alpha > \beta$, then relation (53) is trivial, and relation (52) becomes $([a, b^\varphi]_\theta)^{2^\beta} = ([b, [a, b]^\varphi]_\theta)^{2^{\beta-1}}$. If $\alpha > \beta = \gamma$, then $([a, b^\varphi]_\theta)^{2^\beta}$ is of order 2, and contained in the subgroup generated by $[b, [a, b]^\varphi]_\theta$; so $H_2(G)$ is generated by $[a, [a, b]^\varphi]_\theta$ and $[b, [a, b]^\varphi]_\theta$. These two are independent, each of order $2^\gamma = 2^\sigma$, and we obtain $H_2(G) \cong C_{2^\sigma} \times C_{2^\sigma}$, which agrees with the description in the statement. On the other hand, if we have $\alpha > \beta > \gamma$, then relation (52) is also trivial, $\delta_{\sigma\beta} = 0$, and so we have $H_2(G) \cong C_{2^{\beta-\gamma}} \times C_{2^\sigma} \times C_{2^\sigma}$, as claimed.

Finally, if $\alpha = \beta \geq \gamma = \rho = \sigma$ then relations (52) and (53) become

$$([a, b^\varphi]_\theta)^{2^\beta} = ([b, [a, b]^\varphi]_\theta)^{2^{\beta-1}} \tag{54}$$

$$([a, b^\varphi]_\theta)^{2^\alpha} = ([a, [a, b]^\varphi]_\theta)^{2^{\alpha-1}}. \tag{55}$$

If $\beta > \gamma$ then both relations are trivial and the abelian invariants for $H_2(G)$ are $C_{2^{\beta-\gamma}} \times C_{2^\sigma} \times C_{2^\sigma}$. If $\beta = \gamma$ then both relations (54) and (55) are nontrivial. The element $([a, b^\varphi]_\theta)^{2^\beta}$ is again of order 2, and contained in the cyclic subgroup generated by $[a, [a, b]^\varphi]_\theta$; so $H_2(G)$ is generated by $[a, [a, b]^\varphi]_\theta$ and $[b, [a, b]^\varphi]_\theta$; both are of order $2^\gamma = 2^\sigma$, but by (54) and (55) we have $([a, [a, b]^\varphi]_\theta)^{2^{\sigma-1}} = ([b, [a, b]^\varphi]_\theta)^{2^{\sigma-1}}$. This yields the desired decomposition of $H_2(G)$ as $C_{2^{\beta-\gamma}} \times C_{2^\sigma} \times C_{2^{\sigma-1}}$ when $\alpha = \beta = \gamma = \rho = \sigma$. This establishes the description of $H_2(G)$.

To finish, recall that if A is a finite abelian p-group and $x \in A$ generates a cyclic subgroup of maximal order, then A is isomorphic to $\langle x \rangle \times A/\langle x \rangle$ (see for example the proof in [22, Theorem 8.2]). Applying this successively to the elements $[a, a^\varphi]$, $[b, b^\varphi]$, and $[a, b^\varphi][b, a^\varphi]$, which independently generate $\nabla(G)$, we deduce that we will have $J_2(G) \cong \nabla(G) \times J_2(G)/\nabla(G)$. The diagram in Figure 1 shows that $J_2(G)$ is an extension of $\mathrm{im}(\psi)$ by $H_2(G)$, and that $\mathrm{im}(\psi) \cong \nabla(G)$; therefore, $J_2(G) \cong \nabla(G) \times H_2(G)$, as claimed. □

5. The subgroups $Z^\otimes(G)$ and $Z^\wedge(G)$

The nonabelian tensor square and exterior square of a group H define the central subgroups $Z^\otimes(H)$ and $Z^\wedge(H)$ of H called the nonabelian tensor center and nonabelian exterior center, respectively. These groups are defined as follows:

$$Z^\otimes(H) = \langle h \in Z(H) \mid h \otimes x = 1 \text{ for all } x \in H \rangle$$
$$Z^\wedge(H) = \langle h \in Z(H) \mid h \wedge x = 1 \text{ for all } x \in H \rangle.$$

We interpret $Z^\otimes(H)$ as the largest subgroup of H such that

$$H \otimes H \cong (H/Z^\otimes(H)) \otimes (H/Z^\otimes(H)),$$

and similarly $Z^\wedge(H)$ is the largest subgroup of H such that

$$H \wedge H \cong (H/Z^\wedge(H)) \wedge (H/Z^\wedge(H)).$$

A group H is called capable if there exists a group K such that $K/Z(K) \cong H$. The subgroup $Z^*(H)$, called the epicenter of H [13], is the smallest central subgroup of H such that $H/Z^*(H)$ is capable [6]. Therefore, a group H is capable if and only if $Z^*(H)$ is trivial. The nonabelian exterior center $Z^\wedge(H)$ of H is isomorphic to $Z^*(H)$ [17]. Hence by computing the nonabelian exterior centers of all the groups in Theorem 1 we determine those that are capable. Note then that for any group H, we have the following inclusions:

$$Z^{\otimes}(H) \leq Z^{\wedge}(H) = Z^*(H) \leq Z(H).$$

At the other extreme from the capable groups are the groups H satisfying $Z^*(H) = Z(H)$. Such groups are called unicentral [6]. Again we will be able to determine which 2-generator p-groups of class 2 are unicentral by our description of the exterior centers of these groups.

We can consider the factor group of $\nu(H)/\nabla(H)$ which we label $\tau(H)$. The nonabelian exterior square, $H \wedge H$, isomorphically embeds into $\tau(H)$ via the mapping $g \wedge h \mapsto [g, h^{\varphi}]_{\tau(H)}$ by Proposition 16 in [8]. Theorem 19 in [8] provides a direct way to compute both $Z^{\otimes}(H)$ and $Z^{\wedge}(H)$.

THEOREM 56. *Let H be any group. Then*

$$Z^{\otimes}(H) = H \cap Z(\nu(H)) \qquad \text{and} \qquad Z^{\wedge}(H) = H \cap Z(\tau(H)),$$

where H on the right-hand side of each equation is interpreted as the natural isomorphic copy of H in $\nu(H)$ and $\tau(H)$ respectively.

5.1. The tensor center. Let $G = G_p(\alpha, \beta, \gamma; \rho, \sigma)$ be any 2-generator p-group of class 2 with presentation as in Theorem 1. We claim that the tensor center $Z^{\otimes}(G)$ will always be contained in the commutator subgroup $[G, G]$ of G.

The center of G is generated by a^{p^γ}, b^{p^γ}, and $[a, b]$; the order of a is $p^{\alpha+\gamma-\rho}$, since $a^{p^\alpha} = [a, b]^{p^\rho}$, and $[a, b]$ is of order p^γ. Likewise, the order of b is $p^{\beta+\gamma-\sigma}$. The defining relations of G then give us that:

$$\begin{aligned} Z(G) &= \left\langle a^{p^\gamma}, b^{p^\gamma}, [a,b] \;\middle|\; (a^{p^\gamma})^{p^{\alpha-\gamma}} = [a,b]^{p^\rho}, (b^{p^\gamma})^{p^{\beta-\gamma}} = [a,b]^{p^\sigma}, [a,b]^{p^\gamma} = 1 \right\rangle \\ &\cong \begin{cases} C_{p^{\alpha-\rho}} \times C_{p^{\beta-\gamma}} \times C_{p^\rho} & \text{if } \rho \leq \sigma, \\ C_{p^{\alpha-\gamma}} \times C_{p^{\beta-\sigma}} \times C_{p^\sigma} & \text{if } \sigma < \rho. \end{cases} \end{aligned}$$

As noted above, we claim that the tensor center is contained in the commutator subgroup. Indeed, suppose $a^r b^s [a, b]^t$ lies in $G \cap Z(\nu(G)) = Z^{\otimes}(G)$. Then $[a, (a^r b^s [a, b]^t)^{\varphi}] = 1$ in $\nu(G)$. Expanding this commutator using the rule $[x, yz] = [x, y][y, [x, z]][x, z]$, Lemma 20, and the usual commutator calculus of $\nu(G)$ we have:

$$\begin{aligned} 1 &= [a, (a^r b^s [a,b]^t)^{\varphi}] = [a, a^{\varphi}]^r [a, [a,b]^{\varphi}]^{rs} [a, (b^{\varphi})^s] \\ &= [a, a^{\varphi}]^r [a, [a,b]^{\varphi}]^s [a, b^{\varphi}]^s [b, [a,b]^{\varphi}]^{\binom{s}{2}}. \end{aligned}$$

For this to be trivial, we must have $p^\alpha | r$ (as the order of $[a, a^{\varphi}]$ is at least p^α, and this commutator is independent of the other three); and we know $a^{p^\alpha} \in [G, G]$. Computing $[b, (a^r b^s [a, b]^t)^{\varphi}]$ we obtain a factor $[b, b^{\varphi}]^s$, so we likewise deduce that $p^\beta | s$. Since $b^{p^\beta} \in [G, G]$, we conclude that $Z^{\otimes}(G) \leq \langle [a, b] \rangle$, as claimed.

LEMMA 57. *$G = G_p(\alpha, \beta, \gamma; \rho, \sigma)$ be any 2-generator p-group of class 2 with presentation as in Theorem 1. Then*

$$Z^{\otimes}(G) = \langle [a,b]^m \rangle \cong C_{p^{\gamma-m}}$$

where p^m is the maximum order of $[a, [a, b]^{\varphi}]$ and $[b, [a, b]^{\varphi}]$.

PROOF. By the analysis above $Z^{\otimes}(G) \leq \langle [a,b] \rangle$. Choosing p^m be the maximum of the orders of $[a,[a,b]^{\varphi}]$ and $[b,[a,b]^{\varphi}]$ insures that p^m is the smallest integer power such that both $[a,([a,b]^{p^m})^{\varphi}] = 1$ and $[b,([a,b]^{p^m})^{\varphi}] = 1$, as p^m is the order of one of the commutators. We conclude that $Z^{\otimes}(G)$ is generated by $\langle [a,b]^{p^m} \rangle \cong C_{p^{\gamma-m}}$. □

We use Lemma 57 together with our previous analysis on the orders of $[b,[a,b]^{\varphi}]$ and $[a,[a,b]^{\varphi}]$ for $p > 2$ and $p = 2$. The proofs of the following two propositions now follow immediately from those results:

PROPOSITION 58. *$G = G_p(\alpha,\beta,\gamma;\rho,\sigma)$ be any 2-generator p-group of class 2 with presentation as in Theorem 1 and $p > 2$. Then*

$$Z^{\otimes}(G) \cong \begin{cases} C_{p^{\gamma-\rho}} & \text{if } \rho \leq \sigma \\ C_{p^{\gamma-\sigma-\tau}} & \text{if } \rho > \sigma \end{cases}$$

where $\tau = \min(\alpha-\beta, \rho-\sigma)$.

PROPOSITION 59. *$G = G_p(\alpha,\beta,\gamma;\rho,\sigma)$ be any 2-generator p-group of class 2 with presentation as in Theorem 1 and $p = 2$. Then*

$$Z^{\otimes}(G) \cong \begin{cases} C_{2^{\gamma-\rho-1}} & \text{if } \rho < \sigma = \gamma \\ C_{2^{\gamma-\rho-1}} & \text{if } \sigma \leq \rho < \gamma \\ C_{2^{\gamma-\sigma-\mu}} & \text{if } \sigma < \rho = \gamma \\ C_{2^{\gamma-\sigma-\tau}} & \text{if } \sigma = \rho = \gamma \end{cases}$$

where $\tau = \min(\rho-\sigma, \alpha-\beta)$ and $\mu = \max(\tau, 1)$.

5.2. The exterior center. The group $\tau(G)$ is generated by a, b, $[a,b]$, a^{φ}, b^{φ}, and $[a,b]^{\varphi}$. In $\tau(G)$ we have the additional relations that $[a,a^{\varphi}] = 1$ and $[b,b^{\varphi}] = 1$. To show which central elements of G^{φ} intersect with $Z(\tau(G))$ we need to analyze the following commutators:

$$[b,(a^{p^{\gamma}})^{\varphi}],\ [a,(b^{p^{\gamma}})^{\varphi}],\ [a,[a,b]^{\varphi}],\ [b,[a,b]^{\varphi}],$$

since $[[a,b],(a^{p^{\gamma}})^{\varphi}] = [[a,b]^{p^{\gamma}},a^{\varphi}] = 1$, $[[a,b],(b^{p^{\gamma}})^{\varphi}] = [[a,b]^{p^{\gamma}},b^{\varphi}] = 1$, and $[[a,b],[a,b]^{\varphi}] = 1$.

In this case, we cannot conclude that the exterior center will be contained in $[G,G]$. The reason is that the latter conclusion relied on factors of $[a,a^{\varphi}]$ and $[b,b^{\varphi}]$ that occurred when expanding $a \otimes a^r b^s [a,b]^t$ and $b \otimes a^r b^s [a,b]^t$, and which allowed us to bound r and s from below. When we work in the exterior square, however, we have $a \wedge a = b \wedge b = 1$, so the argument will not hold. In fact, it may be that there is a power of $a^{p^{\gamma}}$ that lies in $Z^{\wedge}(G)$ but not in $[G,G]$, and similarly with $b^{p^{\gamma}}$; even so, the exterior center will be cyclic in all cases. The analysis is similar to that for the tensor center.

LEMMA 60. *$G = G_p(\alpha,\beta,\gamma;\rho,\sigma)$ be any 2-generator p-group of class 2 with presentation as in Theorem 1. Then*

$$Z^{\wedge}(G) = \langle a^{p^{\gamma+m}}, b^{p^{\gamma+n}}, [a,b]^{p^t} \rangle,$$

where p^m and p^n are the orders of $[a^{p^{\gamma}},b^{\varphi}]$ and $[a,(b^{p^{\gamma}})^{\varphi}]$ respectively and p^t is the maximum of the orders of $[a,[a,b]^{\varphi}]$ and $[b,[a,b]^{\varphi}]$.

PROOF. Note that since $\gamma \geq 1$, we have $[a^{p^{\gamma+k}}, b^{\varphi}] = [a^{p^{\gamma}}, b^{\varphi}]^{p^k}$ for all nonnegative integers k and all primes p, and analogously with $[a, (b^{p^{\gamma+k}})^{\varphi}]$. The elements $a^{p^{\gamma}}$ and $b^{p^{\gamma}}$ are the smallest powers of a and b in the center of G. These element commute with $[a,b]^{\varphi}$ since

$$[a^{p^{\gamma}}, [a,b]^{\varphi}] = [a, ([a,b]^{p^{\gamma}})^{\varphi}] = 1$$
$$[b^{p^{\gamma}}, [a,b]^{\varphi}] = [b, ([a,b]^{p^{\gamma}})^{\varphi}] = 1.$$

In $\tau(G)$ we have $[a, a^{\varphi}] = 1$, so if some power of $a^{p^{\gamma}}$ is in $Z^{\wedge}(G)$ it must commute with b^{φ}. The smallest such power is the order of $[a^{p^{\gamma}}, b^{\varphi}]$. The analysis is symmetric for $b^{p^{\gamma}}$. Finally as with the tensor center, the maximum of the orders of $[a, [a,b]^{\varphi}]$ and $[b, [a,b]^{\varphi}]$ gives the minimum order of $[a,b]^{\varphi}$ in $\tau(G)$ that commutes with both a and b. □

With these computations and the orders of the generators in the exterior square of G that were previously computed, the following two propositions follow directly:

PROPOSITION 61. *Let $G = G_p(\alpha,\beta,\gamma;\rho,\sigma)$ be any 2-generator p-group of class 2 with presentation as in Theorem 1 and $p > 2$. Then*

$$Z^{\wedge}(G) \cong \begin{cases} C_{p^{\alpha+\gamma-\beta-\rho}} & \text{if } \rho \leq \sigma \\ C_{p^{-\alpha+\gamma+\beta-\sigma}} & \text{if } \rho > \sigma \text{ and } \tau = \alpha - \beta \\ C_{p^{\alpha+\gamma-\beta-2\rho+\sigma}} & \text{if } \rho > \sigma \text{ and } \tau = \rho - \sigma \end{cases}$$

where $\tau = \min(\alpha-\beta, \rho-\sigma)$.

PROPOSITION 62. *Let $G = G_2(\alpha,\beta,\gamma;\rho,\sigma)$ be any 2-generator 2-group of class 2 with presentation as in Theorem 1. Then*

$$Z^{\wedge}(G) \cong \begin{cases} C_{2^{\alpha+\gamma-\beta-\rho}} & \text{if } \rho < \sigma = \gamma; \\ C_{2^{\alpha+\gamma-\beta-2\rho+\sigma}} & \text{if } \sigma \leq \rho < \gamma; \\ C_{2^{-\alpha+\gamma+\beta-\sigma+\delta_{\beta(\sigma+1)}}} & \text{if } \sigma < \rho = \gamma \text{ and } \tau = \alpha - \beta; \\ C_{2^{\alpha+\gamma-\beta-2\rho+\sigma+\delta_{\beta(\sigma+1)}}} & \text{if } \sigma < \rho = \gamma \text{ and } \tau = \rho - \sigma; \\ C_{2^{\alpha-\beta-\delta_{\sigma\beta}}} & \text{if } \sigma = \rho = \gamma \text{ and } \alpha > \beta; \\ 1 & \text{if } \sigma = \rho = \gamma \text{ and } \alpha = \beta; \end{cases}$$

where $\tau = \min(\alpha-\beta, \rho-\sigma)$ and δ_{ij} is Kronecker's delta.

5.3. Capable and unicentral groups. From our description above, we are able to describe exactly those groups in $G_p(\alpha,\beta,\gamma;\rho,\sigma)$, for $p > 2$ that are capable by characterizing when $Z^{\wedge}(G)$ is trivial, as well as those that are unicentral by characterizing when $Z^{\wedge}(G) = Z(G)$.

The case $p = 2$ is different enough to justify a separate statement and proof.

THEOREM 63. *Let $G = G_p(\alpha,\beta,\gamma;\rho,\sigma)$ be any 2-generator p-groups of class 2 with representation as in Theorem 1 and $p > 2$. Then G is capable if and only if*

$$\alpha - \beta = \rho - \sigma \quad \text{and} \quad \gamma = \rho.$$

When $\rho \leq \sigma$, the conditions exclude all groups from Family 1(a) of Theorem 1, and characterize all of the capable groups from Family 2. When $\rho > \sigma$, the conditions characterize all of the groups from Family 1(b) that are capable, and exclude all the groups of Family 1(c).

PROOF. If G is capable then one of the following condition must hold:

$$\alpha+\gamma-\beta-\rho=0 \qquad \text{if } \rho\leq\sigma \tag{64}$$

$$-\alpha+\gamma+\beta-\sigma=0 \qquad \text{if } \rho>\sigma \text{ and } \tau=\alpha-\beta \tag{65}$$

$$\alpha+\gamma-\beta-2\rho+\sigma=0 \qquad \text{if } \rho>\sigma \text{ and } \tau=\rho-\sigma. \tag{66}$$

Equation (64) can be expressed as $\alpha-\beta=\rho-\gamma$. Since $\rho-\gamma$ must be nonnegative and $\rho\leq\gamma$ it follows that $\rho=\gamma$ and hence $\alpha-\beta=0$. This excludes Family 1(a), where $\alpha>\beta$, and describes the capable groups of Family 2.

Equation (65) can be rewritten $\alpha-\beta=\gamma-\sigma$. The condition $\rho>\sigma$ implies $\alpha-\beta>0$; and if $\tau=\alpha-\beta$, then $\rho=\gamma$. This condition determines all of the capable groups of Family 1(b). All groups in Family 1(c) have $\alpha-\beta>\rho-\sigma$ and are excluded from this condition.

Equation (66) can be rewritten as $\alpha-\beta-(\rho-\sigma)=\rho-\gamma$. By hypothesis $\tau=\rho-\sigma$ so $\alpha-\beta-(\rho-\sigma)\geq 0$. In general $\rho\leq\gamma$ so we can only have $\rho=\gamma$. Then Equation (66) reduces to (65) with $\tau=\rho-\sigma=\alpha-\beta$. □

THEOREM 67. *Let $G=G_2(\alpha,\beta,\gamma;\rho,\sigma)$ be any 2-generator 2-groups of class 2 with representation as in Theorem 1. The group G is capable if and only if it meets one of the following conditions:*

(i) *$\rho\leq\sigma$, $\alpha=\beta$, and $\rho=\gamma$;*
(ii) *$\rho>\sigma$, $(\alpha-\beta)=(\rho-\sigma)>\delta_{\beta\gamma}$, and $\rho=\gamma$;*
(iii) *$\rho=\sigma=\gamma=\beta$, and $\alpha=\gamma+1$;*

where δ_{ij} is Kronecker's delta. All groups from Family 1(c) and Families 3(a) and 3(b) of Theorem 1 are excluded from being capable. The three conditions characterize exactly when groups in Families 1(a), 1(b), 2, and 3(c) are capable.

PROOF. If G is capable then either $\rho=\sigma=\gamma$ and $\alpha=\beta$, which is case (i) in the statement; or one of the following conditions hold:

$$\alpha+\gamma-\beta-\rho=0 \qquad \text{if } \rho<\sigma=\gamma \tag{68}$$

$$\alpha+\gamma-\beta-2\rho+\sigma=0 \qquad \text{if } \sigma\leq\rho<\gamma \tag{69}$$

$$-\alpha+\gamma+\beta-\sigma+\delta_{\beta(\sigma+1)}=0 \qquad \text{if } \sigma<\rho=\gamma \text{ and } \tau=\alpha-\beta \tag{70}$$

$$\alpha+\gamma-\beta-2\rho+\sigma+\delta_{\beta(\sigma+1)}=0 \qquad \text{if } \sigma<\rho=\gamma \text{ and } \tau=\rho-\sigma \tag{71}$$

$$\alpha-\beta-\delta_{\sigma\beta}=0 \qquad \text{if } \rho=\sigma=\gamma. \tag{72}$$

Equation (68) can be rewritten as $\alpha-\beta=\rho-\gamma$; but $\alpha-\beta\geq 0$ and $\rho<\gamma$, so this condition cannot hold.

Consider now equation (69); the canonical values of ρ and σ in Theorem 1 only allow $\sigma\leq\rho<\gamma$ if $\rho-\sigma<\alpha-\beta$, or if $\alpha=\beta=\gamma=\sigma+1=\rho+1$. Now, (69) is equivalent to $(\alpha-\beta)-(\rho-\sigma)=(\rho-\gamma)$. The right hand side is negative under the assumptions of this case, while the left hand side is positive in the first allowable situation, and zero in the second; either way, this condition cannot hold either.

The latter two cases exclude all groups in Families 1(a), 1(c), 3(a), and 3(b).

Equation (70) can be rewritten as $\alpha-\beta=\rho-\sigma+\delta_{\beta(\sigma+1)}$. Since we also have $\tau=\alpha-\beta$, this means that $\alpha-\beta\leq\rho-\sigma$, so we must have $\beta>\sigma+1$. If $\beta=\gamma$, this means that $\tau\geq 2$, so the condition $\alpha-\beta>\delta_{\beta\gamma}$ holds; if $\beta>\gamma$, this condition places no restriction on τ. In any case, we must have $\rho=\gamma$. This condition determines all of the capable groups of Family 1(b), and is condition (ii) given in the statement.

Equation (71) can be expressed as $\alpha - \beta - (\rho - \sigma) + \delta_{\beta(\sigma+1)} = \rho - \gamma$. However, as $\rho = \gamma$, this equation becomes $\alpha - \beta + \delta_{\beta(\sigma+1)} = \rho - \sigma$. If $\beta > \sigma + 1$ we are back to the previous case. If $\beta = \sigma + 1$, then $\beta = \gamma = \sigma + 1$, so $\rho - \sigma = 1$, which would require $\alpha - \beta = 0$ and this cannot hold since $\sigma < \rho$ implies $\alpha > \beta$.

Finally, equation (72) requires $\alpha - \beta = \delta_{\sigma\beta}$. If $\beta = \sigma = \rho = \gamma$, this means that $\alpha = \gamma + 1$, giving part (iii) of the statement, and corresponding to groups in Family 1(a); if $\beta > \gamma$, then this means $\alpha = \beta$, and we are back to condition (i) and Family 2 or 3(c). □

It is worth noting that these results agree with the previous descriptions of the capable groups found in [23,24], as it turns out that no group from Family 1(c) (the family that was missing in the classifications on which the previous work relied) is capable. We also note that for fixed $n \geq 3$, the number of capable 2-generator p-groups of class 2 is independent of p. This follows since a group $G_p(\alpha, \beta, \gamma; \rho, \sigma)$ is capable for $p > 2$ if and only if it is capable for $p = 2$, with two exceptions: $G_p(\gamma + 1, \gamma, \gamma; \gamma, \gamma - 1)$ is capable if and only if $p > 2$, and $G_p(\gamma + 1, \gamma, \gamma; \gamma, \gamma)$ is capable if and only if $p = 2$. They both occur with $n = 3\gamma + 1$, so the total number will be independent of p.

COROLLARY 73. *Let $n \geq 3$ and let p be a prime. The number of capable 2-generator p-groups of class exactly 2 and order p^n is $C(n)$, where:*

$$C(n) = \begin{cases} \displaystyle\sum_{\gamma=1}^{\lfloor \frac{n}{3} \rfloor} \left\lfloor \frac{\min(\gamma + 2, n - 3\gamma + 2)}{2} \right\rfloor & \text{if } n \text{ is even,} \\ \displaystyle\sum_{\gamma=1}^{\lfloor \frac{n}{3} \rfloor} \left\lfloor \frac{\min(\gamma + 1, n - 3\gamma + 2)}{2} \right\rfloor & \text{if } n \text{ is odd,} \end{cases}$$

and $\lfloor m \rfloor$ denotes the greatest integer less than or equal to m.

PROOF. Since, as we noted above, $C(n)$ does not depend on p, we work with $p > 2$ since the capability conditions are simpler. In that situation, for a fixed γ, $0 < \gamma < \lfloor \frac{n}{3} \rfloor$, there is at most one capable group corresponding to each value of k, $0 \leq k \leq \gamma$, namely the group $G_p(\beta + k, \beta, \gamma; \gamma, \gamma - k)$, with $\gamma + 2\beta + k = n$ and $\beta \geq \gamma$. Replacing β with $\gamma + \ell$, $\ell \geq 0$, this is the group $G_p(\gamma + \ell + k, \gamma + \ell, \gamma; \gamma, \gamma - k)$. In particular, $n = 3\gamma + 2\ell + k$ so $n - (3\gamma + k)$ must be even and nonnegative. Conversely, if k is nonnegative, $0 \leq k \leq \min(\gamma, n - 3\gamma)$, and $n - (3\gamma + k)$ is even, then the group above is capable. Thus, for a fixed n and γ, there is exactly one capable group for each k such that $0 \leq k \leq \min(\gamma, n - 3\gamma)$ and $n - 3\gamma \equiv k \pmod 2$.

Suppose first that n is even. If γ is odd then k must be odd as well. When $\gamma \leq n - 3\gamma$, k may take the values $1, 3, \ldots, \gamma$, for a total of $\lfloor \frac{\gamma+1}{2} \rfloor = \lfloor \frac{\gamma+2}{2} \rfloor$ capable groups. When $n - 3\gamma < \gamma$, k may take the values $1, 3, \ldots, n - 3\gamma$, so we have $\lfloor \frac{n-3\gamma+2}{2} \rfloor$ possible choices; the total for odd γ is then $\lfloor \frac{\min(\gamma+2, n-3\gamma+2)}{2} \rfloor$. Similarly, if γ is even then k must be even, so the possible values for k are 0, $2, \ldots, \min(\gamma, n - 3\gamma)$. We conclude the number is independent of the parity of γ, and is given by $\lfloor \frac{\min(\gamma+2, n-3\gamma+2)}{2} \rfloor$.

Next, suppose that n is odd. If γ is odd then k must be even; if $\gamma \leq n - 3\gamma$ then k may take the values $0, 2, \ldots, \gamma - 1$. If $n - 3\gamma < \gamma$, then k may take the values 0, $2, \ldots, n - 3\gamma$. Thus, the total number of choices when γ is odd is $\lfloor \frac{\min(\gamma+1, n-3\gamma+2)}{2} \rfloor$. If γ is even then k must be odd. If $\gamma \leq n - 3\gamma$ then k may take the values 1,

$3,\ldots,\gamma-1$; and when $n-3\gamma<\gamma$, k may take the values $1,3,\ldots,n-3\gamma$. Thus, the total number of choices for even γ is $\lfloor\frac{\min(\gamma+1,n-3\gamma+1)}{2}\rfloor = \lfloor\frac{\min(\gamma+1,n-3\gamma+2)}{2}\rfloor$. Again, the total is independent of the parity of γ, and is given by $\lfloor\frac{\min(\gamma+1,n-3\gamma+2)}{2}\rfloor$.

To obtain $C(n)$, we add over all possible choices of γ; these range from $\gamma=1$ through $\gamma=\lfloor\frac{n}{3}\rfloor$, yielding the formula in the statement. □

We now consider the case of unicentral groups. Recall that the unicentral groups lie at the opposite extreme as the capable groups: a group G is unicentral if and only if the smallest subgroup N of $Z(G)$ such that G/N is capable is $N=Z(G)$; that is, if and only if $Z^\wedge(G)=Z(G)$.

As we saw earlier, the center of G is isomorphic to:

$$Z(G)\cong\begin{cases} C_{p^{\alpha-\rho}}\times C_{p^{\beta-\gamma}}\times C_{p^{\rho}} & \text{if } \rho\leq\sigma,\\ C_{p^{\alpha-\gamma}}\times C_{p^{\beta-\sigma}}\times C_{p^{\sigma}} & \text{if } \sigma<\rho.\end{cases}$$

Note that $Z^\wedge(G)$ is always cyclic. If $0<\min(\rho,\sigma)<\gamma$, then the center is not cyclic; this means that for G to be unicentral we must have $\min(\rho,\sigma)=0$ or $\rho=\sigma=\gamma$. If $\rho=\sigma=\gamma$, then we further need $\alpha=\beta=\gamma$; but in this case, the exterior center $Z^\wedge(G)$ is trivial, and so cannot equal the center. Thus, if G is unicentral then we must have $\min(\rho,\sigma)=0$.

If $\sigma<\rho\leq\gamma$, we would then also require $\alpha-\gamma=0$, but this means $\beta=\gamma=\alpha$, and then $\rho\leq\sigma$; thus, this case is impossible. Hence a necessary condition for G to be unicentral is $\rho=0$ and $\beta=\gamma$, in which case the center is cyclic of order p^α. We deduce that:

THEOREM 74. *Let $G=G_p(\alpha,\beta,\gamma;\rho,\sigma)$ be any 2-generator p-groups of class 2 with representation as in Theorem 1. Then G is unicentral if and only if*

(i) *$\beta=\gamma=\sigma$ and $\rho=0$; or*
(ii) *$p=2$, $\alpha=\beta=\gamma=1$, and $\rho=\sigma=0$.*

In particular, no groups from Families 1(b), 1(c), or 3(c) are unicentral, and the only group from Family 3(b) that is unicentral is the group $G_2(1,1,1;0,0)$.

PROOF. We have already seen that $\beta=\gamma$ and $\rho=0$ are necessary.

Suppose first that $p>2$. Then $\rho\leq\sigma$ means that we must also have $\sigma=\gamma$, so the necessity of (i) follows; to show the sufficiency, it is enough to show that the order of $Z^\wedge(G)$ in this situation will be p^α, since the exterior center is always a subgroup of $Z(G)$. And this is indeed the case, as we see in Proposition 61.

Suppose then that $p=2$. If we have $\beta=\gamma$, $\rho=0$, and $\sigma<\gamma$, then we must have $\sigma=\rho=0$ and $\gamma=\beta=\alpha=1$ (that is, we must be in Family 3(b) from Theorem 1, with $\alpha=\beta=\gamma=1$). For G to be unicentral in this case, the exterior center must be of order $2^\alpha=2$, which is indeed the case as we see in Proposition 62. So we may assume that $\sigma=\gamma=\beta$, $\rho=0$, and we want to show that this is sufficient for unicentrality. Again, Proposition 62 shows that in this case, the exterior center has order 2^α, and hence is equal to the center, as desired. □

It is interesting to note that the number of unicentral groups among the 2-generated p-groups of class exactly 2 and order p^n is the same regardless of p; for in the case with $p=2$, the group $G_2(1,1,1;0,0)$ with $n=3$ "takes the place" of the group $G_p(1,1,1;0,1)$ that we have for $p>2$.

COROLLARY 75. *Let $n \geq 3$ and p a prime. Then the number of unicentral 2-generator p-groups of class exactly 2 and order p^n is $\lfloor \frac{n}{3} \rfloor$, where $\lfloor m \rfloor$ is the greatest integer less than or equal to m.*

PROOF. For each n we obtain exactly one group for each choice of α, β, and γ with $\alpha \geq \beta = \gamma > 0$ and $\alpha + \beta + \gamma = n$; once $\beta = \gamma$ is chosen, α is forced, and the common value of β and γ cannot exceed $\lfloor \frac{n}{3} \rfloor$. □

6. Computational support

Most of the results in this paper were motivated and checked using GAP [19], a computational group theory system. In this section we provide details on how these computations were completed. We start by showing how to create the groups from Theorem 1.

Let $G = G_p(\alpha, \beta, \gamma; \rho, \sigma)$ be any 2-generators p-group of class 2. Then G has the following consistent polycyclic presentation

$$G = \langle a, b, c \mid a^{p^\alpha} = c^{p^\rho},\ b^{p^\beta} = c^{p^\sigma},\ b^a = bc,\ c^{p^\gamma} = 1,\ c^a = c,\ c^b = c \rangle. \tag{76}$$

For a particular prime p and α, β, γ, ρ, and σ it is straightforward to create the finitely presented group (76) in GAP. For illustration let us consider the group $G = G_3(6, 5, 4; 3, 2)$. The following GAP commands create a finitely presented group H isomorphic to G. We suppress most of the output for brevity with the double semicolons.

```
gap> F := FreeGroup("a","b","c");;
gap> a:=F.1;; b:=F.2;; c:=F.3;;
gap> R := [a^(3^6)/c^(3^3), b^(3^5)/c^(3^2), c^(3^4),
> (a^-1*b*a)/(b*c), (a^-1*c*a)/c, (b^-1*c*b)/c];;
gap> H := F/R;
<fp group on the generators [ a, b, c ]>
```

Since the presentation is a consistent polycyclic presentation we can convert H to a polycyclic representation, which allows for effective and efficient computations using the Polycyclic package [15].

```
gap> G := Image(IsomorphismPcpGroup(H));
Pcp-group with orders [ 729, 243, 81 ]
```

As a Pcp-group we can compute directly most of the groups we are considering via the built-in functions found in Polycyclic.

```
gap> ts := NonAbelianTensorSquare(G);
Pcp-group with orders [ 81, 3, 27, 9, 729, 243, 243 ]
gap> IsAbelian(ts);
true
gap> AbelianInvariants(ts);
[ 9, 9, 243, 243, 729, 729 ]

gap> es := NonAbelianExteriorSquare(G);
Pcp-group with orders [ 81, 3, 27, 9 ]
gap> AbelianInvariants(es);
[ 9, 9, 729 ]

gap> wh := WhiteheadQuadraticFunctor(G);
```

```
Pcp-group with orders [ 243, 729, 243 ]
gap> AbelianInvariants(wh);
[ 243, 243, 729 ]

gap> SchurMultiplicator(G);
[ [ 3, 1 ], [ 9, 1 ], [ 27, 1 ] ]
```

The `SchurMultiplicator` command returns the abelian invariants of the Schur multiplier, where the first number in each list of length 2 is the order of the cyclic factor and the second number is the number of factors of that order. In our example,

$$H_2(G) \cong C_3 \times C_9 \times C_{27}.$$

There are no direct commands in `Polycyclic` to compute $\nabla(G)$, $J_2(G)$, $Z^{\otimes}(G)$ and $Z^{\wedge}(G)$. However, these groups can be determined from computations made by `Polycyclic`. The epimorphisms from $\theta : G \otimes G \to G \wedge G$ and $\kappa' : G \wedge G \to G'$ can be obtained via the commands

```
gap> theta := NonAbelianTensorSquareEpimorphism(G);;
gap> kappap := NonAbelianExteriorSquareEpimorphism(G);;
```

respectively. The kernel of θ is equal to $\nabla(G)$, and $J_2(G)$ is the kernel of the composition of θ and κ'. As defined in the GAP above, `theta` has a different image than the source of `kappap`. This can be corrected by computing an isomorphism between the two. However, `kappap` is computed as part of the computation of `theta` in `Polycyclic` setting up the appropriate isomorphism eliminating the need to find one explicitly.

```
gap> nabla := Kernel(theta);;
gap> AbelianInvariants(nabla);
[ 243, 243, 729 ]
gap> # Redefine kappap so we can make the
gap> # composition.
gap> kappap := Range(theta)!.epimorphism;;
gap> J2 := Kernel(theta*kappap);;
gap> AbelianInvariants(J2);
[ 3, 9, 27, 243, 243, 729 ]
```

The groups $\nu(G)$ and $\tau(G)$ can be found using the commands

```
gap> nu := NonAbelianTensorSquarePlus(G);;
gap> tau := NonAbelianExteriorSquarePlus(G);;
```

We need to find the embedding of G in $\nu(G)$ and $\tau(G)$ to compute the tensor center and exterior center of G. There is not a clean way to do this via a `Polycyclic` command; however the generators of $\nu(G)$ and $\tau(G)$ are by construction

$$a, b, c, a^{\varphi}, b^{\varphi}, c^{\varphi}$$

in that order. Hence we use this information to find the embeddings

```
gap> G_nu := Subgroup(nu,Igs(nu){[1,2,3]});;
gap> G_tau := Subgroup(tau,Igs(tau){[1,2,3]};;
```

that are each isomorphic to G. The following GAP code checks that G is isomorphic to G_{ν}.

```
gap> hom:= GroupGeneralMappingByImages(G,G_nu,Igs(G),Igs(G_nu));;
gap> IsPcpGroupHomomorphism(hom);
```

```
true
gap> IsTrivial(Kernel(hom));
true
```

To compute the tensor center and exterior center of G we apply Theorem 56 and find the intersection of G_ν and $Z(\nu(G))$ and G_τ and $Z(\tau(G))$ respectively and record their abelian invariants.

```
gap> tc := Intersection(G_nu,Center(nu));;
gap> AbelianInvariants(tc);
[ 3 ]
gap> ec := Intersection(G_tau,Center(tau));;
gap> AbelianInvariants(ec);
[ 3 ]
```

The computations above line up with our theoretical results. Indeed, the group $G = G_3(6,5,4;3,2)$ has $\rho - \sigma \geq \alpha - \beta$ with p odd, so by Theorem 1(1)(b) we have $G \cong G_3(6,5,4;4,2)$ with $\tau = 1$. Hence the abelian invariants for $G \otimes G$ are

$$(p^\alpha, p^\beta, p^\beta, p^{\beta+\tau}, p^\sigma, p^\sigma) = (3^6, 3^5, 3^5, 3^6, 3^2, 3^2)$$

and the abelian invariants for $G \wedge G$ are

$$(p^{\beta+\tau}, p^\sigma, p^\sigma) = (3^6, 3^2, 3^2)$$

as predicted.

We conclude this section with GAP code that enumerates all the 2-generator p-groups of class 2 of order p^n, denoted by $\mathcal{G}_{p^n}$. Being able to consider all such groups for a fixed prime and n was very helpful in obtaining general information about these groups. We also give timings for creating these groups.

Enumerating all groups in $\mathcal{G}_{p^n}$ is naturally broken up into three functions. The first function generates all 5-tuples $(\alpha, \beta, \gamma; \rho, \sigma)$ for a given n and p that give nonisomorphic groups by Theorem 1. The second function generates a specific group in $\mathcal{G}_{p^n}$ as a `PcpGroup` by parameterizing the polycyclic presentation (76). The third function uses the first two functions to compute a list of all nonisomorphic groups in $\mathcal{G}_{p^n}$ as `Polycyclic` groups. We start with the second function.

```
# Parameterization of G_p(alpha,beta,gamma;rho,sigma) as
# a GAP function
#
2grp :=
  function(alpha,beta,gamma,rho,sigma,p)
    local F, R,     # Free group and relations
          a, b, c; # Generators of F

      F := FreeGroup("a","b","c");
      a := F.1; b := F.2; c := F.3;
      R := [];

      if rho=gamma then
        Add(R,a^(p^alpha));
      else
        Add(R,a^(p^alpha)/c^(p^rho);
      fi;

      if sigma=gamma then
```

```
    Add(R,b^(p^beta));
  else
    Add(R,b^(p^beta)/c^(p^sigma));
  fi;

  Append(R, [(a^-1*b*a)/(b*c), c^(p^gamma),
             (a^-1*c*a)/c, (b^-1*c*b)/c]);

  return Image(IsomorphismPcpGroup(F/R));
end;
```

To enumerate $(\alpha, \beta, \gamma; \rho, \sigma)$ for a given n and p we compute each nonnegative partition of n of length 3 and enumerate over all ρ and σ pairs and find the canonical values for ρ and σ for each Family using Theorem 1. Enumerating over all possible values of ρ and σ pairs for each Family can lead to duplicate 5-tuples. For example suppose $\alpha > \beta$ and $\gamma = 1$. Then the ρ, σ pairs $(0,0)$ and $(0,1)$ both correspond to the canonical 5-tuple $(\alpha, \beta, \gamma; 0, \gamma)$. We allow for this possibility and eliminate possible duplicates by forming a set of tuples. However, by enumerating through all possible ρ and σ pairs we ensure that all isomorphism types are found. The GAP code implementing this is as follows:

```
# Return a list of all five tuples (alpha, beta, gamma; rho, sigma)
# for a given n and p that represent all nonisomorphic 2-generator
# p-groups of order p^n
#
5tuples :=
  function(n,p)
    local P,                    # Partitions of n
          par,                  # Particular partition
          alpha, beta, gamma,   # Partition values in sorted order
          rho, sigma,
          tups;                 # Set of distinct 5-tuples
    tups := [];

    # Each partition is in sorted order.
    P := Partitions(n,3);

    for par in P do
      alpha := par[1]; beta := par[2]; gamma := par[3];

      # Family 1
      if alpha > beta then
        for rho in [0..gamma] do
          for sigma in [0..gamma] do

            # Case (a)
            if rho <= sigma then
              AddSet(tups, [alpha, beta, gamma, rho, gamma]);

            # Case (b)
            elif (sigma < sigma+alpha-beta and sigma+alpha-beta <= rho) or
                 (sigma < rho and rho = gamma) then
              AddSet(tups, [alpha, beta, gamma, gamma, sigma]);
```

```
          # Case (c)
          elif sigma < rho and rho < Minimum(gamma,sigma+alpha-beta) then
            AddSet(tups, [alpha, beta, gamma, rho, sigma]);
          else
            Error("We shouldn't get here");
          fi;
        od;
      od;
    fi;

    # Family 2
    if (alpha = beta and beta > gamma) or
       (alpha = beta and beta = gamma and p > 2) then
      for rho in [0..gamma] do
        for sigma in [0..gamma] do
          AddSet(tups, [alpha, beta, gamma, Minimum(rho, sigma), gamma]);
        od;
      od;
    fi;

    # Family 3
    if (alpha = beta and beta = gamma and p = 2) then
      for rho in [0..gamma] do
        for sigma in [0..gamma] do

          # Case (a)
          if Minimum(rho, sigma) < (gamma-1) then
            AddSet(tups, [alpha, beta, gamma, Minimum(rho,sigma), gamma]);

          # Case (b)
          elif rho = sigma and sigma = gamma-1 then
             AddSet(tups, [alpha, beta, gamma, gamma-1, gamma-1]);

          # Case (c)
          elif Minimum(rho,sigma) >= gamma-1 and Maximum(rho,sigma)=gamma then
            AddSet(tups, [alpha, beta, gamma, gamma, gamma]);
          else
            Error("We shouldn't get here");
          fi;
        od;
      od;
    fi;
  od;
  return tups;
end;
```

To enumerate all tuples $(\alpha, \beta, \gamma; \rho, \sigma)$ for a given p and n that represent all nonisomorphic groups in $\mathcal{G}_{p^n}$ takes little computational time on a modern laptop or desktop. For example

```
gap> 5tuples(35,3);; time;
800
gap> Length(5tuples(50,3));; time;
13497
```

```
gap> Length(5tuples(50,2));; time;
13321
```

where the times are in milliseconds. Creating the group associated with each 5-tuple can take significant time. This has to do with converting the finitely presented group (`FpGroup`) into a polycyclic group (`PcpGroup`) when the powers of the polycyclic generators are very large either because the prime is large or n is large. For example

```
gap> 2grp(6,1,1,1,1,3);time;
Pcp-group with orders [ 729, 3, 3 ]
4
gap> 2grp(6,1,1,1,1,17);time;
Pcp-group with orders [ 24137569, 17, 17 ]
34674
```

However for small primes and n up to about 20 we can create all nonisomorphic groups in $\mathcal{G}_{p^n}$ in GAP. The third function returns a list of all nonisomorphic groups in $\mathcal{G}_{p^n}$ with type `PcpGroup` using the previous two functions.

```
all2gps :=
  function(n,p)
     return List(5tuples(n,p),
       x->2grp(x[1],x[2],x[3],x[4],x[5],p));
  end;
```

We see that for small primes and small n this is quite feasible. For example

```
gap> all2gps(5,3);;time
16
gap> all2gps(15,2);;time;
376
gap> all2gps(15,3);;time;
14189
```

where the times are in milliseconds. As polycyclic groups, the `Polycyclic` package has functionality, as illustrated above, to compute the homological functors we are investigating.

References

[1] Azhana Ahmad, Arturo Magidin, and Robert Fitzgerald Morse, *A new classification of* 2-*generator p-groups of class* 2. To be submitted.

[2] Michael R. Bacon, *On the nonabelian tensor square of a nilpotent group of class two*, Glasgow Math. J. **36** (1994), no. 3, 291–296. MR **1295503 (95i:**20054)

[3] Michael R. Bacon and Luise-Charlotte Kappe, *On capable p-groups of nilpotency class two*, Illinois J. Math. **47** (2003), no. 1-2, 49–62. Special issue in honor of Reinhold Baer (1902–1979). MR **2031305 (2004j:**20036)

[4] Michael R. Bacon and Luise-Charlotte Kappe, *The nonabelian tensor square of a* 2-*generator p-group of class* 2, Arch. Math. (Basel) **61** (1993), no. 6, 508–516. MR **1254062 (95h:**20041)

[5] Michael R. Bacon, Luise-Charlotte Kappe, and Robert Fitzgerald Morse, *On the nonabelian tensor square of a* 2-*Engel group*, Arch. Math. (Basel) **69** (1997), no. 5, 353–364. MR **1473089 (98i:**20042)

[6] F. Rudolf Beyl, Ulrich Felgner, and Peter Schmid, *On groups occurring as center factor groups*, J. Algebra **61** (1979), no. 1, 161–177. MR **554857 (81i:**20034)

[7] Russell D. Blyth, Primož Moravec, and Robert Fitzgerald Morse, *On the nonabelian tensor squares of free nilpotent groups of finite rank*, Computational group theory and the theory of groups, Contemp. Math., vol. 470, Amer. Math. Soc., Providence, RI, 2008, pp. 27–43. MR **2478412 (2009m:**20047)

[8] Russell D. Blyth and Robert Fitzgerald Morse, *Computing the nonabelian tensor squares of polycyclic groups*, J. Algebra **321** (2009), no. 8, 2139–2148. MR **2501513**

[9] Russell D. Blyth, Robert Fitzgerald Morse, and Joanne L. Redden, *On computing the non-abelian tensor squares of the free 2-Engel groups*, Proc. Edinb. Math. Soc. (2) **47** (2004), no. 2, 305–323. MR **2081056 (2005d:**20064)

[10] Russel D. Blyth, Francesco Fumagalli, and Marta Morigi, *Some structural results on the non-abelian tensor square of groups*, J. of Group Theory. To appear.

[11] R. Brown, D. L. Johnson, and E. F. Robertson, *Some computations of nonabelian tensor products of groups*, J. Algebra **111** (1987), no. 1, 177–202. MR **913203 (88m:**20071)

[12] Ronald Brown and Jean-Louis Loday, *Van Kampen theorems for diagrams of spaces*, Topology **26** (1987), no. 3, 311–335. With an appendix by M. Zisman. MR **899052 (88m:**55008)

[13] John Burns and Graham Ellis, *On the nilpotent multipliers of a group*, Math. Z. **226** (1997), no. 3, 405–428. MR **1483540 (98h:**20050)

[14] R. Keith Dennis, *In search of new "Homology" functors having a close relationship to K-theory.* Unpublished preprint (1976).

[15] B. Eick and W. Nickel, *Polycyclic – Computing with polycyclic groups*, 2002. A GAP Package, see [19].

[16] Bettina Eick and Werner Nickel, *Computing the Schur multiplicator and the nonabelian tensor square of a polycyclic group*, J. Algebra **320** (2008), no. 2, 927–944. MR **2422322 (2009d:**20123)

[17] Graham Ellis, *On the capability of groups*, Proc. Edinburgh Math. Soc. (2) **41** (1998), no. 3, 487–495. MR **1697585 (2000e:**20053)

[18] Graham Ellis and Frank Leonard, *Computing Schur multipliers and tensor products of finite groups*, Proc. Roy. Irish Acad. Sect. A **95** (1995), no. 2, 137–147. MR **1660373 (99h:**20084)

[19] The GAP Group, GAP – *Groups, Algorithms and Programming, Version 4.4* (2005), available at http://www.gap-system.org.

[20] P. Hall, *The classification of prime-power groups*, J. Reine Angew. Math. **182** (1940), 130–141. MR **0003389** (**2**,211b)

[21] Luise-Charlotte Kappe, Matthew P. Visscher, and Nor Haniza Sarmin, *Two-generator two-groups of class two and their nonabelian tensor squares*, Glasg. Math. J. **41** (1999), no. 3, 417–430. MR **1720422 (2000k:**20037)

[22] Serge Lang, *Algebra*, 3rd ed., Graduate Texts in Mathematics, vol. 211, Springer-Verlag, New York, 2002. MR **1878556 (2003e:**00003)

[23] Arturo Magidin, *Capable 2-generator 2-groups of class two*, Comm. Algebra **34** (2006), no. 6, 2183–2193. MR **2236108 (2007b:**20037)

[24] Arturo Magidin, *Capability of nilpotent products of cyclic groups*, J. Group Theory **8** (2005), no. 4, 431–452. MR **2152690 (2006c:**20073)

[25] Clair Miller, *The second homology group of a group; relations among commutators*, Proc. Amer. Math. Soc. **3** (1952), 588–595. MR **0049191** (**14**,133c)

[26] W. Nickel, *nq – Nilpotent Quotients of Finitely Presented Groups*, 2003. A GAP Package, see [19].

[27] N. R. Rocco, *On a construction related to the nonabelian tensor square of a group*, Bol. Soc. Brasil. Mat. (N.S.) **22** (1991), no. 1, 63–79. MR **1159385 (93b:**20060)

[28] Urs Stammbach, *Über die ganzzahlige Homologie von Gruppen*, Exposition. Math. **3** (1985), no. 4, 359–372 (German, with English summary). MR **844416 (87h:**20094)

[29] D. Ya. Trebenko, *Nilpotent groups of class two with two generators*, Current analysis and its applications (Russian), "Naukova Dumka", Kiev, 1989, pp. 201–208, 228 (Russian). MR **1054426 (91f:**20044)

Mathematics Dept. University of Louisiana at Lafayette, 217 Maxim Doucet Hall, P.O. Box 41010, Lafayette LA 70504-1010

E-mail address: magidin@member.ams.org

Dept. of Electrical Engineering and Computer Science, University of Evansville, Evansville IN 47722 USA

E-mail address: rfmorse@evansville.edu

Contemporary Mathematics
Volume **511**, 2010

Geometric Algorithms for Resolutions for Bieberbach Groups

Marc Röder

ABSTRACT. We present a geometric method for calculating free resolutions for Bieberbach groups. An implementation is compared to a more standard algebraic method and, as an example, we calculate the additive integral homology of all 7-dimensional orientable Hantzsche-Wendt manifolds.

1. Outline

By an *n-dimensional crystallographic group* we mean a topologically discrete subgroup of the group $O(n) \ltimes \mathbb{R}^n$ of isometries of n-dimensional Euclidean space with a compact fundamental domain. We say that a crystallographic group G is *Bieberbach* if it is torsion free (i.e. it contains no non-trivial elements of finite order). Bieberbach [Bie11, Bie12] proved that an n-dimensional crystallographic group is torsion free if and only if it is the fundamental group of an n-dimensional, flat and compact, Riemannian manifold.

There is a long tradition in studying cohomological properties of Bieberbach groups and the associated manifolds [CV65]. The cohomology of Bieberbach groups—particularly the calculation of cohomology rings—has recently attracted some attention [MR99, DM05, AGPP08].

In this paper, we compute the additive integral homology of all orientable Hantzsche-Wendt manifolds of dimension 7 (sometimes called "generalised orientable Hantzsche-Wendt" manifolds or just "Hantzsche-Wendt" manifolds; see [MR99] or [Szc06] for a definition). We achieve this by constructing, for the fundamental group G of the manifold, a free $\mathbb{Z}G$-resolution of $\mathbb{Z}$ (with trivial action of G) from the natural action of G on $\mathbb{R}^n$. The calculations are made on a computer using a GAP implementation [GAP07, Röd08a] of the algorithms presented in Sections 3–5.

2000 *Mathematics Subject Classification.* 18G10, 20H15, 20J06.

Key words and phrases. Bierberbach groups, crystallographic groups, flat manifolds, chain complexes, homology.

Supported by Marie Curie Transfer of Knowledge grant MTKD-CT-2006-042685.

Section 3 provides methods for dealing with orbits and stabilisers of the natural representation of G on $\mathbb{R}^n$. In Section 4 we give a method for finding a Dirichlet domain for G and Section 5 contains an algorithm that generates a free $\mathbb{Z}G$-resolution from such a domain.

The performance of the algorithm is illustrated in Section 6 and is compared to the algebraic methods for computing free resolutions of crystallographic groups implemented in the GAP package HAP [Ell08]. Furthermore, this Section (6.3) contains the main results of this paper, namely the additive integral homology of the orientable Hantzsche-Wendt manifolds of dimension 7 mentioned above.

2. Preliminaries

Let G be a group of isometries (i.e. affine transformations) of Euclidean space $X = \mathbb{R}^n$. Any element $g \in G \leq \mathrm{O}(n) \ltimes \mathbb{R}^n$ can be written $g = At$ with $a \in \mathrm{O}(n)$ and t a translation represented by a vector $t \in \mathbb{R}^n$. We refer to A as the *linear part* of g and to t as the *translational part* of g.

We denote by x^g the image of $x \in X$ under the action of $g \in G$. We write $h^g = g^{-1}hg$ for $g, h \in G$. For $x \in X$ we consider the *stabiliser* $G_x = \{g \in G \colon x^g = x\}$ and *orbit* $x^G = \{x^g \colon g \in G\}$. For a function $d \colon X \to X$ and for $x \in X$ we write xd instead of $d(x)$.

THEOREM 1 (Bieberbach, [Bie11, Bie12]). *An n-dimensional Bieberbach group G contains a free abelian normal subgroup T of rank n and finite index (this is the subgroup of pure translations).*

The finite group G/T is called the *point group* of G.

DEFINITION 1 (Dirichlet-Voronoi construction). Let $x \neq y \in \mathbb{R}^n$ and let $\mathrm{H}(x, y) := \{a \in \mathbb{R}^n \mid \|x - a\| \leq \|y - a\|\}$. Let G be a crystallographic group acting naturally on $\mathbb{R}^n$. Define the *Dirichlet domain* to be

$$\mathrm{D}_G(x) := \bigcap_{y \in x^G} \mathrm{H}(x, y).$$

THEOREM 2. *Let G be an n-dimensional crystallographic group and $x \in \mathbb{R}^n$ a point with trivial stabiliser G_x. Then $D_G(x)$ is a convex, bounded polyhedron (i.e. a convex polytope) and contains a dense system of representatives of G-orbits (i.e. it is a fundamental domain).*

Let $\mathfrak{P} = \mathrm{D}_G(x) \subseteq \mathbb{R}^n$ be the Dirichlet domain of the Bieberbach group G with respect to $x \in \mathbb{R}^n$. The orbit of $\mathfrak{P}$ under G is a tessellation of $\mathbb{R}^n$. This tessellation endows $X = \mathbb{R}^n$ with the structure of a CW-space. The resulting chain complex $C_*(X)$ is, by definition, a chain complex of free abelian groups

$$0 \to (\mathbb{Z}G)^{n_n} \to \cdots \to (\mathbb{Z}G)^{n_1} \to (\mathbb{Z}G)^{n_0}$$

where $\mathfrak{P}_i$ has a basis in one-to-one correspondence with the i-dimensional cells in the tessellation. Since the tessellation arises from the fundamental domain $\mathrm{D}_G(X)$, the action of G on X induces the structure of a $\mathbb{Z}G$-module of $\mathbb{Z}G$-rank equal to the number of orbits of i-cells under the action of G.

In the following section we discuss algorithms for calculating $\mathrm{D}_G(X)$ and for calculating the boundary homomorphisms in the free $\mathbb{Z}G$-resolution $C_*(X)$.

3. Orbit-Stabiliser Type Algorithms

In this section we look at algorithms that will be needed during the calculation of Dirichlet domains and the resolutions defined by the associated tessellation. Algorithm 1 calculates the point group of an n-dimensional crystallographic group G and a system of representatives for it in G. We assume the existence of a function LIN that returns the linear part of an element from G.

Algorithm 1 Calculating The Point Group And Representatives For It

```
Require: n-dimensional crystallographic group G
procedure POINTGROUPREPRESENTATIVES(G)
    Set γ = R = ∅
    function RECURSION(h, Γ, P, R)
        for g ∈ Γ do
            if LIN(hg) ∉ P then
                add hg to R
                add LIN(hg) to P
                RECURSION(hg, Γ, P, R)
            end if
        end for
    end function
    Let Γ be a set of generators of G
    Let P := ∅; R := ∅
    RECURSION(1, Γ, P, R)
    return point group P with representatives R
end procedure
```

NOTATION. The orbits of an n-dimensional crystallographic group G are infinite. So for computational purposes, the orbit of $x \in \mathbb{R}^n$ is represented by the set $[0,1)^n \cap x^G$ (the intersection with a half-open unit cube centred at $(1/2)^n$). We call this set the *orbit-part* of x under G.

Note that two vectors are conjugate under G if and only if their orbit-parts are equal.

Algorithm 2 below calculates the orbit-part of x under G. Note that the algorithm requires the vector x to be contained in $[0,1)^n$. Of course, this restriction can be removed by applying an appropriate translation. The reason for keeping this restriction is that otherwise the orbit representatives of x do not necessarily contain x. An algorithm for orbit and stabiliser on point sets is just a slight variation of Algorithm 2. In this case, the orbit representatives are all images that contain a point of $[0,1)^n$.

Algorithm 3 calculates all elements of a crystallographic group G that map a vector x into a given (not necessarily convex) polytope $\mathfrak{P}$.

4. Calculating a Dirichlet Domain

Algorithm 4 calculates the Dirichlet domain $\mathrm{D}_G(x)$ of the n-dimensional Bieberbach group G with translation subgroup $T \trianglelefteq G$ with respect to a given starting point $x \in \mathbb{R}^n$.

Algorithm 2 Orbit-Part and Stabiliser

```
Require: crystallographic group G, vector x ∈ [0,1)^n
Let P be a set of point group representatives of G (Alg. 1)
Let O := ∅, S := ∅
for g ∈ P do
    i := x^g mod 1
    if i ∉ O then
        Add i to O
    else if x^g = x then
        Add i to S
    else if i = x then
        Let t be the translation mapping x^g to x
        Add gt to S
    end if
end for
return orbit-part O, stabiliser ⟨S⟩
```

Algorithm 3 Mapping a Point Into a Polytope

```
Require: G crystallographic group, x ∈ ℝ^n, 𝔓 ⊆ ℝ^n polytope
procedure MapsToPolytope(G,x,𝔓)
    M := ∅
    Let P be a set of point group representatives (Alg. 1)
    Choose a point v in the interior of 𝔓
    Let d := max { |v − y| | y ∈ vert(𝔓) }
    for g ∈ P do
        i := x^g
        Let 𝔗 be the set of translations mapping i into v + [−d/2, d/2]^n
        for t ∈ 𝔗 do
            if i^t ∈ 𝔓 then
                Add gt to M
            end if
        end for
    end for
    return M
end procedure
```

NOTE. We make the following remarks on Algorithm 4 and its implementation.

(1) The algorithm is designed to work with a computer algebra system where group computation and convex-hull computations are done separately and there is some overhead involved in changing from one sort of computation to the other. The algorithm was implemented in GAP using an interface to polymake for the convex hull computations [GAP07, Röd08b, Röd08a, GJ]. In [Eng81], Engel describes an algorithm for calculating Dirichlet domains in dimension 3 which exploits the fact that $\mathrm{D}_G(x)$ is "centred in x." The obvious generalisation to n dimensions may be faster than our method, but it comes at the cost of calculating and storing the full face lattice of

Algorithm 4 Calculating a Dirichlet domain

1: calculate the initial polytope $\mathfrak{P} = \mathrm{D}_T(x)$
2: Set $I = V_n = \emptyset$
3: **repeat**
4: Calculate vertices vert($\mathfrak{P}$) and facets fac($\mathfrak{P}$) of $\mathfrak{P}$
5: $V := \mathrm{vert}(\mathfrak{P}) \setminus V_n$
6: **if** $V = \emptyset$ **then**
7: return $\mathfrak{P}$
8: **end if**
9: Set $N = I := \emptyset$
10: **repeat**
11: choose random $v \in V$ and remove it from V
12: Add v to V_n
13: $N := \{g \in G \mid v^g \in \mathfrak{P} - \mathrm{vert}(P)\}$ ▷ Algorithm 3
14: Append $\{ H_x(x^g) \mid g \in N \cup N^{-1} \}$ to I
15: **until** $V = \emptyset$ or $|I| > b$ for some suitable bound b
16: Redefine $\mathfrak{P}$ by the inequalities fac($\mathfrak{P}$) $\cup\, I$
17: **until** $I = \emptyset$
18: return $\mathfrak{P}$

approximating polytopes. These face lattices tend to be very large and hence Engel's method is probably very memory-demanding.

(2) In line 1 of Algorithm 4, the initial polytope can be chosen to be any polytope of the form $\mathrm{D}_S(x)$ with suitable $S \subseteq G$. Here "suitable" just means that S has to be chosen so that $\mathrm{D}_S(x)$ is a polytope. All subgroups of G which are n-dimensional crystallographic groups are suitable.

(3) For the calculation of vertices and facets in line 4 of Algorithm 4 a convex hull computation is performed. The implementation of this part uses the computational geometry package polymake [GJ] as a black box.

(4) In line 15, the value of b is chosen by experiment. It balances the load between convex hull computations and the translation calculations of line 13

(5) The same algorithm can also be used to produce a Dirichlet domain for a crystallographic (non-Bieberbach) group. For this, x has to be a point with trivial stabiliser in G.

LEMMA 3. *Let G be an n-dimensional Bieberbach group and $x \in \mathbb{R}^n$. Then Algorithm 4 terminates in finitely many steps and the returned object is the Dirichlet domain with respect to x.*

PROOF. Let $\mathfrak{P} = \mathrm{D}_T(x)$ be the initial approximating polytope of Algorithm 4. Denote by vert($\mathfrak{P}$) the vertices of $\mathfrak{P}$ and by fac($\mathfrak{P}$) the facets of $\mathfrak{P}$.

Suppose that $N = \{g \in G \mid v^g \in \mathfrak{P} - \mathrm{vert}(\mathfrak{P})\} \neq \emptyset$. Then $\mathfrak{P} \cap \mathfrak{P}^g \neq \emptyset$ either contains a non-trivial open set for some $g \in N$ or $\mathfrak{P}$ is already a fundamental domain, as it contains one. So let's suppose we are in the first case.

Then $\mathfrak{P}' \subsetneq \mathfrak{P}$, where $\mathfrak{P}'$ is the polytope defined by the inequalities

$$\mathrm{fac}(\mathfrak{P}) \cup \{H_x(x^g), H_x(x^{g^{-1}})\}.$$

In particular, $\mathfrak{P}'$ is smaller than $\mathfrak{P}$ and we still have $\mathfrak{P} \supseteq D_G(x)$. But as $|G/T| < \infty$, this process stops after finitely many steps.

The polytope $\tilde{\mathfrak{P}}$ in the final step has the property that $\tilde{\mathfrak{P}} \cap \tilde{\mathfrak{P}}^g$ has Lebesgue measure 0 for all $g \in G$. Hence $\tilde{\mathfrak{P}} = \mathrm{D}_G(x)$. $\square$

5. Building a Resolution

Let G be a crystallographic group acting naturally on $\mathbb{R}^n$ and $\mathfrak{P} \subseteq \mathbb{R}^n$ a convex polytope that is a fundamental domain for G (the Dirichlet domain $\mathrm{D}_G(x)$, for example, as it is convex by definition). From the natural inclusion of $(i-1)$-faces into i-faces of $\mathfrak{P}$, we get mappings

$$\begin{aligned} \partial_i\colon \quad & \mathfrak{P}_i \to \mathrm{Pow}(\mathfrak{P}_{i-1}) \\ & p \mapsto \{q \in \mathfrak{P}_{i-1} \mid q \in p\}, \end{aligned} \tag{1}$$

where $\mathrm{Pow}(\mathfrak{P}_{i-1})$ denotes the power set of $\mathfrak{P}_{i-1}$. These mappings extend naturally to mappings of free $\mathbb{Z}$-modules using the injection $\mathrm{Pow}(\mathfrak{P}_i) \to \mathbb{Z}\mathfrak{P}_i$, $S \mapsto \sum_{p \in S} p$ into the free $\mathbb{Z}$-module $\mathbb{Z}\mathfrak{P}_i$ generated by the i-faces of $\mathfrak{P}$.

Now let G be a Bieberbach group. This group acts freely on the faces of the tessellation $\mathfrak{P}^G$ (of any given dimension). So by identifying faces with group elements, the mappings ∂_i give rise to a sequence of mappings

$$\mathfrak{P}_n^G \xrightarrow{\partial_n} \mathfrak{P}_{n-1}^G \to \cdots \xrightarrow{\partial_1} \mathfrak{P}_0^G. \tag{2}$$

Note that the $\mathfrak{P}_i^G$ are free $\mathbb{Z}G$-modules whose $\mathbb{Z}G$-ranks correspond to the numbers of G-orbits on $\mathfrak{P}_i$. Note also that we did not use any information about the orientation of the i-faces of $\mathfrak{P}$. So the homomorphisms in sequence (2) do not satisfy $\partial_i \partial_{i-1} = 0$.

The aim of this section is to obtain an explicit description of $\mathbb{Z}G$-homomorphisms d_i from the homomorphisms ∂_i which satisfy $d_i d_{i-1} = 0$ and hence define a free $\mathbb{Z}G$-resolution $C_*(\mathfrak{P}_i, d_i)$. It suffices to describe d_i on the free $\mathbb{Z}G$-generators of $\mathfrak{P}_i$.

Notation. Let $f \in (\mathbb{Z}G)^{n_k}$ and $B = \{g_1, \dots, g_{n_k}\}$ be a basis of $(\mathbb{Z}G)^{n_k}$. We say that f is a *face* if its expression in B is of the form $f = g_i h$ for some $h \in G$ and $1 \le i \le n_k$.

For a face f in dimension k, we call the $(k-1)$-dimensional faces contained in f the *lines* of f and the $(k-2)$-dimensional faces in f will be called the *points* of f.

Observe that every face $f \in (\mathbb{Z}G)^{n_k}$ corresponds to a face of a polytope, so that each face in dimension $k-2$ is contained in exactly two faces in dimention $k-1$, so talking of lines and points does make sense.

In Algorithm 5, we calculate the image of a generator in dimension k under the boundary d_k from the boundary map in dimension $k-1$.

Also, by abuse of notation, we will identify the images of a face under the mappings ∂_i (from (2)) with a set of faces. Whenever there is no risk of confusion, we will write ∂ instead of ∂_i and d instead of d_i.

The general idea of Algorithm 5 is as follows:

(1) The boundary of the 1-faces is chosen such that for each 1-face one of the 0-faces contained in it has $+1$ as a sign and the other one has -1.
(2) For a given face b_i^k, calculate its points and lines. For every point of b_i^k find the lines of b_i^k incident with it (exactly two).

(3) Now assign signs to the lines $b_i^k\partial$ to get a boundary map d satisfying $d^2 = 0$. Start with any point p_0 and a line l containing p_0. Set the sign of l to $+1$.
(4) Let l_2 be the second line containing p. If the sign of p in $l_2 d$ is the same as in ld, set the sign of l_2 in $b_i^k d$ to -1. Note that d is already known on the lines by induction.
(5) Do the same with all other points in $l, l_1, \ldots$ Because of connectedness, all lines will eventually be processed.

The algorithm assumes an ordering on the generators of every module of the sequence (2). This is used to determine the orientation of the faces in dimension 1 and again to find the line l from which the sign- calculation is started. This is just done for convenience as it ensures the same output in every run and makes debugging easier.

6. Performance of the Implementation and Examples

The algorithms described in the previous sections are implemented in GAP as part of a package called HAPcryst [GAP07, Röd08a]. The GAP package HAP [Ell08] has methods for the calculation of resolutions that can be applied to crystallographic groups as well, so it is natural to compare the two.

It is, however, worth noting that the algorithms used by HAP are very different from the ones in HAPcryst and the returned resolutions can be quite different. The most obvious difference is that HAPcryst only calculates resolutions for Bieberbach groups whereas HAP handles a much larger class of groups. On the other hand, HAP calculates only finitely many terms of a resolution with infinitely many non-trivial terms while the resolution obtained from HAPcryst has only finitely many non-trivial terms (all of which are calculated).

We consider a couple of example cases to illustrate the performance of the two algorithms in different areas.

The resolutions calculated by HAPcryst depend on the calculation of a Dirichlet domain with respect to some point. In all cases, this point was chosen to be the origin.

Note that the times for HAPcryst are not CPU time, but "wall time" while the times for HAP are CPU time. It is also woth noting that HAPcryst does some calculations in random order which can lead to a slight variation in runtimes. All tests were done on an ordinary desktop computer.

All tests were done on a computer running SuSE Linux 10.3, GAP 4.4.10, HAP 1.8.8, HAPcryst 0.1.8, polymake 2.3, using one core of an Intel Core2 Duo CPU at 2.66GHz and 2GB of memory.

6.1. Low Dimensional Groups. In the low dimensional cases, HAPcryst produces quite some overhead because, in its current implementation, it calls polymake for convex hull computations. For the group `SpaceGroup(3,165)`, which has point group $\mathbb{Z}_6$, calculating a resolution with HAPcryst takes almost 2 seconds. `ResolutionAlmostCrystGroup` calculates a free resolution with 4 terms in 0.2 seconds and a resolution with 10 terms in about 1.2 seconds.

6.2. Small Point Groups: Hyperkähler Manifolds. As an example for groups with small point group, we use Hyperkähler manifolds of dimension 8. A Bieberbach group is called "Hyperkähler" if the point group is contained in

Algorithm 5 Boundary of Generator

```
Require: Bieberbach group G, for every module (ZG)^{n_k} from (2): a basis B^k =
    { b_i^k | 1 ≤ i ≤ n_k }, mappings ∂ from (2)
  procedure BoundaryOfGenerator(b_i^k, ∂)
    if k=1 then
      write b_i^1 ∂ := b_{ℓ1}^0 g_1 + b_{ℓ2}^0 g_2 with b_{ℓ1}^0, b_{ℓ2}^0 ∈ B^0 and g_1, g_2 ∈ G
      if ℓ1 > ℓ2 then
        Define b_i^1 d := b_{ℓ1}^0 g_1 − b_{ℓ2}^0 g_2
      else
        Define b_i^1 d := b_{ℓ1}^0 g_1 + b_{ℓ2}^0 g_2
      end if
    else
      Let L be the set of faces in b_i^k ∂                ▷ The "lines" to consider
      Choose l ∈ L and define λ := {l}
      π := ∅                                    ▷ The "points" that are already done
      repeat
        λ_n := ∅
        for l ∈ λ do
          P := l∂ − π
          for p ∈ P do
            Add p to π
            Let s be the sign of p in ld
            Let l_2 ≠ l be the second (k − 1)-face in b_i^k ∂ containing p
            if l_2 ∈ L then
              Remove l_2 from L
              Let s_2 be the sign of p in l_2 d
              if s = s_2 then
                Define the sign of l_2 in b_i^k d to be −1
                Add −l_2 to λ_n
              else
                Define the sign of l_2 in b_i^k d to be +1
                Add l_2 to λ_n
              end if
            end if
          end for
        end for
        overwrite λ with the value of λ_n
      until λ = ∅
    end if
    return b_i^k d
  end procedure
```

$\mathrm{Sp}(n)$ (see [DM05]). In dimension 8, the point groups of these Bieberbach groups are cyclic of order $2, 4, 3$, and 6 (in cases 1–4, 5–7, 8–10, and 11, enumeration as in [DM05]).

In [DM05], Dotti and Miatello calculated the Poincaré polynomials and the real cohomolgy rings of these manifolds.

	timing h:mm:ss			
G	HAPcryst	HAP(9)	HAP(5)	dmin – dmax
1	0:00:07	0:00:38	0:00:04	1 – 70
2	0:03:55	0:01:22	0:00:07	18 – 601
3	0:06:00	0:01:43	0:00:09	27 – 1 031
4	0:14:21	0:01:51	0:00:12	34 – 1 750
5	0:00:08	0:01:01	0:00:05	1 – 70
6	0:19:16	0:10:57	0:00:26	45 – 1 937
7	28 hours	12:33:10	0:05:34	142 – 22 122
8	0:41:10	0:48:39	0:00:59	44 – 2 689
9	1:39:57	1:34:19	0:01:38	68 – 4 473
10	0:00:26	0:01:42	0:00:06	4 – 155
11	0:00:31	0:06:18	0:00:12	4 – 155

TABLE 1. Results for Resolutions of 8-dimensional Hyperkähler Manifolds

We calculate nine terms of a resolution using HAP (Table 1). As Hyperkähler manifolds are orientable, the corresponding groups are Poincaré duality groups and hence five terms of a resolution are enough to obtain all (additive) homology and cohomology groups. So the timing for calculating five terms in HAP is added as well. The minimum and maximum of the dimensions of the modules of the resolution HAPcryst calculated are given in the last column (where the minimum is taken over the terms 0 to 7, the 8th term is one-dimensional by construction). The dimensions of HAP's resolutions are 1, 9, 37, 93, 163, 219, 247, 255, 256 in all cases.

6.3. Large Point Groups in Dimension 7: Hantzsche-Wendt Groups. We look at the orientable Hantzsche-Wendt groups in dimension 7 for an example of groups with large point group. These are Bieberbach groups with an elementary abelian point group of order 2^6. Up to isomorphism, there are 64 groups of this type [MR99].

Calculating resolutions with HAPcryst for all of these 62 groups takes about 41 minutes in total. The pointwise minimum and maximum of the module-dimensions in the resulting resolutions are (3, 33, 113, 183, 157, 71, 15) and (3, 142, 490, 742, 630, 280, 42), and for both bounds there is actually a group with a resolution having these dimensions.

From these resolutions, the integral homology groups can be calculated by HAP. The results of this calculation are listed in Table 2 (enumeration as in [MR99]).

HAP is able to calculate resolutions of length 3 for all 62 groups in 26 minutes while calculating four terms takes many hours for a single group.

6.4. The General Case in Dimensions 5 and 6. The Bieberbach groups up to dimension 6 are known [CS01] and available in various electronic formats from the website of the CARAT project [CAR].

There are 1060 Bieberbach groups in dimension 5 and calculating free resolutions for all of them with HAPcryst takes just under one hour in total. On the other hand, HAP needs 41 hours (median: 2.5 seconds) to calculate resolutions with only four terms for all groups.

Now consider an example of a Bieberbach group in dimension 6 with large non-abelian point group. Up to isomorphism, there is exactly one 6-dimensional

Group number [MR99]	H_0	H_1	H_2	H_3	H_4	H_5	H_6	H_7
1, 30, 44, 60	$\mathbb{Z}$	$\mathbb{Z}_2^6$	$\mathbb{Z}_2^8$	$\mathbb{Z}_4^6$	$\mathbb{Z}_2^8$	$\mathbb{Z}_2^6$	0	$\mathbb{Z}$
2, 3, 5, 6, 11, 13, 15, 17, 23, 24, 35, 39	$\mathbb{Z}$	$\mathbb{Z}_2^6$	$\mathbb{Z}_2^9$	$\mathbb{Z}_4^2 \oplus \mathbb{Z}_2^{10}$	$\mathbb{Z}_2^9$	$\mathbb{Z}_2^6$	0	$\mathbb{Z}$
4, 7, 10, 16, 19, 20, 22, 25, 27, 31, 32, 33, 34, 36, 37, 38, 40, 42, 43, 47, 48, 50, 52, 53, 55, 56, 57, 58, 59, 61, 62	$\mathbb{Z}$	$\mathbb{Z}_2^6$	$\mathbb{Z}_2^8$	$\mathbb{Z}_4^4 \oplus \mathbb{Z}_2^4$	$\mathbb{Z}_2^8$	$\mathbb{Z}_2^6$	0	$\mathbb{Z}$
8, 9, 12, 14, 18, 21, 26, 28, 29, 41, 45, 46, 49, 51, 54	$\mathbb{Z}$	$\mathbb{Z}_2^6$	$\mathbb{Z}_2^8$	$\mathbb{Z}_4^2 \oplus \mathbb{Z}_2^8$	$\mathbb{Z}_2^8$	$\mathbb{Z}_2^6$	0	$\mathbb{Z}$

TABLE 2. Integral Homology of 7-Dimensional Hantzsche-Wendt Manifolds

Bieberbach group with point group $\mathbb{Z}_6 \times \mathrm{Alt}_4$. Using HAPcryst, we get a free resolution with modules of dimension 20, 102, 194, 176, 79, 16, 1 in 30 seconds of which 5 seconds are spent for convex hull computations. HAP calculates 3 terms of a free resolution in 6 minutes and 4 terms in 11 days (dimensions $1, 8, 31, 81, 169$).

Calculating free resolutions for the 176 Bieberbach groups in dimension 6 which have a point group of size at least 36 takes a little more than 12 hours (average per group 4 minutes, median 14 seconds).

7. Concluding Remarks

The geometric approach described in this paper provides a generic method for calculating free resolutions for Bieberbach groups. In contrast to existing algebraic methods, the geometric method finds a resolution with only finitely many non-trivial terms and calculates all of them.

We have shown in Section 6.4 that the method can be used to calculate free resolutions of groups of dimension 5 (and probably 6) in reasonable time. Experiments suggest that calculations in higher dimensional groups are possible at least in cases where the group is "well-behaved" (i.e. has a Dirichlet domain with not too many faces). As seen in 6.1 and 6.2, the use of the algebraic methods of HAP can, however, be much faster than the geometric approach for groups of low dimension or small point group when only few terms of a resolution are needed. This is the case when calculating the homology of groups for Bieberbach groups with Poincaré duality, for example.

7.1. Possible Extensions and Further Work. The construction of a free resolution as a geometric object suggests a number of further applications. It seems possible to calculate a contracting homotopy on this resolution in a geometric way. This could then be used to calculate the cohomology rings for Bieberbach groups. Another possibility is the construction of resolutions of manifolds from others, if the new manifold is constructed in terms of fundamental domains (this applies to direct sums, for example).

While experiments suggest that for groups given in the notation of GAP's crystallographic groups library and the CARAT catalogue [CAR], the origin is a good choice, it would be very desirable to have a method to find a "suitable" starting point for resolution calculations. Suitability here seems to be mainly determined

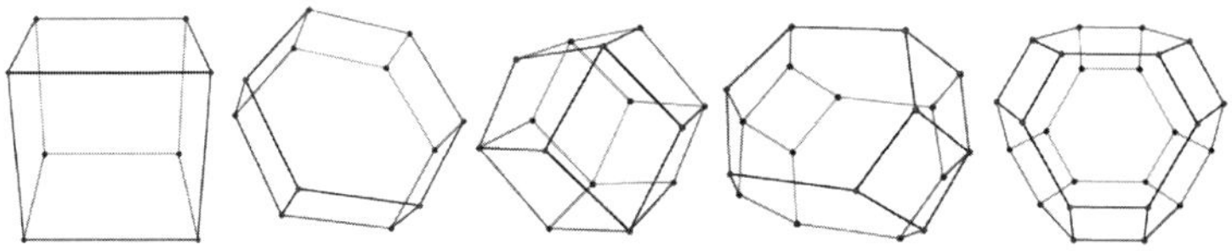

FIGURE 1. Combinatorial types of Dirichlet domains for group IT_4

by the number of faces of the corresponding Dirichlet domain. As the complexity of the algorithms we described depends mainly on the number of faces of the Dirichlet domain, starting points which induce Dirichlet domains with few faces are more suitable than those for which the Dirichlet domain has many faces.

EXAMPLE (Some Dirichlet domains in dimension 3). Experiments show that there are at least 136 combinatorial types of Dirichlet domains for the 10 Bieberbach groups in dimension 3. Consider the following small example:

The group

$$IT_4 := \left\langle \begin{pmatrix} -1 & 0 & 0 & 0 \\ 0 & 1 & 0 & 0 \\ 0 & 0 & -1 & 0 \\ 0 & \frac{1}{2} & 0 & 1 \end{pmatrix}, \begin{pmatrix} 1 & 0 & 0 & 0 \\ 0 & 1 & 0 & 0 \\ 0 & 0 & 1 & 0 \\ 1 & 0 & 0 & 1 \end{pmatrix}, \begin{pmatrix} 1 & 0 & 0 & 0 \\ 0 & 1 & 0 & 0 \\ 0 & 0 & 1 & 0 \\ 0 & 0 & 1 & 1 \end{pmatrix} \right\rangle$$

acts as a 3-dimensional Bieberbach group on the space of affine row-vectors of the form $(x_1, x_2, x_3, 1)$ by multiplication from the right. Experiments show that this group admits at least five different combinatorial types of Dirichlet domains (see Figure 1 for illustrations).

Figure 2 shows the five combinatorial type of $\mathrm{D}_G((x, 0, y))$ for $0 \leq x, y < 1$. Every shade of gray represents one combinatorial type. There are dark-grey horizontal and vertical lines, middle-grey and white areas and very light gray points where the dark-grey lines intersect with each other or the black circles – this is probably hard to see in print. Note that it is enough to know $\mathrm{D}_G(v)$ in this region because of symmetry. For more geometric models of Dirichlet domains for 3-dimensional Bieberbach groups see [Röd].

8. Acknowledgements

The author would like to thank Graham Ellis for helpful suggestions, discussions, and for pointing out this problem.

References

[AGPP08] Alejandro Adem, Jianquan Ge, Jianzhong Pan, and Nansen Petrosyan, *Compatible actions and cohomology of crystallographic groups*, J. Algebra **320** (2008), no. 1, 341–353. MR 2417992

[Bie11] L. Bieberbach, *Über die Bewegungsgruppen der euklidischen Räume. (Erste Abh.)*, Math. Ann. **70** (1911), 297–336 (German).

[Bie12] L. Bieberbach, *Über die Bewegungsgruppen der euklidischen Räume. (Zweite Abh.) Die Gruppen mit einem endlichen Fundamentalbereich.*, Math. Ann. **72** (1912), 400–412 (German).

[CAR] W. Plesken, *Torsion free space group database.* `http://wwwb.math.rwth-aachen.de/carat/bieberbach.html`.

[CS01] Carlos Cid and Tilman Schulz, *Computation of five- and six-dimensional Bieberbach groups*, Experiment. Math. **10** (2001), no. 1, 109–115. MR **1822856 (2002m:**20082)

[CV65] L. S. Charlap and A. T. Vasquez, *Compact flat Riemannian manifolds. II. The cohomology of Z_p-manifolds*, Amer. J. Math. **87** (1965), 551–563. MR 0189064 (32 #6491)

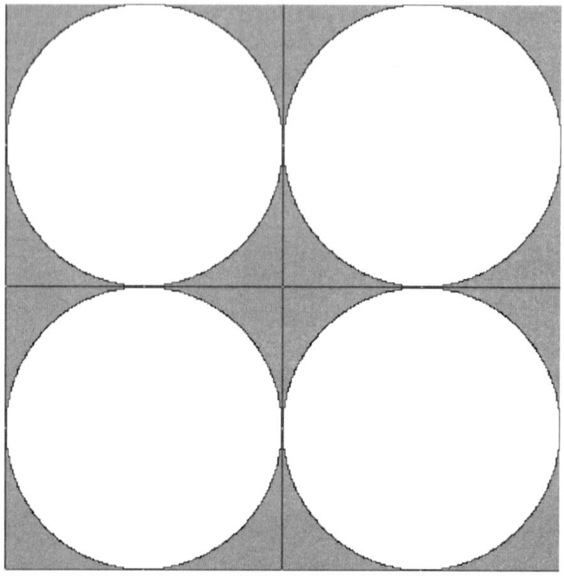

FIGURE 2. Location of combinatorial types in $(x, 0, z)$, $0 \leq x, z < 1$

[DM05] I. G. Dotti and R. J. Miatello, *On the cohomology ring of flat manifolds with a special structure*, Rev. Un. Mat. Argentina **46** (2005), no. 2, 133–147 (2006). MR **2281677 (2008g:**53048)

[Ell08] Graham Ellis, *HAP, version 1.8.6*, 2008. available at `http://hamilton.nuigalway.ie/Hap/www/index.html`.

[Eng81] Peter Engel, *Über Wirkungsbereichsteilungen von kubischer Symmetrie*, Z. Krist. **154** (1981), no. 3-4, 199–215 (German, with English summary). MR **598811 (82i:**52025)

[GAP07] The GAP Group, *GAP – Groups, Algorithms, and Programming, Version 4.4.10*, 2007. `http://www.gap-system.org`.

[GJ] Ewgenij Gawrilow and Michael Joswig, *polymake*. `http://www.math.tu-berlin.de/polymake/`.

[MR99] R. J. Miatello and J. P. Rossetti, *Hantzsche-Wendt manifolds of dimension* 7, Differential geometry and applications (Brno, 1998), Masaryk Univ., Brno, 1999, pp. 379–390. MR **1708926 (2001c:**58032)

[Röd] Marc Röder, *Geometric models for Dirichlet domains of* 3*-dimensional Bieberbach groups*. Submitted to `http://www.eg-models.de`.

[Röd08a] Marc Röder, *HAPcryst, version 0.1.6*, 2008. `http://www.gap-system.org/Packages/hapcryst.html`.

[Röd08b] Marc Röder, *polymaking, version 0.7.3*, 2008. `http://www.gap-system.org/Packages/polymaking.html`.

[Szc06] Andrzej Szczepański, *Problems on Bieberbach groups and flat manifolds*, Geom. Dedicata **120** (2006), 111–118. MR **2252897 (2007k:**20109)

E-mail address: `roeder.marc@gmail.com`

DEPARTMENT OF MATHEMATICS, NATIONAL UNIVERSITY OF IRELAND, GALWAY

Contemporary Mathematics
Volume **511**, 2010

Nonabelian tensor product of soluble minimax groups

Francesco Russo

ABSTRACT. In this paper we prove two results. The first is that the nonabelian tensor product of two Chernikov groups is a Chernikov group. The second is that the nonabelian tensor product of two soluble minimax groups is a soluble minimax group.

1. Introduction and Statement of Results

Let M and N be groups acting upon each other in a compatible way, that is,

$$ {}^{{}^{m}n}m' = {}^{m}({}^{n}({}^{m^{-1}}m')), \tag{1.1} $$

$$ {}^{{}^{n}m}n' = {}^{n}({}^{m}({}^{n^{-1}}n')), \tag{1.2} $$

for $m, m' \in M$ and $n, n' \in N$, and acting upon themselves by conjugation. The *nonabelian tensor product* $M \otimes N$ of M and N is a group generated by the symbols $m \otimes n$, where $m \in M$ and $n \in N$, with defining relations

$$ mm' \otimes n = ({}^{m}m' \otimes {}^{m}n)(m \otimes n), \tag{1.3} $$

$$ m \otimes nn' = (m \otimes n)({}^{n}m \otimes {}^{n}n'), \tag{1.4} $$

where $m, m' \in M$ and $n, n' \in N$. When $N = M$ and all actions are conjugations, $M \otimes M$ is called the *nonabelian tensor square* of M. Classical results on the nonabelian tensor product can be found in [1, 2, 4, 10]. Some of the structural results on the nonabelian tensor products of groups can be found in [6–8, 12].

As it turns out, many abstract group theoretical properties are closed with respect to forming the nonabelian tensor product. For instance, this is true for finite groups [4], nilpotent groups [12], soluble groups [8, 12], polycyclic groups [6], and locally finite groups [7]. Some topological properties are also closed with respect to forming the nonabelian tensor product. See [2].

We recall two notions respectively from [5, (1), p.85] and [3, §1.6 and 1.7]. A group G is said to have *finite torsion-free rank* if it has a finite subnormal series such that each non-torsion factor is infinite cyclic. The number of infinite cyclic

---

2000 *Mathematics Subject Classification.* Primary: 20J99; Secondary: 20F18.

Key words and phrases. Nonabelian tensor product, soluble minimax groups.

The author thanks GNSAGA of Indam for the financial support, but specially the editor and the referee for their excellent work.

factors is independent of the series. It is called the torsion-free rank of G and denoted by $r_0(G)$. Polycyclic groups have finite torsion-free rank.

A group G is said to have *finite Prüfer rank* $r(G)$ if every finitely generated subgroup can be generated by $r(G)$ elements and $r(G)$ is the least such a integer. Chernikov groups have finite Prüfer rank. See [3, Corollary 1.6.24]. In general, $r_0(G) \leq r(G)$. Obviously, a Chernikov group G has $r_0(G) = 0$. But, the infinite dihedral group D_∞ is a polycyclic group with $r_0(D_\infty) = 1$ and $r(D_\infty) = 2$.

A group is called *minimax* if it has a series (of finite length) for which each factor satisfies *max* or *min*. A *soluble minimax* group has both finite torsion-free rank and finite Prüfer rank. Recall that each torsion soluble minimax group is a Chernikov group. Furthermore, the class of Chernikov groups is closed with respect to forming subgroups, quotients and finite extensions of its members. The same happens for the class of soluble minimax groups. See [5, Chapter 5] or [3].

We prove the following two results.

THEOREM 1.1. *If M and N are two Chernikov groups acting compatibly upon each other, then the group $M \otimes N$ is a Chernikov group. Furthermore, $r(M \otimes N)$ is bounded by a function depending only on $r(M)$ and $r(N)$.*

THEOREM 1.2. *If M and N are two soluble minimax groups acting compatibly upon each other, then the group $M \otimes N$ is a soluble minimax group. Furthermore,*

(i) *$r_0(M \otimes N)$ is bounded by a function depending only on $r_0(M)$ and $r_0(N)$;*
(ii) *$r(M \otimes N)$ is bounded by a function depending only on $r(M)$ and $r(N)$.*

Obviously, (i) and (ii) of Theorem 1.2 are equivalent for polycyclic groups. Note that Theorem 1.1 agrees with [7, Theorem 1], because Chernikov groups are locally finite groups of infinite exponent, when they are not finite. Theorem 1.2 extends [6, Theorem] and [7, Theorem 1] to the wider class of soluble minimax groups. Let $m \neq \pm 1$ be a non-zero integer and write π for the set of prime divisors of m. Consider the group $G = \langle x, a \ : \ a^m = a^x \rangle = \langle x \rangle \ltimes \mathbb{Q}_\pi$, where $\langle a \rangle^G \simeq \mathbb{Q}_\pi$ is the additive group of π-adic rationals and x has infinite order, inducing in $\mathbb{Q}_\pi$ the automorphism $b \mapsto mb$. Then G is a metabelian group, finitely presented, but neither a polycyclic nor a Chernikov group. Of course, G is a soluble minimax group. Note that $r_0(G) = r(G) = 2$. See [5, pp. 240–244] for details. We have a concrete situation in which Theorem 1.2 generalizes Theorem 1.1 and [6, Theorem].

2. Proofs

We mention some properties of the nonabelian tensor product of groups which can be found in [1,2]. Whenever the groups M and N act trivially on each other, their tensor product $M \otimes N$ is isomorphic to the usual tensor product $M^{\mathrm{ab}} \otimes_{\mathbb{Z}} N^{\mathrm{ab}}$. If M and N act compatibly on each other, then these actions induce an action of the free product $M * N$ on $M \otimes N$ given by ${}^p(m \otimes n) = {}^pm \otimes {}^pn$ for all $m \in M$, $n \in N$, and $p \in M * N$.

Let G and H be groups. A *crossed module* is a group homomorphism $\mu : G \longrightarrow H$ together with an action of H on G satisfying $\mu({}^hg) = h\mu(g)h^{-1}$ and ${}^{\mu(g)}g' = gg'g^{-1}$ for all $g, g' \in G$ and $h \in H$. It follows easily from the definitions that im i is normal in H and $\ker \mu$ is central in G.

In the beginning, we quote some homological results that will be used in our proof of Theorem 1.1.

By [5, 5.1.4], an abelian group A is minimax if and only if it has a finitely generated subgroup X such that A/X satisfies *min*. Because of the structure of abelian groups with *min*, the subgroup X may be chosen so that A/X is divisible, and hence a direct sum of finitely many quasicyclic groups. See [5, pp. 86–87]. The finite set $\pi(A/X)$ of the prime divisors of the orders of the elements in A/X is independent of X. This set of primes is called the *spectrum* of A and denoted by $Sp(A)$. Having a soluble minimax group Q, we may consider the spectrum of each factor of the derived series of Q. The (finite) set of all these primes is called the *spectrum* of Q and denoted by $Sp(Q)$.

If p is a prime, the *finite p-rank* of an abelian group A is the finite Prüfer rank of the Sylow p-subgroup of A. See [5, Chapter 5] or [3, Defintion 1.6.22].

LEMMA 2.1. *Let Q be a soluble minimax group and A be a Q-module. Assume that A is a torsion abelian group of finite p-rank with no nontrivial p-elements for any $p \in Sp(Q)$. Then*

(i) *$H^n(Q, A)$ is a torsion abelian group of finite p-rank for all $p \in Sp(Q)$ and for all $n \geq 1$;*

(ii) *$H_n(Q, A)$ is a torsion abelian group of finite p-rank for all $p \in Sp(Q)$ and for all $n \geq 0$.*

We will use the following consequence of Lemma 2.1.

COROLLARY 2.2. *Let Q be a soluble minimax group and A be a Q-module. Assume that A is an abelian p-group of finite p-rank with no nontrivial p-elements for any $p \in Sp(Q)$. Then*

(i) *$H^n(Q, A)$ is an abelian p-group of finite p-rank for all $p \in Sp(Q)$ and for all $n \geq 1$;*

(ii) *$H_n(Q, A)$ is an abelian p-group of finite p-rank for all $p \in Sp(Q)$ and for all $n \geq 0$.*

PROOF OF THEOREM 1.1. This proof follows closely the proof of the main result of [6].

Assume that M and N are Chernikov groups, acting compatibly upon each other. Let G be the Peiffer product of M and N. See [13]. More precisely, let $G = M * N/IJ$, where I and J are the normal closures in $M * N$ of

$$\langle {}^n m n m^{-1} n^{-1} \ : m \in M, n \in N\rangle \tag{2.1}$$

and

$$\langle {}^m n m n^{-1} m^{-1} \ : m \in M, n \in N\rangle, \tag{2.2}$$

respectively. Note that G is an image of $N \ltimes M$, hence G is a Chernikov group. Let $\mu\colon M \to G$ and $\nu\colon N \to G$ be the natural maps and denote $\overline{M} = \mu(M)$ and $\overline{N} = \nu(N)$. Then $\overline{M}$ and $\overline{N}$ are normal subgroups of G and $G = \overline{M}\,\overline{N}$. Since $\mu\colon M \to G$ and $\nu\colon N \to G$ are crossed modules, $\ker\mu$ is central in M and $\ker\nu$ is central in N. An argument as in [1, Proposition 9] shows that the following sequence is exact:

$$(M \otimes \ker\nu) \times (\ker\mu \otimes N) \xrightarrow{\ i\ } M \otimes N \longrightarrow \overline{M} \otimes \overline{N} \longrightarrow 1, \tag{2.3}$$

where i is induced by $(m \otimes n', m' \otimes n) \mapsto (m \otimes n')(m' \otimes n)$. It can be seen that $\operatorname{im} i$ is a central subgroup of $M \otimes N$. Since ${}^n m = {}^{\nu(m)} m$ and ${}^m n = {}^{\mu(m)} n$ for all $m \in M$ and $n \in N$, $\ker\mu$ and $\ker\nu$ act trivially on N and M, respectively.

Therefore,

$$M \otimes \ker \nu \simeq M^{\mathrm{ab}} \otimes_{\mathbb{Z}} \ker \nu^{\mathrm{ab}} = M^{\mathrm{ab}} \otimes_{\mathbb{Z}} \ker \nu \tag{2.4}$$

and

$$\ker \mu \otimes N \simeq \ker \mu^{\mathrm{ab}} \otimes_{\mathbb{Z}} N^{\mathrm{ab}} = \ker \mu \otimes_{\mathbb{Z}} N^{\mathrm{ab}}. \tag{2.5}$$

In particular, $M \otimes \ker \nu$ is a finite group, because it is isomorphic to the usual tensor product of two abelian groups with *min*. This is an elementary fact, which can be found in [11]. Analogously, $\ker \mu \otimes N$ is a finite group. It follows that $\mathrm{im}\, i$ is a finite group, because it is a quotient of the finite group $(M \otimes \ker \nu) \times (\ker \mu \otimes N)$.

Since $\overline{M} \otimes \overline{N} \simeq (M \otimes N)/\mathrm{im}\, i$, it is enough to prove that $\overline{M} \otimes \overline{N}$ is a Chernikov group. We may work with $\overline{M}$ instead of M and with $\overline{N}$ instead of N in order to get our result. Then there is no loss of generality in assuming that M and N are normal subgroups of G, $G = MN$, and all actions are induced by conjugation in G. Define

$$M \wedge N = (M \otimes N)/D, \tag{2.6}$$

where $D = \langle x \otimes x : x \in M \cap N \rangle$, and let K be the kernel of the commutator map $M \wedge N \to [M, N]$. In order to finish our proof, it suffices to prove that K and D are Chernikov groups. By [2, Theorem 4.5], we have an exact sequence:

$$\cdots \longrightarrow H_3(G/M) \oplus H_3(G/N) \longrightarrow K \longrightarrow H_2(G) \longrightarrow \cdots . \tag{2.7}$$

Since G, G/M, and G/N are Chernikov groups, Corollary 2.2 implies that $H_2(G)$, $H_3(G/M)$, and $H_3(G/N)$ are abelian p-groups of finite p-rank. Then they satisfy [3, Lemma 1.6.25] and so they are abelian p-groups with *min*. On the other hand, K is an extension of the abelian p-group $H_3(G/M) \oplus H_3(G/N)$ with *min* by the abelian p-group $H_2(G)$ with *min*. Therefore, K is an abelian p-group with *min*. Then K is a Chernikov group. See [5, Chapter 5] or [3, Theorem 6.1.15].

There exists a well-defined homomorphism of groups

$$\psi : \Gamma((M \cap N)^{\mathrm{ab}}) \to (M \cap N) \otimes (M \cap N), \tag{2.8}$$

where ψ is the Whitehead function and Γ is the Whitehead quadratic functor. See [1, p. 181] or [2]. Furthermore, $\mathrm{im}\, \psi = D$. Since $(M \cap N)^{\mathrm{ab}}$ is an abelian group with *min*, it is the direct product of finitely many (finite) cyclic groups and (periodic) quasicyclic groups. But $\Gamma((M \cap N)^{\mathrm{ab}})$ is an abelian group with *min*, combining [13, (B) and (C), pp. 62–63] and [1, (a), p. 181]. We conclude that D is an abelian group with *min*, since it is a homomorphic image of $\Gamma((M \cap N)^{\mathrm{ab}})$ via ψ.

More generally, the results in [13, Chapters 5 and 9] show that $\Gamma((M \cap N)^{\mathrm{ab}})$ is an abelian minimax group whenever $(M \cap N)^{\mathrm{ab}}$ is an abelian minimax group.

The first part of the result follows.

We obtained $M \otimes N$ as an extension of K by D. Since the finite Prüfer rank is additive on the extensions [3, Lemma 1.6.23], $r(M \otimes N) \leq r(K) + r(D)$. The construction of K shows that $r(K)$ depends only on $r(M)$ and $r(N)$. The same is true for $r(D)$. Therefore, the same is true for $r(K) + r(D)$. The result follows completely. □

Proof of Theorem 1.2. We may use the closure properties of the class of soluble minimax groups and follow the proof of Theorem 1.1. □

References

[1] R. Brown, D.L. Johnson, and E.F. Robertson, *Some computations of non-abelian tensor products of groups*, J. Algebra **111** (1987), 177–202.
[2] R. Brown and J.L. Loday, *Van Kampen theorems for diagrams of spaces*, Topology **26** (1987), 311–335.
[3] M.D. Dixon, *Sylow theory, formations and Fitting classes in locally finite groups*, World Scientific, 1994.
[4] G. Ellis, *The nonabelian tensor product of finite groups is finite*, J. Algebra **111** (1987), 203–205.
[5] J.C. Lennox and D.J.S. Robinson, *The Theory of Infinite Soluble Groups*, Clarendon Press, Oxford, 2004.
[6] P. Moravec, *The nonabelian tensor product of polycyclic groups is polycyclic*, J. Group Theory **10** (2007), 795–798.
[7] P. Moravec, *The exponents of nonabelian tensor products of groups*, J. Pure Appl. Algebra **212** (2008), 1840–1848.
[8] I. Nokaoka, *Non-abelian tensor product of solvable groups*, J. Group Theory **3** (2000), 157–167.
[9] D.J.S. Robinson, *Applications of Cohomology to the Theory of Groups*, Proceedings, Groups St. Andrews 1981, LMS Lecture Note Series, vol. 71, Cambridge University Press, 1982, pp. 46–80.
[10] N.R. Rocco, *On a construction related to the nonabelian tensor square of a group*, Bol. Soc. Brasil Mat. **22** (1991), 63–79.
[11] J. Rotman, *An Introduction to Homological Algebra*, Academic Press, 1979.
[12] M.P. Visscher, *On the nilpotency class and solvability length of a nonabelian tensor product of groups*, Arch. Math. (Basel) **73** (1999), 161–171.
[13] J.H.C. Whitehead, *A certain exact sequence*, Ann. of Math. **52** (1950), 51–110.

Current address: Department of Mathematics, Faculty of Mathematical Sciences, University of Naples Federico II, via Cinthia, I-80126, Naples, Italy

E-mail address: francesco.russo@dma.unina.it

Contemporary Mathematics
Volume **511**, 2010

Finite groups have short rewriting systems

Jack Schmidt

ABSTRACT. All but finitely many finite groups have a rewriting system whose number of rules is less than the square root of the group order. A natural rewriting system is given for finite groups of Lie type.

Rewriting systems are widely used in commutative algebra and more general algebras over fields under the name of Gröbner bases. Their use in infinite group theory flourished briefly, especially their applications to homological algebra [Ani86, Squ87, Gro90, Gro97], but gradually gave way to finite automata and automatic groups. Their use for finite groups has been hampered by the lack of a practical method to find complete rewriting systems for finite groups, and the worry that small complete rewriting systems may simply not exist. The purpose of this work is to outline a general procedure for finding rewriting systems with relatively few rules compared to the group order. The number of rules proposed for a finite group will be roughly the sum of the number of rules proposed for its composition factors, and so the simple groups will be the main concern.

DEFINITION 1. For a finite set X, the free monoid on X is denoted X^*. A **complete rewriting system** for a group G is a pair (X, R) where $X \subseteq G$ and $R \subseteq X^* \times X^*$ subject to a list of conditions. For the following auxiliary definitions:

- $f : X^* \to G$ the unique monoid homomorphism fixing X
- $\to$ the relation $\{(axb, ayb) \in X^* \times X^* : (x, y) \in R, a, b \in X^*\}$,
- $\twoheadrightarrow$ the reflexive, transitive closure of $\to$, and
- $\equiv$ the reflexive, symmetric, transitive closure of $\to$

one must have:

- f is surjective
- $f(a) = f(b)$ if and only if $a \equiv b$
- $\to$ is noetherian; that is, every chain $a_1 \to a_2 \to \dots$ is finite
- if $a \equiv b$, then there is some $c \in X^*$ with $a \twoheadrightarrow c$ and $b \twoheadrightarrow c$

The elements of X are called **generators** and the elements of R are called **rules**. If $a \in X^*$ is such that $a \twoheadrightarrow b$ implies $a = b$, then a is called **final**. A complete rewriting system is **strongly reduced** if for $x \in X$, $a, b \in X^*$, x is final, and if $(ax, b) \in R$ or $(xa, b) \in R$ then both a and b are final. "Strongly reduced, complete

2010 *Mathematics Subject Classification.* Primary 20F05, 20D06.
Key words and phrases. Rewriting systems, finite groups, groups of Lie type.

rewriting system" is abbreviated as **rws**. A rws of a finite group is called **short** if $|R|^2 \leq |G|$. A rws is said to **exhibit** a subgroup H of G if $H = \langle Y \rangle$ for some subset $Y \subseteq X$, and to exhibit a subquotient H/K if it exhibits both H and K.

The main result is:

THEOREM A. *All but finitely many isomorphism classes of finite groups have a short rws exhibiting a composition series.*

The proof follows by reducing to simple groups in Proposition 8, which are then handled in Proposition 9. All groups considered will be finite unless specified otherwise.

Rewriting systems in group theory are discussed in [Coh93], in Chapter 8 of [HEO05], and in Chapter 9 of [Sim94].

1. Rewriting systems exhibiting a composition series

A rewriting system for a quotient group and the corresponding normal subgroup can be used to form a rewriting system for the group. This can be done effectively if elements of the group that happen to lie in the normal subgroup can be effectively rewritten as words in the generators of the normal subgroup. In particular, in the context of a composition tree [Ser06, NS06], rewriting systems can be lifted from leaves to the original group. Proofs of the results quoted here are deferred until section 1.1, but are usually elementary or well known.

The following is well known [Gro90, Gro97]:

LEMMA 1. *Suppose G is a group with normal subgroup N. If N has a rws on g_1 generators and r_1 rules, and if G/N has a rws on g_2 generators and r_2 rules, then G has a rws on $g_1 + g_2$ generators and $r_1 + r_2 + g_1 g_2$ rules.*

The lemma can be used inductively on a composition series to obtain a rws that refines a composition series and has simple formulas for its number of generators and number of rules in terms of its composition factors. In particular, notice that the number of generators and rules produced for G and for $G/N \times N$ are the same, so that without loss of generality, one may assume each group is a direct product of its composition factors. When this lemma is applied to finite soluble groups and a composition series, one gets the well known polycyclic rewriting systems:

LEMMA 2. *If G is a finite soluble group with composition series of length n, then G has a rws on n generators and $\binom{n+1}{2}$ rules. In particular, every finite soluble group of order at least 2^{14} has a short rws.*

The polycyclic rws are extremal in how few rules they have for a given number of generators:

LEMMA 3. *If G has a rws on r rules and n generators, then $n^2 \leq 2r$.*

This gives a convenient bound for the number of rules of a rws of a direct product:

LEMMA 4. *Let G_1, G_2 be groups and set $G_3 = G_1 \times G_2$. If the rws of G_i has g_i generators and r_i rules then $r_3 = r_1 + r_2 + g_1 g_2 \leq r_1 + r_2 + 2\sqrt{r_1 r_2} = (\sqrt{r_1} + \sqrt{r_2})^2$.*

This can then be used to maintain or produce short rws, as in the following three lemmas:

LEMMA 5. *If for* $i = 1, 2$, *the finite groups* G_i *have short rws and* $|G_2| \geq 16$, *then* $G_1 \times G_2$ *has a short rws.*

LEMMA 6. *If* G_1 *is solvable and* $|G_2| \geq 300$ *and* G_2 *has a short rws, then* $G_1 \times G_2$ *has a short rws.*

LEMMA 7. *For each non-abelian finite simple group* G, *there is a rws on two generators and with at most* $2|G|$ *rules. Hence* G^k *has a short rws for every* $k \geq 3$.

DEFINITION 2. Let $\mathcal{X}$ be a set of isomorphism class representatives of the finite non-abelian simple groups without a short rws.

Proposition 9 will show that $\mathcal{X}$ is finite, detailed calculations have shown it contains at most 22 groups, and conjecturally it is empty. Combining the previous lemmas, it is possible to avoid any bad rws from $\mathcal{X}$ by requiring that the group be large enough:

PROPOSITION 8. *If* G *is a finite group of order at least* $\max(300 \cdot 2^{14}, 4N_1^2) \cdot N_2^2$ *where* $N_1 = \sum\{|X_i| : X_i \in \mathcal{X}\}$ *and* $N_2 = \prod\{|X_i| : X_i \in \mathcal{X}\}$, *then* G *has a short rws.*

Assuming $\mathcal{X}$ is finite, this shows that all but finitely many finite groups have short rws.

1.1. Proofs. The proofs of the lemmas and propositions that are stated in the first Section are collected here.

LEMMA 1. *Suppose* G *is a group with normal subgroup* N. *If* N *has a rws on* g_1 *generators and* r_1 *rules, and if* G/N *has a rws on* g_2 *generators and* r_2 *rules, then* G *has a rws on* $g_1 + g_2$ *generators and* $r_1 + r_2 + g_1 g_2$ *rules.*

PROOF. Suppose G is a group with an epimorphism $\pi : G \to K$ and that (X, R) is a rws for K and that (Y, S) is a rws for the kernel of π. By renaming elements so that a generating set of K is identified with a set of preimages in G under π, one can assume G is chosen such that $X \subset G$, π fixes X, and $X \cap Y = \{\}$. Make the following definitions:

- Set $f_1 : Y^* \to \ker(\pi)$ to be the unique monoid homomorphism fixing Y
- Set $f_2 : X^* \to K$ and $f_3 : X^* \to G$ to be the unique monoid homomorphisms that fix X
- Set $Z = X \cup Y$ and consider X^* and Y^* as submonoids of Z^*
- Set $\tilde{R} = \{(g, hv) \in Z^* \times Z^* : g, h \in X^*, v \in Y^*, (g, h) \in R,$ $v = f_1^{-1}(f_3(h)^{-1} \cdot f_3(g))\}$
- Set $T = \{(yx, xv) \in Z^* \times Z^* : x \in X, y \in Y, v \in Y^*,$ $v = f_1^{-1}(f_3(x)^{-1} \cdot f_1(y) \cdot f_3(x))\}$

Then $(Z, \tilde{R} \cup S \cup T)$ is a rws for G, as is easily verified. □

LEMMA 2. *If* G *is a finite soluble group with composition series of length* n, *then* G *has a rws on* n *generators and* $\binom{n+1}{2}$ *rules. In particular, every finite soluble group of order at least* 2^{14} *has a short rws.*

PROOF. The previous lemma can be applied inductively to a composition series to get if each composition factor G_i has a rws with g_i generators and r_i rules, then G has a rws with $\sum g_i$ generators and $\sum r_i + \sum_{i \neq j} g_i g_j$ rules. In a composition series of the finite solvable group G, each composition factor is cyclic of some order p, so has a rws $(\{x\}, \{(x^p, 1)\})$ where x is a generator of the factor. One can thus take $g_i = r_i = 1$ for all i, to get a rws with n generators and $n + \binom{n}{2} = \binom{n+1}{2}$ rules. The last claim follows by comparing 2^n and $\binom{n+1}{2}^2$, with the former being greater for $n \geq 14$. □

LEMMA 3. *If G has a rws on r rules and n generators, then $n^2 \leq 2r$.*

PROOF. A complete rewriting system of a finite group must have certain rules for each generator and pair of generators. Let (X, R) be a rws for a finite group G. It suffices to show that $|R| \geq \binom{|X|+1}{2}$:

For every $x \in X$ there is a positive integer n, an element $a \in X^*$, and a rule $(x^n, a) \in R$: Otherwise the set of words $\{x, x^2, x^3, \dots\}$ is an infinite set of final words. Since G is finite, there are two distinct words, x^i and x^j that are equal in G. Hence there is some $c \in X^*$ with $x^i \twoheadrightarrow c$ and $x^j \twoheadrightarrow c$. However, x^i and x^j are final, so $x^i = c = x^j$, a contradiction.

For every distinct pair $x, y \in X$, there must be $i, j \in \{0, 1\}$, a non-negative integer n, an element $a \in X^*$, and $(y^i(xy)^n x^j, a) \in R$: Otherwise the set of words $\{xy, (xy)^2, (xy)^3, \dots\}$ is an infinite set of final words.

Note that a polycyclic rewriting system has only these $\binom{|X|+1}{2}$ rules, so the bound $|R| \geq \binom{|X|+1}{2}$ is sharp. It is more convenient to use the looser bound $|X|^2 \leq |R|$. □

LEMMA 4. *Let G_1, G_2 be groups and set $G_3 = G_1 \times G_2$. If the rws of G_i has g_i generators and r_i rules then $r_3 = r_1 + r_2 + g_1 g_2 \leq r_1 + r_2 + 2\sqrt{r_1 r_2} = (\sqrt{r_1} + \sqrt{r_2})^2$.*

PROOF. Set $G_3 = G_1 \times G_2$. Let G_i have a rws with g_i generators and r_i rules, choosing short rws for $i = 1, 2$ and the rws of Lemma 1 for $i = 3$. Then $r_3 = r_1 + r_2 + g_1 g_2$ by Lemma 1. Lemma 3 gives $g_i \leq \sqrt{2r_i}$ and so we have $r_3 \leq r_1 + r_2 + 2\sqrt{r_1 r_2} = (\sqrt{r_1} + \sqrt{r_2})^2$. □

LEMMA 5. *If for $i = 1, 2$, the finite groups G_i have short rws and $|G_2| \geq 16$, then $G_1 \times G_2$ has a short rws.*

PROOF. Set $G_3 = G_1 \times G_2$. Let G_i have a rws with g_i generators and r_i rules, choosing short rws for $i = 1, 2$ and the rws of Lemma 1 for $i = 3$. Let G_i be of order o_i. Then $r_3 = (\sqrt{r_1} + \sqrt{r_2})^2$ by Lemma 4.

First assume that $|G_1| \geq 16$, so that $\sqrt[4]{o_i} \geq 2$. Set $x = \max(2, \sqrt{r_1})$ and $y = \max(2, \sqrt{r_2})$. Then $x^4 \leq o_1$ and $y^4 \leq o_2$. Also $x + y \leq 2\max(x, y) \leq xy$ and so $r_3^2 \leq (x + y)^4 \leq (xy)^4 \leq o_1 o_2 = o_3$, as was to be shown.

Now assume that $o_1 < 16$, so $r_1 < 4$. Since $\binom{g_1+1}{2} \leq r_1$ and $\binom{3+1}{2} = 6 > r_1$, either $g_1 = 1$ and G_1 is cyclic with $r_1 = 1$, or $g_1 = 2$ and $r_1 = 3$. Both cases are now easily verified: Firstly, $r_3 = 1 + r_2 + g_2 \leq 1 + \sqrt{2}x + x^2$ where $x = \sqrt[4]{o_2}$, and one checks that $x^4 > 1 + x + x^2$ for $x \geq 2$. Secondly, $r_3 = 3 + r_2 + 2g_2 \leq 3 + 2\sqrt{2}x + x^2$, and again $x^4 > 3 + 2\sqrt{2}x + x^2$ for $x \geq 2$. □

LEMMA 6. *If G_1 is solvable and $|G_2| \geq 300$ and G_2 has a short rws, then $G_1 \times G_2$ has a short rws.*

PROOF. Set $G_3 = G_1 \times G_2$. Let G_i have a rws with g_i generators and r_i rules, choosing a polycyclic rws for G_1, a short rws for G_2, and the rws of Lemma 1 for G_3. Let G_i be of order o_i. Then

$$r_3 = r_1 + r_2 + g_1 g_2 = \frac{k(k+1)}{2} + r_2 + k g_2 \leq \binom{k+1}{2} + k\sqrt{2}x + x^2,$$

where k is the composition length of G_1 and $x = \sqrt[4]{o_2}$. We want to show that $\binom{k+1}{2} + k\sqrt{2}x + x^2 \leq 2^{k/2}x^2$, which is a quadratic inequality in x. One checks that for $k \geq 1$, the minimum positive x for which the inequality holds is decreasing in k, and that for $k = 1$, $x \geq \sqrt[4]{300}$ suffices. □

LEMMA 7. *For each non-abelian finite simple group G, there is a rws on two generators and with at most $2|G|$ rules. Hence G^k has a short rws for every $k \geq 3$.*

PROOF. Every finite simple group is two-generated, and if G has a rws (X, R), then $|R| \leq |X| \cdot |G|$, since there are $|G|$ final elements of X^* and so at most $|X| \cdot |G|$ left hand sides of rules.

Let G have a rws with $g = 2$ generators and $r = r_1 \leq 2|G|$ rules. We set $o = |G| \geq 60$, and set r_k to be the number of rules of G^k, using Lemma 1 inductively. One gets that $r_k = kr_1 + 4\binom{k}{2} \leq 2k|G| + 4\binom{k}{2}$.

Set $a_k = 2k|G|$ $b_k = 4\binom{k}{2}$ and $c_k = |G|^{k/2}$. Then for $k \geq 3$: $a_{k+1}/a_k = 1 + \frac{1}{k} \leq 2$, $b_{k+1}/b_k = 1 + \frac{2}{k-1} \leq 2$, $c_{k+1}/c_k = \sqrt{|G|}$. Hence $a_k + b_k \leq c_k$ for all $k \geq K$ follows from $a_K + b_K \leq c_K$ if $|G| \geq 4$.

However $r_3 = 3r + 12 \leq 6|G| + 12$ which must be less than $|G|^{3/2}$. Consider $r_3^2 = 36o^2 + 144o + 144$ which is less then o^3 when $o \geq 40$. □

PROPOSITION 8. *If G is a finite group of order at least* $\max(300 \cdot 2^{14}, 4N_1^2) \cdot N_2^2$ *where* $N_1 = \sum\{|X_i| : X_i \in \mathcal{X}\}$ *and* $N_2 = \prod\{|X_i| : X_i \in \mathcal{X}\}$, *then G has a short rws.*

PROOF. Without loss of generality, G is the direct product of its composition factors. Write $G = A \times B \times C \times D$ where A is the product of the abelian composition factors, B is the product of the composition factors that appear in X but have multiplicity at most 2 in G, C is the product of the composition factors appearing in X and with multiplicity at least 3 in G, and D is the product of the non-abelian composition factors not in X.

D has a short rws: If $D = 1$, then there is nothing to prove, so assume D is non-trivial. Write $D = \prod_{i=1}^n D_i$ where each D_i is simple. Note that each D_i has a short rws by definition of D and $\mathcal{X}$. If $\prod_{i=1}^k D_i$ has a short rws, then $\prod_{i=1}^{k+1} D_i = \prod_{i=1}^k D_i \times D_{k+1}$ has a short rws by Lemma 5. Hence by induction D has a short rws.

C has a short rws: If $C = 1$, then there is nothing to prove, so assume C is non-trivial. Write $C = \prod_{i=1}^n C_i^{k_i}$ where C_i are distinct simple groups and each $k_i \geq 3$. Each $C_i^{k_i}$ has a short rws by Lemma 7. As before induction and Lemma 5 gives that C has a short rws.

$C \times D$ has a short rws: If either or both are trivial, then there is nothing to prove so assume $C \neq 1 \neq D$. Then $|C|, |D| \geq 60$, so Lemma 5 shows $C \times D$ has a short rws.

$A \times C \times D$ has a short rws: If both C and D are trivial, then $|G| = |A| \cdot |B|$ and $|B| \leq N_2^2$, so $|A| \geq 300 \cdot 2^{14} \geq 2^14$ and by Lemma 2 has a short rws. If $|C \times D| \geq 300$, then Lemma 6 shows $A \times C \times D$ has a short rws. Finally, if $1 < |C \times D| < 300$, then as before $|A| \geq 2^{14}$ so has a short rws, and $|C \times D| \geq 60 \geq 16$, so Lemma 5 again shows $A \times C \times D$ has a short rws.

$A \times B \times C \times D$ has a short rws: If B is trivial, then there is nothing to prove, so assume B is non-trivial. Set $G_1 = A \times C \times D$, $G_2 = B$, and $G_3 = G_1 \times G_2$. As usual let g_i be the number of generators, r_i the number of rules, and o_i the order, where a short rws is chosen for G_1, and the rewriting system of Lemma 7 is chosen for the simple factors of G_2 and combined using Lemma 1, and the rws for G_3 according to Lemma 1. Since $o_2 \leq N_2^2$, $o_1 \geq 4N_1^2$. Also $r_2 \leq \sum\{4|X_i| : X_i \in \mathcal{X}\} + 4\binom{2|\mathcal{X}|}{2}$, since B has at most $2|\mathcal{X}|$ factors, each $X_i \in \mathcal{X}$ occurring at most twice. Hence $r_2 \leq 5N_1$. Then $r_3 = r_1 + r_2 + g_1 g_2 \leq (\sqrt{r_1} + \sqrt{r_2})^2$ and

$$r_3^3 \leq (\sqrt{r_1} + \sqrt{r_2})^4 \leq 8(r_1^2 + r_2^2) \leq 8o_1 + 200N_1^2 \leq 8o_1 + 50o_1 \leq o_1 o_2,$$

since $o_2 \geq 60$. Hence $r_3 \leq o_3$ and G has a short rws, as was to be shown. □

2. Finite simple groups

To prove Theorem A it remains to prove:

PROPOSITION 9. *All but finitely many isomorphism classes of finite simple groups have a short rws.*

The proof considers the various types of simple groups separately: cyclic, alternating, low rank Lie type, high rank Lie type, and sporadic. The cases of cyclic and sporadic types are handled trivially: cyclic simple groups have a short rws with a single rule, and the sporadics are considered to be part of the finitely many exceptions. A uniform rewriting system for groups of Lie type is described. The rws is shown to be short for all groups of Lie type over sufficiently large fields, and over all fields if the group has sufficiently large rank. The latter requires somewhat more detailed information on rws for Coxeter groups of type A, B, and D. The case of alternating groups is similar to the case of the Coxeter groups of type A.

2.1. Bruhat rewriting system. This and the next few sections use the theory and notation laid out in [Car85, Car72, Ste68] for finite simple groups of Lie type. To uniformly construct rws for these groups, it is more convenient to uniformly describe the final forms:

DEFINITION 3. A **reduction ordering** on X is a well-ordering on the set of strings X^* such that if $x \leq y$ then $axb \leq ayb$ for all $x, y, a, b \in X^*$.

The following lemma is well known:

LEMMA 10. *Let G be a group with generating set X not containing the identity of G. For a reduction ordering $\leq$ on X, there is a unique rws (X, R) such that the final forms are exactly the set of $\leq$-least pre-images of the elements of G; that is $w \in X^*$ is a final form iff w is the $\leq$-least element of $f^{-1}(f(w))$. The rules of this rws are exactly the pairs $(x, y) \in X^* \times X^*$ where every proper subword of x is final, but the final form of x is $y \neq x$.*

Thus in the special case of rws induced from a reduction ordering, the rules of the rws can be read directly from the final forms. Groups of Lie type come with an extremely natural set of "normal forms" called the Bruhat decomposition. The algorithmic properties of these normal forms were studied in [CMT04]. However these normal forms are not the set of final forms of any complete rws since they are not closed under subwords unless the reduced rank of the Lie group is 1. To fix this, one simply splits part of the Bruhat decomposition into smaller pieces whose subwords are of the same form. Mostly to set notation, we give this definition which closely follows [Car85].

DEFINITION 4. A group G is said to have a **BN-pair** (or Tits system) if it has subgroups B and N satisfying the following axioms:

(1) $G = \langle B, N\rangle$
(2) $H := B \cap N$ is normal in N
(3) $W := N/H$ is generated by elements s_i, $i \in I$, of order 2
(4) If $n_iH = s_i$, then $n_iBn_i \neq B$
(5) For each $n \in N$ and n_i with $n_iH = s_i$, one has $n_iBn \subseteq Bn_inB \cup BnB$

The **length** of an element of W is the length of the shortlex minimal expression for it as a product of the s_i. Denote a fixed set of pre-images of the s_i in N as n_i. There is a unique element of maximal length in W; denote the **longest element** by w_0. Denote a preimage of w_0 by $n_0 \in N$.

The BN-pair is said to be **split of characteristic** p if $U := O_p(B)$ is a normal complement to the abelian p'-group H, and $H = \cap_{n\in N} B^n$ is the intersection of the N-conjugates of B.

The **simple root subgroups** X_i are defined as $U \cap U^{n_0 n_i}$. The **root subgroups** X_i^n are the N-conjugates of the simple root subgroups. The **positive root subgroups** are those root subgroups contained in U.

The split BN-pair is said to **satisfy the weak commutator relations** if $[X_i, Y] \leq Z$ whenever X_i is a simple root subgroup, Y is a positive root subgroup distinct from X_i, and Z is the subgroup of U generated by all positive root subgroups that are not simple root subgroups.

A finite simple group of Lie type (other than the Tits group) has a split BN-pair of characteristic p satisfying the weak commutator relations. Since B is nilpotent by abelian and finite it is finite solvable, and so has a polycyclic rws. Indeed, $B = H \ltimes U$ so one can choose a polycyclic rws that exhibits H. Since N/H is a Coxeter group, it comes with a very efficient rewriting system. One can use this to give a rewriting system for N, re-using the generators for H. Using Lemma 1 as inspiration, one treats G as if it were an extension of the "quotient" W by the "normal" subgroup B even though W is merely a subquotient and B a non-normal subgroup. The resulting rewriting system has a simple form and a small number of rules.

LEMMA 11. *Let G be a group with a split BN-pair of characteristic p satisfying the weak commutator relations. Set $U_i = U \cap U^{n_i}$. Then $U = U_iX_i$, $U_i \cap X_i = 1$, $U_i^{n_i} = U_i$. If Y is a positive root subgroup distinct from X_i, then $Y \leq U_i$. In particular, $[X_i, U_i] \leq Z \leq U_i$.*

LEMMA 12. *Let G be a group with a split BN-pair of characteristic p satisfying the weak commutator relations. If the simple root subgroups are labeled with integers 1 to d, then B has the following subnormal series:*

$$1 \trianglelefteq Z \trianglelefteq ZX_d \trianglelefteq ZX_dX_{d-1} \trianglelefteq \ldots \trianglelefteq ZX_d \cdots X_2 \trianglelefteq ZX_d \cdots X_1 = U \trianglelefteq B.$$

In particular, B has a polycyclic rws exhibiting the simple root subgroups.

LEMMA 13. *Let G have a split BN-pair satisfying the weak commutator relations. Then $n_iU = U_in_iX_i$ and writing $u = abc$ with $a, c \in U_i$ and $b \in X_i$, there is some $d \in U_i$ with $n_iabc = dn_ib$.*

LEMMA 14. [Car85, Prop 2.5.13] *Let G be a group with a split BN-pair of characteristic p. Every element of G has a unique expression as $b\dot{w}u$ for $b \in B$, $w \in W$, and $u \in U_w$, where $U_w = U \cap U^{\dot{w}}$.*

Unfortunately this cannot be used to define final forms for a rewriting system as $s_is_jx_i$ is in the form specified but its subword s_jx_i is not whenever $i \neq j$. This is because $X_i \leq U_{s_is_j}$ but $X_i \cap U_{s_j} = X_i \cap X_j = 1$. Luckily, this is easily fixed:

LEMMA 15. *Let G be a group with a split BN-pair of characteristic p satisfying the weak commutator relations. If ws_i is reduced in W, then $\dot{w}n_iU_{ws_i} = \dot{w}U_wn_iX_i$. Indeed, if $w = \prod s_{a_i}$ is reduced, and $\dot{w} = \prod n_{a_i}$, then $\dot{w}U_w = \prod s_{a_i}X_{a_i}$.*

This allows one to find the final forms given the polycyclic rewriting systems of B and W, using the fact that the X_i are expressed as "subsystems" of B by Lemma 12. The generating set $X = X_B \cup X_W$ for the rewriting system are the generators for B and the n_i. The ordering is the wreath product of the orderings on W and B: Define a function from X^* to $X_W^* \times (X_B^*)^*$ by sending a product $\prod b_iw_i \cdot b_n$ to $[\prod w_i, b_n, b_{n-1}, \ldots, b_1]$, where $w_i \in X_W$ and $b_i \in X_B^*$. Compare elements in X^* by applying the function and then comparing lexicographically left to right, using the shortlex ordering on X_W, and the polycyclic ordering on each coordinate taking values in X_B^*.

LEMMA 16. *The final forms of G under this ordering are precisely the $b\prod w_iu_i$ where $w_i \in X_W$, b is the final form for an element of B, u_i is a final form for an element of X_i, and $\prod w_i$ is a final form for W.*

LEMMA 17. *Let G be a finite group with a split BN-pair of characteristic p satisfying the weak commutator relations. The left hand sides of the rules of G with respect to the previously indicated generating set are a subset of the following:*

1. *The polycyclic rules of B*
2. *The B,N action: For every generator $n_i \in N$, every element $x_i \in X_i$, and every generator $b \in B$ such that x_ib is reduced, the word n_ix_ib is a left hand side*
3. *The decorated W rules: for every left hand side $s_{i_1} \cdots s_{i_n}$ of a rule of W, and every sequence $x_k \in X_{i_k}$, the words $(\prod_{k=1}^{n-1} s_{i_k}x_k) \cdot s_{i_n}$ are left hand sides*

2.2. Low rank groups of Lie type. In this section we consider the finite simple groups of Lie type of some fixed rank. Since there are only finitely many families of such groups, each indexed by a parameter q, it will suffice to confine attention to large q. Simple numerical facts about these groups suffice to show that such groups have short rws.

Let G be a group of Lie type over an algebraic closed field, B be a Borel subgroup, N the normalizer of a maximal torus T of B and U the unipotent subgroup of B. $W = B/N$ is a finite Coxeter group, so let n be the number of positive roots, and r be the number of simple (positive) roots. Let F be a Frobenius map on G whose corresponding group G^F is finite, and assume B, T, and U are F-stable, and T is maximally F-split. Let $q = p^k$ be the number defined in [Car85, p. 35],

[Car72, p. 251], and [Ste68, p. 189]. F induces an action on the roots of W; let $\omega = (\omega_1, \ldots, \omega_{r_0})$ be the sizes of the r_0 orbits of simple roots. Let W^1 be the induced Coxeter group, and let ℓ_W be the length function inherited from W. The simple root subgroups X_i have order $q^{\ell_W(s_i)}$. If F induces the identity graph automorphism, then q is the sized of the fixed field, $W = W^1$, $r = r_0$, and $\omega_i = 1$. There is a simple bound on the order of the group that does not depend on the twisting:

$$|G^F| = |T^F||U^F| \sum_{w \in W^1} |U_w^F| = \prod_{i=1}^{r_0} (q^{\omega_i} - 1) \cdot q^n \cdot \sum_{w \in W^1} q^{\ell_W(w)} \geq (q-1)^r \cdot q^n \cdot q^n$$

Now a rws for the reduced Weyl group W^1 on the standard Coxeter generators has $r_0 \leq r$ generators and some finite set of rules not depending on q (but depending on which of the finitely many diagram automorphisms F induces). The left hand sides $w \cdot s_i$ of the rules of W^1 have the property that $\ell_W(w) \leq n$. The group H has a polycyclic rws with r_0 generators, and the group U has a polycyclic rws with kn generators. Totalling the number of generators from Lemma 17 one gets that number of rules is:

$$\binom{kn + r_0 + 1}{2} + \left(r_0(r_0 + k(n - r_0)) + \sum_{i=2}^{r_0} kiq^{\ell_W(s_i)} \right) + \sum_{(w \cdot s, v) \in R} q^{\ell_W(w)}$$

Note that the only terms with q dependence are the second half of the middle term and the last term. This is then sufficient to conclude that the Bruhat rws is short for all but finitely many q: the number of rules is $O(q^n)$ and the order is $\Omega(q^{2n+r})$ where n, r are fixed. Here the standard "Big O" notations are used: $O(f)$ is the class of functions g such that $\limsup_{q \to \infty} \frac{g}{f}$ is finite and $\Omega(f)$ is the class of functions g such that $\liminf_{q \to \infty} \frac{g}{f}$ is greater than zero, and $\Theta(f) = O(f) \cap \Omega(f)$.

A more careful analysis reveals that for A_1 the number of rules is $\Theta(q)$ while the order is $\Theta(q^3)$, for G_2 the number of rules is $\Theta(q^5)$ while the order is $\Theta(q^{14})$, for ${}^2F_4$ the number of rules is $\Theta(q^{20})$ while the order is $\Theta(q^{52})$, for ${}^2G_2$ the number of rules is $\Theta(q^6)$ while the order is $\Theta(q^{14})$, and that amongst all simple groups of Lie type, these are the four largest ratios of the logarithm of the number of rules to the logarithm of the order: $\frac{1}{3}$, $\frac{5}{14}$, $\frac{5}{13}$, and $\frac{3}{7}$. For this reason, the current methods cannot do significantly better than bounding the number of rules by the square root of the group order.

LEMMA 18. *There is a sequence of positive integers q_r such that if G is a finite simple group of Lie type and rank r with parameter q satisfying $q \geq q_r$, then G has a short rws.*

2.3. High rank groups of Lie type. The groups of Lie type that are simple as algebraic groups and have $r \geq 9$ fall into six families: A_r, B_r, C_r, D_r, ${}^2A_r$, and ${}^2D_r$. To prove Proposition 9, it is necessary to show that for all sufficiently large r and for all q, the finite simple groups of Lie type of rank r with given q have a short rws. For this a slightly better estimate on the number of rules of the corresponding Weyl groups is needed.

LEMMA 19. *The polynomial $\Psi(q)$ for the Coxeter group of type X_r satisfies $\Psi(q) \leq f(q)$ with equality at $q = 1$, where X and f come from the following table that also includes the number of positive roots n in terms of the number of simple (positive) roots r:*

X	n	f		
A	$\binom{r+1}{2}$	q^r	$+$	$r(r-1)q^{r-1}$
B	r^2	q^{2r-1}	$+$	$r(r-1)q^{2r-3}$
D	$r(r-1)$	q^{2r-2}	$+$	$(3\binom{r}{2}-n+1)q^{2r-4}$

For large r, the term corresponding to the rules of W^1 clearly dominates the number of rules. It is not difficult to make this precise: the number of rules is bounded above by $r^4p^{6kr} = r^4q^{6r}$. Similarly one can bound the order below by $\frac{1}{r+1}p^{kr^2} = \frac{1}{r+1}q^{r^2}$. Since $r^{10} \leq q^{r^2-12r}$ for all $q \geq 2$ and $r \geq 15$, one gets:

LEMMA 20. *If G is a finite simple group of Lie type whose corresponding simple algebraic group has rank $r \geq 15$, then G has a short rws.*

2.4. Alternating groups. An efficient rewriting system on $n-2$ generators and $(18n^2 - 13n + 159 + (-1)^n)/8$ rules is given by a slight modification of the Coxeter rewriting system for $\mathrm{Sym}(n)$. The generators are $(i, i+1) \cdot (n-1, n)$ for $1 \leq i \leq n-2$, and the reduction ordering is the wreath product ordering which compares the number of occurrences of the last generator first, and then uses lexicographic comparison of each of the subwords between those occurrences, using shortlex comparison on the individual subwords. The maximum length of a left hand side is $2n-4$ with multiplicity 1 for n odd, and $2n+1$ with multiplicity 2 for n even.

This rewriting system views an embedded copy of $\mathrm{Sym}(n-2)$ in $\mathrm{Alt}(n)$ as a sort of "normal subgroup", and the last generator $(n-2, n, n-1)$ as some sort of "quotient group". Since the rewriting system for $\mathrm{Sym}(n-2)$ is extremely efficient, and the wreath product ordering attempts to collect all occurrences of the "bad" generator as quickly as possible, the result is a rewriting system not much more complicated than the one for $\mathrm{Sym}(n)$. The list of left hand sides is regular, but divides into a large number of cases.

LEMMA 21. *All but finitely many alternating groups have a short rws.*

2.5. Proofs. The proofs of the lemmas and propositions stated in Section 2 are collected here. The proof of the main proposition is simply gathering up the lemmas:

PROPOSITION 9. *All but finitely many isomorphism classes of finite simple groups have a short rws.*

PROOF. Lemma 21 shows all but finitely many alternating groups have a short rws. Lemma 20 shows that finite simple groups of Lie type have a short rws for all but finitely many ranks. Lemma 18 shows amongst those finitely many ranks, all but finitely many groups have a short rws. Cyclic groups obviously have a short rws. This leaves the sporadics (including the Tits group), and we appeal to the CFSG to conclude there are only finitely many sporadics, completing the proof. □

LEMMA 10. *Let G be a group with generating set X not containing the identity of G. For a reduction ordering $\leq$ on X, there is a unique rws (X, R) such that the final forms are exactly the set of $\leq$-least pre-images of the elements of G; that is*

$w \in X^*$ is a final form iff w is the $\leq$-least element of $f^{-1}(f(w))$. The rules of this rws are exactly the pairs $(x, y) \in X^* \times X^*$ where every proper subword of x is final, but the final form of x is $y \neq x$.

PROOF. This is the sum of Propositions 2.7, 2.8, and 2.9 from [Sim94]. □

LEMMA 11. *Let G be a group with a split BN-pair of characteristic p satisfying the weak commutator relations. Set $U_i = U \cap U^{n_i}$. Then $U = U_i X_i$, $U_i \cap X_i = 1$, $U_i^{n_i} = U_i$. If Y is a positive root subgroup distinct from X_i, then $Y \leq U_i$. In particular, $[X_i, U_i] \leq Z \leq U_i$.*

PROOF. This is just gathered from [Car85, p. 50, Prop 2.5.11, pp. 61,62]. Much of this follows directly from the axioms and [Car85, Prop 2.5.16], which gives a formula for U_i as the subgroup generated by the positive root subgroups distinct from the subgroup X_i. □

LEMMA 12. *Let G be a group with a split BN-pair of characteristic p satisfying the weak commutator relations. If the simple root subgroups are labeled with integers 1 to d, then B has the following subnormal series:*

$$1 \trianglelefteq Z \trianglelefteq ZX_d \trianglelefteq ZX_dX_{d-1} \trianglelefteq \ldots \trianglelefteq ZX_d \cdots X_2 \trianglelefteq ZX_d \cdots X_1 = U \trianglelefteq B.$$

In particular, B has a polycyclic rws exhibiting the simple root subgroups.

PROOF. Since $[X_d, Z] \leq Z$, X_d normalizes Z, and X_dZ is a subgroup of G, indeed of U_{d-1}. Since $[X_{d-1}, ZX_d] \leq [X_{d-1}, U_{d-1}] \leq Z \leq ZX_d$, X_{d-1} normalizes ZX_d. Continuing in this manner, we get that each term is a semidirect product, and in particular defines a subnormal series of B. Since the series consists of semidirect products, there is no difficulty in repeating generators. Hence one can choose any polycyclic generating set of Z and the X_i, and the concatenation is a polycyclic generating sequence for U. One then extends to B by finding a polycyclic generating sequence for the abelian group H. Again, because it is a semidirect product, there are no restrictions on how the two polycyclic generating sequences (for U and for H) are chosen. □

LEMMA 13. *Let G have a split BN-pair satisfying the weak commutator relations. Then $n_iU = U_in_iX_i$ and writing $u = abc$ with $a, c \in U_i$ and $b \in X_i$, there is some $d \in U_i$ with $n_iabc = dn_ib$.*

PROOF. The set equality follows simply from $U_i^{n_i} = U_i$ (clear from the definition since $n_i^2 \in N_G(U)$), and $U = U_iX_i$, [Car85, Prop 2.5.11]. By Lemma 11, $[X_i, U_i] \leq U_i$, so $n_iu = n_iabc = n_ia[b^{-1}, c^{-1}]cb = n_ivb$ with $v \in U_i$. Since n_i normalizes U_i, $d = v^{n_i} \in U_i$ satisfies $n_iabc = dn_ib$. □

Lemma 14 is the Bruhat decomposition, [Car85, Prop 2.5.13].

LEMMA 15. *Let G be a group with a split BN-pair of characteristic p satisfying the weak commutator relations. If ws_i is reduced in W, then $\dot{w}n_iU_{ws_i} = \dot{w}U_wn_iX_i$. Indeed, if $w = \prod s_{a_i}$ is reduced, and $\dot{w} = \prod n_{a_i}$, then $\dot{w}U_w = \prod s_{a_i}X_{a_i}$.*

PROOF. By [Car85, Prop 2.5.10.i], $U_{ws_i} = U_w^{n_i}X_i$, and so

$$\dot{w}n_iU_{ws_i} = \dot{w}n_iU_w^{n_i}X_i = \dot{w}U_wn_iX_i.$$

The final statement follows by induction. □

LEMMA 16. *The final forms of G under this ordering are precisely the $b\prod w_i u_i$ where $w_i \in X_W$, b is the final form for an element of B, u_i is a final form for an element of X_i, and $\prod w_i$ is a final form for W.*

PROOF. Every element of G can be written in such a form by Lemmas 14 and 15. Suppose $b_0 \cdot \prod \dot{w}_j b_j$ is some final form for an element $g \in G$ with $b_j \in X_B^*$ and $w_j X_W$.

b_0: The subword b_0 must also be a final form, and it clearly represents an element of B.

$\prod w_j$: The element g lies in some double coset $B\dot{w}B$, so write $g = b\dot{w}b'$ where $b, b' \in B$ and $\dot{w} \in N$. Choose words $c, c' \in X_B^*$ corresponding to b, b', and let $s_{i_1} \cdots s_{i_k}$ be the shortlex final form of $w \in W$. Then g is also represented by $c\prod w_j c_j$, where $c_j = 1$ except for the last index j where $c_j = c'$, and where $c_j = s_{i_j}$. Comparing the final form to this form involves comparing $[\prod w_j, b_n, b_{n-1}, \ldots]$ to $[\prod s_{i_j}, c', 1, 1, \ldots, c]$. Since we have a final form in G, we must have $\prod w_j \leq \prod s_{j_k}$.

Now we have written $g \in Bw_1Bw_2B \cdots Bw_nB$, but using the fourth axiom we see that g is contained in one of the cosets BvB where v is the product of w_{i_k} where i_k is a strictly increasing sequence ending in n (of length between 1 and n). In particular, unless $v = \prod w_j$, the length of the final form of v is strictly less than the length of the formal product $\prod w_j$. Since $\prod w_j$ is minimal, we have $v = \prod w_j$ as formal products, and so they represent the same element of W. Hence $\prod w_j = \prod s_{j_k}$ as formal products since they represent the same element of W, $\prod w_j \leq \prod s_{j_k}$, and $\prod s_{j_k}$ is a final form. Hence $\prod w_j$, too, is the shortlex final form of an element in W.

$w_j b_j$: Since each $\dot{w}_i$ normalizes H and each h_i normalizes U, the only b_i with an H-component is b_0. In other words, the final form must have each b_i the final form of an element of U for $i > 0$.

Now the subwords $w_j b_j$ are reduced, so factor $b_j \in X_B^*$ into rst with r, t final forms of elements in U_i and s the final form of an element in X_i where w_j represents n_i. By Lemma 13, $n_i rst = un_i s$ for some $u \in U_i$. Hence letting u also denote its final form in X_B^*, the words $un_i s$ and $n_i rst$ are equivalent, but $[n_i, s, u] \leq [n_i, rst]$. Since $w_j b_j$ is a final form, we must have $b_j = s \in X_i$.

This shows that final forms are all of the form $b_0 \prod n_{i_k} x_k$ where $\prod s_{i_k}$ is a shortlex final form of an element of W, $b_0 \in X_B^*$ is a final form of a word in B, and each x_k is the final form of a word in X_{i_k}.

Since every element of G has a unique expression in this form and every element's final form is in this form, that unique expression must be the final form, and so each such expression occurs as a final form. □

LEMMA 17. *Let G be a finite group with a split BN-pair of characteristic p satisfying the weak commutator relations. The left hand sides of the rules of G with respect to the previously indicated generating set are a subset of the following:*

(1) *The polycyclic rules of B*
(2) *The B,N action: For every generator $n_i \in N$, every element $x_i \in X_i$, and every generator $b \in B$ such that $x_i b$ is reduced, the word $n_i x_i b$ is a left hand side*
(3) *The decorated W rules: for every left hand side $s_{i_1} \cdots s_{i_n}$ of a rule of W, and every sequence $x_k \in X_{i_k}$, the words $(\prod_{k=1}^{n-1} s_{i_k} x_k) \cdot s_{i_n}$ are left hand sides*

PROOF. The left hand sides are precisely the final forms appended by generators such that the result is not a final form, but every proper subword is a final form. From another perspective, take a final form, append a generator so that the result is not a final form, and then take the shortest suffix of the word that is not a final form to be a left hand side.

To show that the list above is exhaustive, we just show that the final forms appended by generators that do not appear are either final forms or have proper suffixes that are not final forms. If the final form is b_0 and the generator is from B, then this is just a rule of B, which we already know. If the final form is b_0 and the generator is from N, then the result is a final form. If the final form is $b_0 \cdot w$ where w is nontrivial, then the suffix wg will already not be a final form. If the final form is $ws_a x_a$ and the generator is from B but not from X_a, then $s_a x_a g$ is a proper suffix that is not a final form. If the generator is from X_a, then either $ws_a x_a g$ is a final form or the proper suffix $x_a g$ is not a final form. If the generator is from N, then this is a rule listed above. □

LEMMA 18. *There is a sequence of positive integers q_r such that if G is a finite simple group of Lie type and rank r with parameter q satisfying $q \geq q_r$, then G has a short rws.*

PROOF. The proof is already given in the section. The number of rules is bounded above by q^n and the order is bounded below by $q^{2n} q^r 2^{-r}$, and for large q, $q^n \leq \sqrt{q^{2n+r}/2^r}$. □

LEMMA 19. *The polynomial $\Psi(q)$ for the Coxeter group of type X_r satisfies $\Psi(q) \leq f(q)$ with equality at $q = 1$, where X and f come from the following table that also includes the number of positive roots n in terms of the number of simple (positive) roots r:*

X	n	f
A	$\binom{r+1}{2}$	$q^r + r(r-1)q^{r-1}$
B	r^2	$q^{2r-1} + r(r-1)q^{2r-3}$
D	$r(r-1)$	$q^{2r-2} + (3\binom{r}{2} - n + 1)q^{2r-4}$

PROOF. The values of the number of positive roots are well known, for instance, [Car72, p. 43]. Suitable rws for the Coxeter groups of type A_r and B_r are given in [LC86a, LC86b]. These are simply the rws induced by the shortlex reduction order on fairly standard orderings of the Coxeter generators (taking the double bond to occur last for B_r). For reference the rws for A_r is:

$$w_i \cdot w_i \mapsto 1 \qquad 1 \leq i \leq r$$

$$w_i \cdot \boldsymbol{w_j} \mapsto \boldsymbol{w_j} \cdot w_i \qquad 1 \leq j < i - 1 < r - 1$$

$$(w_i \cdot w_{i-1} \cdots w_{i-j}) \cdot \boldsymbol{w_i} \mapsto \boldsymbol{w_{i-1}} \cdot (w_i \cdot w_{i-1} \cdots w_{i-j}) \qquad 1 \leq j < i$$

The rws for B_r is:

$$\begin{array}{rcll} w_i \cdot w_i & \mapsto & 1 & 1 \le i \le r \\ w_i \cdot \boldsymbol{w_j} & \mapsto & \boldsymbol{w_j} \cdot w_i & 1 \le j < i-1 < r-1 \\ (w_i \cdot w_{i-1} \cdots w_{i-j}) \cdot \boldsymbol{w_i} & \mapsto & \boldsymbol{w_{i-1}} \cdot (w_i \cdot w_{i-1} \cdots w_{i-j}) & 1 \le j < i < r \\ (w_i \cdot w_{i-1} \cdots \boldsymbol{w_{i-j}})^2 & \mapsto & \boldsymbol{w_{i-1}} \cdot (w_i \cdot w_{i-1} \cdots w_{i-j+1} \cdot w_{i-j}) & 1 \le j < i = r \\ & & \cdot (w_i \cdot w_{i-1} \cdots w_{i-j+1}) & \end{array}$$

For somewhat unclear reasons, Le Chenadec did not give the rws for D_r, but one can easily be given along the same lines:

$$\begin{array}{rcl} w_i^2 & \mapsto & 1 \\ & & 1 \le i \le r-2, \text{ or } i \in \{a,b\} \\ w_i \cdot \boldsymbol{w_j} & \mapsto & \boldsymbol{w_j} \cdot w_i \\ & & 1 \le j \le i-2 < i \le r-2, \\ & & \text{or } (j \in \{a,b\}, 2 \le i \le r-2) \\ (w_i \cdot w_{i-1} \cdots w_j) \cdot \boldsymbol{w_i} & \mapsto & \boldsymbol{w_{i-1}} \cdot (w_i \cdot w_{i-1} \cdots w_j) \\ & & 1 \le j < i \le r-2 \\ (w_i \cdot w_{i-1} \cdots w_1) \cdot w_k \cdot \boldsymbol{w_i} & \mapsto & \boldsymbol{w_{i-1}} \cdot (w_i \cdot w_{i-1} \cdots w_1) \cdot w_k \\ & & 1 \le i \le r-2, k \in \{a,b\} \\ (w_i \cdot w_{i-1} \cdots w_1) \cdot w_a \cdot w_b \cdot (w_1 \cdots w_j) \cdot \boldsymbol{w_i} & \mapsto & \boldsymbol{w_{i-1}} \cdot (w_i \cdot w_{i-1} \cdots w_1) \cdot w_a \cdot w_b \cdot (w_1 \cdots w_j) \\ & & 0 \le j \le i-2, 2 \le i \le r-2 \\ (w_i \cdot w_{i-1} \cdots w_1) \cdot w_a \cdot w_b \cdot (w_1 \cdots w_i) \cdot \boldsymbol{w_{i-1}} & \mapsto & \boldsymbol{w_{i-1}} \cdot (w_i \cdot w_{i-1} \cdots w_1) \cdot w_a \cdot w_b \cdot (w_1 \cdots w_i) \\ & & 2 \le i \le r-2 \\ w_1 \cdot w_a \cdot w_b \cdot w_1 \cdot \boldsymbol{w_a} & \mapsto & \boldsymbol{w_b} \cdot w_1 \cdot w_a \cdot w_b \cdot w_1 \\ w_1 \cdot w_a \cdot w_b \cdot w_1 \cdot \boldsymbol{w_b} & \mapsto & \boldsymbol{w_a} \cdot w_1 \cdot w_a \cdot w_b \cdot w_1 \end{array}$$

□

LEMMA 20. *If G is a finite simple group of Lie type whose corresponding simple algebraic group has rank $r \ge 15$, then G has a short rws.*

PROOF. The proof of this is outlined in the corresponding section. It is tedious but straightforward to factor out $r^4 p^{6kr}$ from the expression for the number of rules, leaving only fractions with constant numerators. The denominators are minimized at $p = 2$, $k = 1$, $r = 15$, and their sum is significantly less than 1, showing that the number of rules is less than $r^4 p^{6kr}$. If $r^{10} \le p^{kr^2 - 12kr}$ then

$$(r^4 p^{6kr})^2 = r^8 p^{12kr} \le \tfrac{1}{r^2} p^{kr^2} \le \tfrac{1}{r+1} p^{kr^2},$$

as required. □

LEMMA 21. *All but finitely many alternating groups have a short rws.*

PROOF. Much like the proof of Lemma 19, one writes down the rws generically, and then checks confluence, perhaps most conveniently using Lemma 10. The resulting rws has less than $3n^2$ rules and for large n, $r(\mathrm{Alt}(n)) \le 3n^2 \ll \sqrt{\frac{n!}{2}} = \sqrt{o(\mathrm{Alt}(n))}$. □

3. Concluding remarks

The presentation in this paper and at the conference mostly follows a theoretical and asymptotic line of reasoning to make the results more easily digestible. The work itself was done very concretely with explicit rational maps on the algebraic groups which when evaluated at the obvious generating set give the rws. The number of rules was therefore easily calculated exactly and in only a few cases was the rws not short (such as for $A_1(4)$ and $A_1(5)$). Similarly, the rws proposed for the alternating groups is short for all but a few cases. However in every case a short rws was easily found, even if not on the standard generators. Indeed, the first five sporadic groups also have short rws, and there is no indication that the other (22, including the Tits group) sporadic groups do not have short rws.

In terms of practical implementation, the Borel subgroup should be handled a bit peculiarly if one wants to minimize the number of rules. One uses a wreath product of polycyclic orderings, so that final forms are $h \cdot x_n \cdots x_2 \cdot x_1$ where the x_i are in the positive root subgroups, with $1 \leq i \leq r_0$ the simple root subgroups, and where $h \in H$. The rules still describe a "collection" process since H normalizes each X_i, and the resulting rules are even quite sparse. Twisted groups also pose an implementation challenge, as the best rws on the Weyl group for asymptotically large q differ from the corresponding rws for the untwisted case, indeed, often using the reverse order. In general for all groups of Lie type, the best rewriting system on the Weyl group may depend on the field size for small fields, so that one may wish to keep a table of small solutions.

The rational functions describing the rules can be determined automatically using the methods of [CMT04] for untwisted groups, and was done in practice using the implementation in Magma, [BCP97]. For twisted groups of low reduced rank, the calculations are somewhat trivialized, and for very large rank can be done generically by hand, but an automatic method that can handle ${}^2F_4$, ${}^3D_4$, and ${}^2E_6$ is being developed. It is hoped that success there will make it reasonable to reimagine the work on rws to systems of rules on algebraic groups. The current implementation of these rws in GAP, [Gap08], uses a primitive version of these algebraic rws.

Future work also includes eliminating the remaining sporadic exceptions and development of ideas to enumerate all rws of a finite group (which was non-trivial even for groups of order 16). It is hoped that these ideas may lead to a proof that certain families, such as $PSL(2,p)$ or ${}^2G(3^{2n+1})$ do not have even shorter rewriting systems; in particular one wants to eliminate the possibility of rewriting systems whose number of rules is logarithmic in the group order.

The current work would not have been possible without the collegial encouragement and support of the GAP, MAGMA, and Group Pub communities, and I would like to especially thank Alexander Hulpke, Frank Lübeck, and Scott Murray. The conference organizers Luise-Charlotte Kappe, Arturo Magidin, and Robert Morse have as usual provided a welcoming atmosphere, and I thank them again. Kathryn Lybarger and Michael Slone also provided insight into the large sea of inequalities and combinatorial data in which this work swam for quite some time, and I thank them for helping me keep this work afloat.

References

[Ani86] David J. Anick, *On the homology of associative algebras*, Trans. Amer. Math. Soc. **296** (1986), no. 2, 641–659. MR **846601 (87i:**16046)

[BCP97] Wieb Bosma, John Cannon, and Catherine Playoust, *The Magma algebra system. I. The user language*, J. Symbolic Comput. **24** (1997), no. 3-4, 235–265. Computational algebra and number theory (London, 1993). MR 1484478

[Car72] Roger W. Carter, *Simple groups of Lie type*, John Wiley and Sons, London-New York-Sydney, 1972. Pure and Applied Mathematics, Vol. 28. MR 0407163 (53 #10946)

[Car85] ———, *Finite groups of Lie type*, Pure and Applied Mathematics (New York), John Wiley and Sons Inc., New York, 1985. Conjugacy classes and complex characters; A Wiley-Interscience Publication. MR **794307 (87d:**20060)

[CMT04] Arjeh M. Cohen, Scott H. Murray, and D. E. Taylor, *Computing in groups of Lie type*, Math. Comp. **73** (2004), no. 247, 1477–1498 (electronic). MR **2047097 (2005a:**20068)

[Coh93] Daniel E. Cohen, *String rewriting—a survey for group theorists*, Geometric group theory, Vol. 1 (Sussex, 1991), London Math. Soc. Lecture Note Ser., vol. 181, Cambridge Univ. Press, Cambridge, 1993, pp. 37–47. MR **1238514 (94i:**20051)

[Gap08] The GAP Group, *GAP – Groups, Algorithms, and Programming, Version 4.4.12*, 2008.

[Gro90] J. R. J. Groves, *Rewriting systems and homology of groups*, Groups—Canberra 1989, Lecture Notes in Math., vol. 1456, Springer, Berlin, 1990, pp. 114–141. MR **1092227** (**92e**:20022)

[Gro97] ———, *An algorithm for computing homology groups*, J. Algebra **194** (1997), no. 2, 331–361. MR **1467155** (**98g**:20082)

[HEO05] Derek F. Holt, Bettina Eick, and Eamonn A. O'Brien, *Handbook of computational group theory*, Discrete Mathematics and its Applications (Boca Raton), Chapman and Hall/CRC, Boca Raton, FL, 2005. MR **2129747** (**2006f**:20001)

[LC86a] Philippe Le Chenadec, *Canonical forms in finitely presented algebras*, Research Notes in Theoretical Computer Science, Pitman Publishing Ltd., London, 1986. MR **840218** (**88b**:68117)

[LC86b] ———, *A catalogue of complete group presentations*, J. Symbolic Comput. **2** (1986), no. 4, 363–381. MR **872786** (**88f**:68067)

[NS06] Max Neunhöffer and Ákos Seress, *A data structure for a uniform approach to computations with finite groups*, ISSAC 2006, ACM, New York, 2006, pp. 254–261. MR 2289128

[Ser06] Ákos Seress, *A unified approach to computations with permutation and matrix groups*, International Congress of Mathematicians. Vol. II, Eur. Math. Soc., Zürich, 2006, pp. 245–258. MR **2275596** (**2008a**:20006)

[Sim94] Charles C. Sims, *Computation with finitely presented groups*, Encyclopedia of Mathematics and its Applications, vol. 48, Cambridge University Press, Cambridge, 1994. MR **1267733** (**95f**:20053)

[Squ87] Craig C. Squier, *Word problems and a homological finiteness condition for monoids*, J. Pure Appl. Algebra **49** (1987), no. 1-2, 201–217. MR **920522** (**89a**:20059)

[Ste68] Robert Steinberg, *Lectures on Chevalley groups*, Yale University, New Haven, Conn., 1968. Notes prepared by John Faulkner and Robert Wilson. MR 0466335 (57 #6215)

University of Kentucky, 715 Patterson Office Tower, Lexington, KY 40506-0027
E-mail address: jack@ms.uky.edu